ENGINEERING MATERIALS 3

Materials Failure Analysis:
Case Studies and Design Implications

THE UNIVERSITY

HAR

For co

Pergamon Titles of Related Interest

ASHBY
Materials Selection in Mechanical Design

ASHBY & JONES
Engineering Materials 1: An Introduction to their Properties and Applications
Engineering Materials 2: An Introduction to Microstructures, Processing and Design

GIBSON & ASHBY
Cellular Solids

HEARN
Mechanics of Materials 1
Mechanics of Materials 2

HULL & BACON
Introduction to Dislocations, 3rd Edition

JONO & INOUE
Mechanical Behaviour of Materials—VI (ICM6)

SALAMA et al.
Advances in Fracture Research (ICF7)

TAYA & ARSENAULT
Metal Matrix Composites: Thermomechanical Behaviour

Pergamon Related Journals

Acta Metallurgica et Materialia

Computers and Structures

Engineering Failure Analysis

Engineering Fracture Mechanics

International Journal of Mechanical Sciences

International Journal of Solids and Structures

Materials Research Bulletin

Mechanics Research Communications

Progresss in Materials Science

Scripta Metallurgica et Materialia

Structural Engineering Review

Free sample copies of journals are gladly sent on request

ENGINEERING MATERIALS 3

Materials Failure Analysis:
Case Studies and Design Implications

BY

DAVID R. H. JONES

Engineering Department, Cambridge University, UK

PERGAMON PRESS

OXFORD · NEW YORK · SEOUL · TOKYO

UK   Pergamon Press Ltd, Headington Hill Hall
Oxford OX3 0BW, England

USA   Pergamon Press, Inc., 660 White Plains Road,
Tarrytown, New York 10591-5153, USA

KOREA   Pergamon Press Korea, KPO Box 315,
Seoul 110-603, Korea

JAPAN   Pergamon Press Japan, Tsunashima Building
Annex, 3-20-12 Yushima, Bunkyo-ku,
Tokyo 113, Japan

First edition 1993

Library of Congress Cataloging in Publication Data
Jones, David Rayner Hunkin, 1945–
Engineering materials 3 : materials failure analysis—case studies and design implications / David R. H. Jones—1st ed.
p. cm. — (International series on materials science & technology)
Includes index.
1. Fracture mechanics. 2. Materials—Fatigue. I. Title.
II. Title: Engineering materials three. III. Series: International series on materials science and technology.
TA409.J67   1993
620.1′126—dc20   93-20538

British Library Cataloguing in Publication Data
A catalogue record for this book is available from the British Library

ISBN 0 08 041904 6 (Hardcover)

ISBN 0 08 041905 4 (Flexicover)

Printed in Great Britain by BPCC Wheatons Ltd, Exeter

CONTENTS

B. PLASTIC DEFORMATION

C. CREEP

D. FAST FRACTURE

E. BRITTLE FRACTURE

F. FATIGUE

G. ENVIRONMENTAL FAILURES

H. GREAT ENGINEERING DISASTERS

FOREWORD

MANY examples of modern engineering—from suspension bridges to aero engines—are masterpieces of efficient and elegant design. But when one considers the astonishing range of engineering artifacts around us it is not surprising that occasionally failures occur. Perhaps a limitation of the material has been overlooked; or a particular loading case has been ignored; or integrity has been lost through exposure to the environment. Though the consequences can be tragic, an engineering failure often provides us with an experimental test of a realism which it is rarely possible to achieve in the laboratory or on the computer. In fact engineers often learn their most important lessons when things go wrong.

The analysis of engineering failures is a vital element in the design process: something useful can be learned from even the most ordinary failures; the most major incidents can lead to radical changes in the way in which things are done. Since most failures are traced ultimately to failures of the materials themselves, they also provide graphic illustrations of the way in which engineering function depends critically on materials properties.

This book presents a series of case studies drawn from real situations involving materials failures. It is suitable for people who have a working knowledge of engineering materials equivalent to the content of the books *Engineering Materials 1* and *Engineering Materials 2* by M. F. Ashby and D. R. H. Jones (Pergamon, 1980 and 1986). It is intended to be of interest to students taking taught courses in a wide range of engineering subjects (including mechanical, civil, aeronautical, manufacturing, chemical, materials and design engineering) and to people interested in failure analysis and prevention.

As you can see from the Contents, the case studies are arranged in groups, with each group describing failures that were linked primarily to a particular material property. Thus there is a section on failures related to elastic deformation, a section on failures related to plastic deformation, and so on. The case studies provide specific examples of the use of engineering materials in real applications and are meant to illustrate and extend the basic information given in introductory texts on engineering materials, structures, mechanics and design.

The treatment has been kept concise so that the content of each chapter can be summarised in one fifty-minute lecture. Most of the case studies involve a detailed analysis, and numerical solutions are obtained wherever this is appropriate. But the level of mathematics used is basic and standard results from related branches of engineering, such as structures or mechanics, have been quoted as necessary.

A special feature of the book is Appendix 1 which gives a basic collection of the formulae and data which keep on cropping up in failure analysis. The appendix is divided into separate sections, each of which deals with a different topic (Stress and Strain, Elastic Deformation, Plastic Deformation and so on). Full details are given in the Contents Summary at the start of Appendix 1. The appendix is intended both to service the material in the main chapters and to serve as a basic "toolkit" for analysing failures and solving design problems. However, it should be remembered that engineering design is a matter for experienced professionals only and is generally carried out through the medium of established codes and procedures rather than basic analytical results of the type presented in Appendix 1.

In Appendix 2 you will find a set of examples to help consolidate or develop the material of the case studies. The examples have been numbered to show when they can be done: for instance, you should be able to attempt Example 10.3 when you have read Chapter 10, although the example might also draw on material covered earlier in that block of chapters. Appendix 3 contains fully-worked solutions to the examples. Try not to turn to these unless you are sure you have solved the problem or are totally stuck.

The principal symbols used in the book are defined in Appendix 4. In order not to be tedious we have only defined symbols in the main text where these are not obvious: you are unlikely to find E defined in the text, but it is listed in Appendix 4 if you cannot remember what it stands for!

Acknowledgements

The author wishes to thank Professor C. Andrew and Professor D. E. Newland at Cambridge, and Mr. J. E. Gilgunn-Jones and Mr. K. Lambert at Pergamon for their help and encouragement throughout the course of this project.

A. Elastic deformation

CHAPTER 1

FAILURE OF MOUNTAINEERING ROPES

Introduction

At its most advanced level, mountaineering is one of the great tests of human mental and physical endurance. The sport as we know it began towards the end of the nineteenth century when the first classic routes were established in the Swiss, Italian and French Alps and on the rock slabs of North Wales and the English Lake District. In the 1920s and 1930s the first attempts were made on Everest, and many of the great routes like the North Face of the Eiger were conquered. Today, Everest is climbed by the "easy" South Col route almost as a matter of routine and the huge 1000-m walls of Yosemite and the Dolomites have bred a generation of outstanding rock experts.

Modern rock climbers generally adopt a standard set of safety procedures. They usually climb as a pair, and are always roped together with a 50-m length of nylon climbing rope. One climber (the "second") ties or belays onto the rock, while the other (the "leader") climbs up the face. The second feeds out rope so that there is no slack between the two climbers. Were the leader to fall off the face from a point say 10 m above the second, then he or she would fall past the second, and would then have to fall another 10 m before the rope could come under load. At this stage the second would have to grip the rope tightly and would need to wear canvas gloves to avoid rope burn.

At convenient intervals the leader assembles running belays, or "runners". A small spike of hardened steel (a piton) is hammered into a fine crack in the rock, and the rope is run through a snap-link (a karabiner) fixed to the piton. The leader then moves up the face past the running belay, and the rope is automatically pulled up through the karabiner behind the leader. Were the leader to fall off the face say 10 m above a runner he or she would, as before, fall 20 m before the rope could take the load, but the second would experience an *upward* pull during the arrest.

On technically difficult and exposed climbs the leader will put runners in as often as possible in order to minimise the dangers of a fall. Indeed, where there are no cracks into which pitons can be jammed securely, it has been known for climbers to drill holes in the rock with a portable electric drill, securing runners using steel expansion bolts.

When cracks are too wide to take pitons, one can instead jam aluminium "nuts", spring-steel "bongs" or wooden wedges into them. Nuts and wedges are usually fitted with short slings of 5-mm steel wire rope which are led out of the crack to the karabiner.

3

When the leader runs out of rope he or she belays to the rock face. The second then unties from the rock and climbs up the face, removing all the ironmongery from the rock on the way up. The leader continually takes in the slack from the rope so that, were the second to slip, it would be easy for the leader to arrest the fall. When the second arrives alongside the leader the cycle is repeated for the next pitch of the climb. On long and serious climbs, where time is important, the second sometimes "leads through", going straight up the face past the previous leader and becoming the new leader in the process.

The set-up when climbing steep snow or ice is the same, except that the belays are made using snow anchors, ice screws, ice pitons or snow stakes. Figure 1.1 shows a typical snow anchor, complete with wire-rope sling and karabiner. Also shown in the figure is a length of multi-stranded nylon climbing rope.

These modern techniques are not infallible. For example, pitons can pull out of small cracks and ice can break up under the pressure of an ice screw. But it is unusual for the rope to snap unless it becomes badly frayed by contact with a sharp edge of rock. Modern protection is good enough that a leader can contemplate falling off when making a difficult move. Indeed, rock gymnasts may fall two or three times before they succeed in overcoming the crux of a climb.

Before the 1940s things were very different. Techniques for running belays were much more rudimentary and it was quite usual for the leader to use only one or two runners on a pitch. A fall under such circumstances could take the climber down by as much as 50 m, putting enormous strain on both leader, second and rope. Even with a nylon rope such a fall could result in severe injury to the leader and would be a nasty experience for the second. However, the climbing ropes of the time were made from natural fibres such as hemp or manila. As we shall see in this case study such ropes are greatly inferior to those made from modern synthetic fibres like nylon or polypropylene. This meant that even a moderate fall

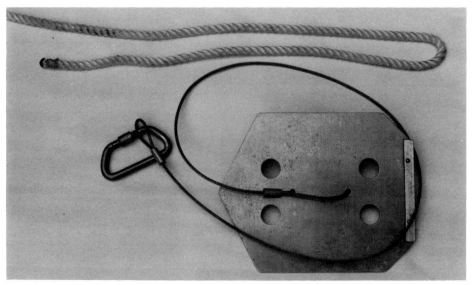

FIG. 1.1. The upper part of the photograph shows a piece of nylon climbing rope. The lower part shows a wire-rope sling and a karabiner, in this case attached to a snow anchor.

carried with it a definite risk that the rope would break. A bad fall was virtually certain to end in disaster unless the rope was in excellent condition and the second was adept at using the progressive "friction brake" method of cushioning the shock of the fall. A leader could not contemplate falling off, and the rope was there very much as a last resort. Routes that are straightforward and perfectly safe with modern protection methods were very serious undertakings before the days of nylon, and many climbers fell to their deaths as a result. The North Face of the Eiger, still an awe-inspiring and dangerous situation because of objective dangers like avalanches and falling stones, could be a gruesome place in the 1930s, with fragments of broken rope and even bits of people frozen in place on the face. So what is it that makes modern synthetic polymers so superior to natural fibres for climbing ropes?

Mechanical Properties of Ropes

Figure 1.2 shows a typical stress–strain curve for a multi-strand rope. At low stress the curve is shallow. This is because the twisted fibres in the rope straighten out a little as they are forced into close contact with one another. Then the curve steepens, and for most of the loading range the behaviour is linear elastic. The area under the curve is the elastic energy required to take the rope to the point of fracture and is equal to the maximum energy that can be absorbed by the rope. Neglecting the behaviour at low stress the energy absorption capability of the rope is given by

$$U_{max}^{el} = \frac{\sigma_{TS}^2}{2E}. \tag{1.1}$$

Figure 1.3 shows the stress–strain curve for the drag line of a spider. This is interesting because it is nature's evolutionary answer to the climbing-rope problem.

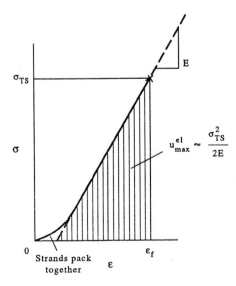

FIG 1.2   Typical stress–strain curve for a multi-strand rope illustrating the maximum energy absorption capability.

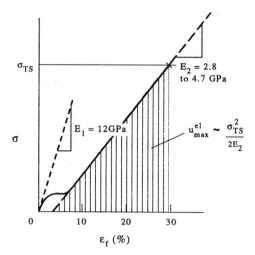

FIG 1.3   Stress–strain curve for spider drag line showing maximum energy absorption capability. The drag line is the thread which the spider uses to raise or lower itself during its normal movements. An adult female spider weighing 0.65 g can support a dead load of 1.0 g.

The drag line is a single filament of part amorphous, part crystalline organic polymers manufactured by the spider and consisting mainly of amino acids. The shape of the stress–strain curve at low stress is quite different from that of a multi-strand rope. However, the behaviour is still linear elastic over most of the loading range and the energy absorption capability is again given quite well by Eqn. (1.1).

Table 1.1 lists the properties of rope made from natural fibres, synthetic fibres and steel wire. It also gives data for spider drag line.

Because the drag line is a monofilament, its properties (σ_{TS}, ε_f, E, ρ) are those of

TABLE 1.1   *Properties of New Ropes (Data from Blackshaw, Lloyds Register and Wainwright et al.)*

Material	σ_{TS} (MPa)	ε_f (%)	E (GPa)	ρ (Mg m^{-3})
Natural fibres				
Manila grade 1	100			0.7
Sisal	90			0.7
Hemp	85	10	0.85	0.7
Synthetic fibres				
Nylon	264	40	0.66	0.6
Polyester	204			0.6
Polypropylene	174			0.45
Steel wire				
Rope	≈ 900	≈ 1–2	≈ 100	≈ 4
Single strand	≈ 1800	≈ 0.85	212	7.9
Spider drag line				
Single strand	870–1420	30	2.8–4.7(E_2)	1.26

the spun polymer itself. However, the properties of a multi-strand rope are not the same as those of the individual fibres. As the data for steel illustrate, the values of σ_{TS}, E and ρ for a rope are about half those for the fibres from which the rope is spun. This is due to two factors:

(a) The fibre material occupies only about 70% of the gross cross-section of the rope.
(b) The strands are wound helically and do not deliver a straight pull along the axis of the rope.

The table immediately points to one remarkable fact: spider drag line is nearly as strong as steel in its strongest, most work-hardened form.

Energy Absorption Capability

The data listed in Table 1.1 can be put into Eqn. (1.1) to find the values of U^{el}_{max} for the different ropes. Table 1.2 lists them in units of J cm^{-3} of rope (not material) volume.

As one might have expected, Darwin gets it right, with drag line having an impressive 135–216 J cm^{-3}. Nylon is a reasonable second with 52.8 J cm^{-3}. Hemp and steel are very poor, with only about 4 J cm^{-3}. Already we see why hemp climbing ropes are so dangerous, and why wire rope is only used for slings.

Climbers are concerned not only with the size of their ropes but also with their weight. Table 1.2 lists values for U^{el}_{max} in units of kJ kg^{-1}. On a weight basis steel drops to bottom position, being six times less efficient than hemp, and eighty-eight times less efficient than nylon.

Peak Loads

As well as being able to absorb the energy of a fall the rope must not exert an unacceptably high load on the body of the falling climber. It is no good if a rope absorbs the energy successfully if in the process it pulls the climber up so suddenly that it causes injury. If the rope successfully absorbs the energy of the fall then we can write

$$U^{el} = \frac{\sigma^2_{peak}}{2E},\tag{1.2}$$

where σ_{peak} is the maximum stress experienced by the rope during the arrest process.

TABLE 1.2 *Energy Absorption and Peak Loads*

Material	U^{el}_{max} (J cm^{-3})	U^{el}_{max} (kJ kg^{-1})	$\sigma_{peak}/\sigma_{peak}$ (nylon)
Hemp	4.25	6.1	1.1
Nylon	52.8	88	1
Steel	4.05	1	12
Spider drag line	135–216	107–171	2.1–2.7

TABLE 1.3 *Environmental Properties of Climbing Ropes*

Natural fibres	Nylon
Absorbs water	*Easily cut by rock*
Stiff and brittle when frozen	Protect rope slings with a sleeve of polyethylene or similar material.
Rots quickly if left damp	Inspect regularly for fraying and cuts.
Breaking strength can be reduced by six times.	Discard after 100 days use.
Rotting often invisible since occurs inside rope.	*Softens at 200°C and melts at 250°C*
Rope must be dried out after use and kept in a dry, well-ventilated space.	Do not use where frictional heat is concentrated on one part of the rope.
Recommended lifetime is only 6 months.	Forbidden for slings except with karabiners.
	Minimal water uptake
	Flexible when frozen.
	Does not rot.

Table 1.2 lists the values of σ_{peak} for the different ropes, all normalised by the peak stress for a nylon rope. Hemp and nylon give essentially the same peak stress. Spider drag line gives a peak stress that is about twice this. But steel rope is awful, with a peak stress twelve times that of hemp or nylon: it is totally unacceptable as a main climbing rope on this basis alone.

Environmental Behaviour

In addition to their poor mechanical performance, natural fibre ropes suffer from environmental degradation. Their environmental properties are compared with those of nylon in Table 1.3. Because natural fibres absorb water readily they are prone to rotting unless they are dried out immediately after use and kept in a dry well-aired place. Rotting usually takes place inside the rope and is not easily visible. A rope may lose five-sixths of its strength with age and the recommended lifetime is only 6 months.

Mechanics of the Arrest Process

Table 1.4 gives details of the ropes actually used in climbing. The standard modern nylon rope is 12 mm in diameter and has a breaking load T_f of 3 tonnef. When hemp was used, 12-mm ropes were often doubled up for extra strength, giving a total T_f of 2 tonnef. Wire rope slings have a breaking load of about 1.8 tonnef.

The mechanics of the arrest process are summarised in Figs. 1.4 and 1.5. By equating energy terms we can write

$$M_f g(a + u_f) = \frac{T_f u_f}{2},$$ (1.3)

TABLE 1.4   *Strengths of Climbing Ropes. Values of the safety factor M_f/M. Human body mass $M = 90$ kg; spider body mass $M = 0.65 \times 10^{-3}$ kg*

Rope type	T_f (kgf)	$a/l = 0$	$a/l = 0.1$	$a/l = 0.5$	$a/l = 2$
Single 12 mm nylon	3000	16	13	7.0	2.6
Double 12 mm hemp	2×1000	11	5.5	1.9	0.6
Old double 12 mm hemp	$\approx 2 \times 170$	1.9	0.9	0.3	0.1
Single 5 mm steel (slings only)	1800	10	0.9	0.2	0.05
Spider drag line	1.0×10^{-3}	0.75	0.56	0.28	0.10

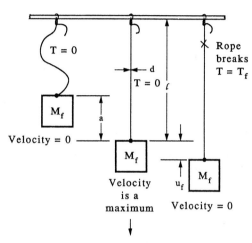

FIG 1.4   Sequence showing how a rope arrests the free fall of a body of mass M. At a critical value of mass equal to M_f the energy released by the falling body is just enough to break the rope.

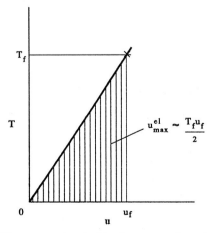

FIG 1.5   The energy absorbed in the rope at the point of failure.

where M_f is the falling mass which is just sufficient to break the rope. Now

$$u_f = \varepsilon_f l. \tag{1.4}$$

Thus

$$M_f = \left(\frac{1}{2g}\right)T_f\left\{\frac{1}{\varepsilon_f}\left(\frac{a}{l}\right) + 1\right\}^{-1}, \tag{1.5}$$

where

$$T_f = A\sigma_{TS} = \left(\frac{\pi d^2}{4}\right)\sigma_{TS}. \tag{1.6}$$

Figure 1.6 plots the term

$$\left\{\frac{1}{\varepsilon_f}\left(\frac{a}{l}\right) + 1\right\}^{-1} = \frac{2M_f g}{T_f} \tag{1.7}$$

as a function of a/l for nylon, spider drag line, hemp and wire rope. The plots are truncated at $a = 2l$ which is the maximum allowable value. They show that, for a given breaking load T_f, the falling mass required to break the rope decreases as a/l increases. The decrease in M_f with a/l is most pronounced for materials with a small ε_f, like wire rope (1%), and least pronounced for materials with a large value of ε_f like nylon (40%). Hemp does not perform very well because of its rather small extensibility (10%). What is interesting is that, as a/l tends to zero, the term

$$\left\{\frac{1}{\varepsilon_f}\left(\frac{a}{l}\right) + 1\right\}^{-1} \to 1 \tag{1.8}$$

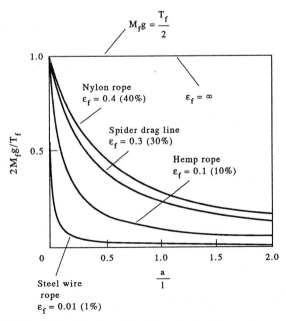

FIG 1.6   Plot of the term $2M_f g/T_f$ against a/l for various ropes.

for all materials. This means that

$$\frac{2M_f g}{T_f} \to 1 \tag{1.9}$$

and

$$M_f g \to \frac{T_f}{2}. \tag{1.10}$$

In other words, a shock loading, no matter how small the value of a, loads the rope to twice the dead load of the falling mass. This unexpected result occurs in a variety of dynamic loading situations and is the reason why grab cranes have a "duty factor" of 2 incorporated into the rules for their design.

Safety Factors of Climbing Ropes

We are now in a position to find out what every climber wants to know: what is the margin of safety in a fall? This can be quantified as the ratio M_f/M, where M is the body weight of the climber. If $M_f/M > 1$, there is a margin of safety on body weight; if $M_f/M < 1$, the rope will break. Values of M_f/M are listed in Table 1.4 for various values of a/l.

The assumed body weight is 90 kg for a climber, and 0.65 g for a spider. Nylon has a factor of safety of 2.6 in the worst fall of $a = 2l$. Under the same conditions hemp will break, and only an expertly executed friction brake can save the leader. The situation with old hemp is even worse, and a small fall by the second (1 m on a 10-m length) can result in breakage. 5-mm wire rope is also unsafe for values of $a/l \geq 0.1$; its use in slings is possible because the nylon rope absorbs the shock loading.

What is surprising is that the spider drag line seems to be unsafe for all values of a/l. The reason is probably that the spider never falls freely as such but lets out and takes in thread in a more controlled manner. Even so, spiders can make quite sudden movements; but they obviously stop short of breaking the thread.

Modern Uses of Natural Ropes

One would have imagined that, with the drawbacks explained above, natural ropes would by now have been made obsolete. This is not the case, however, and they are still used quite widely, for example on ships and in dockyards. Presumably they are cheaper than synthetic ropes. It is therefore not surprising that accidents involving hemp and manila ropes continue to occur even though the materials and the knowledge to prevent them have been in place for 50 years.

Devices for Absorbing Energy

The linear spring is not a very good device for absorbing energy. As Fig. 1.7 shows, the energy that is absorbed by a spring at a peak force F_{peak} can be absorbed by a constant retarding force that is only $0.5 \times F_{peak}$.

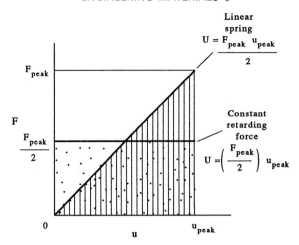

FIG 1.7   Energy absorption by a constant retarding force compared to energy absorption by a linear spring.

There is, therefore, less risk of damage to vehicles or people if they can be slowed down by a constant force.

One can think of various examples drawn from the engineering world. The spring buffers between railway wagons and carriages are obviously not optimal devices. For this reason the buffer stops at the end of the platforms in a main-line terminus usually employ hydraulic cylinders instead. These can be controlled to give constant-force characteristics. If the brakes on a car are applied with a constant force (and if the friction coefficient is independent of speed and temperature) then this, too, is a constant-force situation and the car will slow down with a constant deceleration. The sand drags in the escape lanes on steep mountain roads give a constant retarding force: the resistance that the sand offers to being furrowed by the wheels of the runaway vehicle. The crumple zones at the front and rear of a car use the *plastic* load–deflection curve of the structure to absorb a large amount of energy at roughly constant force. The advantage of a spring, of course, is that it springs back once it has done its job and is automatically ready for the next impact; hydraulic buffers have to be re-set, sand drags re-levelled and crashed cars scrapped.

References

A. Blackshaw, *Mountaineering*, Penguin, 1970.
Lloyds Register of Shipping: "Code for Lifting Appliances in a Marine Environment", 1984.
S. A. Wainwright, W. D. Biggs, J. D. Currey and J. M. Gosline, *Mechanical Design in Organisms*, Arnold, 1976.

CHAPTER 2

WHIRLING FAILURE IN A WOODWORKING LATHE

Background

Figure 2.1 shows a woodworking lathe which is set up for drilling a $\frac{1}{4}$-inch hole all the way along the axis of a wooden cylinder. The wood is sycamore, and the drilling of the $\frac{1}{4}$-inch hole is the first stage in producing the chanter of a set of bagpipes.

The type of drill that is being used is known as a D-bit. The cutting action of the D-bit is shown schematically in Fig. 2.2. Now a D-bit, although an excellent tool for use with metal, is not really suitable for drilling long holes in wood. Cutting speeds for wood are much higher than they are for metal, and the rubbing friction between the wood and the shank of the D-bit warms everything up. As a result, the D-bit tends to stick in the hole and it can be difficult to pull it out again.

The drill is held in a drill chuck which has a tapered arbor, as shown in Fig. 2.3. The arbor is inserted into the tapered socket of the tailstock barrel and is held in place by friction. The drill is fed into the rotating wood by turning the handwheel on the end of the tailstock. This is the standard arrangement for drilling small holes with either wood or metal lathes. Naturally, wood shavings build up in the space between the D-bit and the side of the hole as shown in Fig. 2.2, and the drill has to be retracted frequently in order to get rid of them. When drilling a long hole this is tedious and the temptation when drilling wood, where the cutting forces are

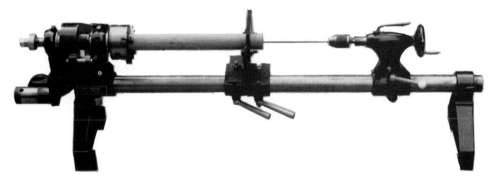

FIG 2.1   Photograph of the woodworking lathe ready for the drilling operation.

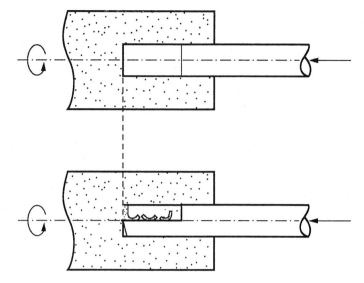

FIG 2.2   How the D-bit cuts the pilot hole.

FIG 2.3   Photograph of the D-bit and the drill chuck used to grip it.

low, is simply to push the whole tailstock bodily along the lathe bed and not to use the handwheel at all.

This procedure was being used at the time of the accident. Unfortunately, the D-bit had started to seize in the hole, and when the tailstock was pulled away in order to retract the drill, the tapered arbor of the drill chuck pulled out of the tailstock socket leaving the drill chuck unsupported. The drill and chuck immediately began to rotate at the speed of the workpiece, the drill bent as shown in Fig. 2.4, and the chuck shot off the end of the drill at high speed, narrowly missing the lathe operator.

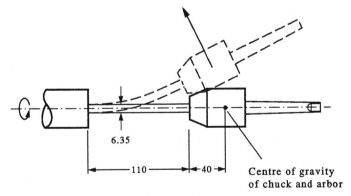

6.35

110 40

Centre of gravity
of chuck and arbor

FIG 2.4   Schematic of the failure. Dimensions in mm.

Failure Mechanism

The accident seems to be a classic example of a "whirling" failure. In order to understand why whirling occurs it is a good idea to run through the theory first. Figure 2.5(a) shows the present situation modelled by a round rod of length l and diameter $2r$, built-in at one end and having a mass M attached to the other. The rod rotates about its axis with angular velocity ω. Provided ω is not too large, the rod will not want to move away from the axis of rotation and the system will be stable. If the angular velocity is high enough, however, the rod will become unstable with respect to a small deflection δ from the axis of rotation (see Fig. 2.5(b)). The rod will fly out sideways and become permanently bent.

The angular velocity at which this instability, or "whirling", takes place is called the critical angular velocity, ω_{cr}. As shown in Appendix 1B, the sideways force that must be applied to the end of the beam to make it deflect by amount δ is given by

$$F = \frac{3EI\delta}{l^3}. \tag{2.1}$$

At the critical angular velocity this force is just provided by the centrifugal force on the mass, i.e.

$$F = M\delta\omega^2. \tag{2.2}$$

Thus

$$\frac{3EI\delta}{l^3} = M\delta\omega_{cr}^2. \tag{2.3}$$

This gives

$$\omega_{cr} = \left(\frac{3EI}{Ml^3}\right)^{1/2}. \tag{2.4}$$

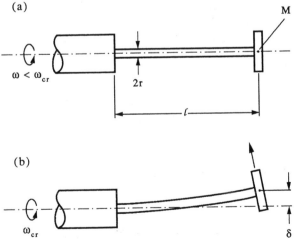

FIG 2.5   Whirling of a weightless rod of diameter $2r$ with a mass M attached to the free end. ω_{cr} is the critical angular velocity for whirling.

The critical number of revolutions per second, f_{cr}, is then given by

$$f_{cr} = \frac{\omega_{cr}}{2\pi} = \frac{1}{2\pi}\left(\frac{3EI}{Ml^3}\right)^{1/2}. \tag{2.5}$$

Appendix 1B gives the value of I for a round bar as

$$I = \frac{\pi r^4}{4}. \tag{2.6}$$

Substituting this result in Eqn. (2.5) finally gives us

$$f_{cr} = \left(\frac{3}{16\pi} \times \frac{Er^4}{Ml^3}\right)^{1/2}. \tag{2.7}$$

Calculations

We begin by assembling information for the parameters in Eqn. (2.7). The D-bit was made from medium carbon steel, for which $E = 212$ GPa (data from Smithells). The mass of the drill chuck was found to be 0.42 kg. The centre of gravity of the chuck was 40 mm from the nose (see Fig. 2.4). The exposed length of drill was about 110 mm, giving an effective value for l of about 150 mm (see Fig. 2.4). Finally, $r = 6.35$ mm$/2 = 3.17$ mm (see Fig. 2.4). Thus

$$f_{cr} = \left\{\frac{3}{16\pi} \times \frac{212 \times 10^9 \text{ N m}^{-2}}{0.42 \text{ kg}} \left(\frac{3.17 \text{ mm}}{150 \text{ mm}}\right)^3 3.17 \times 10^{-3} \text{ m}\right\}^{1/2}. \tag{2.8}$$

Since Eqn. (2.2) is a version of Newton's second law, the units of newtons in Eqn. (2.8) are equivalent to kg m s^{-2}. The units of f_{cr} are thus s^{-1}, with

$$f_{cr} = 30 \text{ revolutions/second} = 1800 \text{ rpm}. \tag{2.9}$$

The lathe motor has a speed of 2850 rpm. The lathe spindle is driven from the motor by a V-belt and a pair of cone pulleys giving three different spindle speeds. At the time of the accident, the pulleys in use were the smallest one on the motor (mean diameter 58 mm) and the largest one on the spindle (mean diameter 83 mm) so the spindle speed was

$$f = 2850 \times \frac{58}{83} = 1992 \text{ rpm}. \tag{2.10}$$

This is just a little higher than f_{cr}, which gives quantitative proof that whirling was indeed responsible for the failure.

Development of the Instability

Although the instability began as an elastic event, δ eventually increased to the point at which the rod began to deform plastically. The instability was then irreversible and a plastic run-away ensued, leading to gross permanent bending of the rod.

Yielding first took place where the drill emerged from the workpiece. The deflection δ_y at which yielding began can be found from the following equations (see Appendix 1B).

$$\sigma_y = \frac{M_{Ay} r}{I}. \tag{2.11}$$

$$M_{Ay} = F_y l. \tag{2.12}$$

$$\delta_y = \frac{F_y l^3}{3EI}. \tag{2.13}$$

These combine to give

$$\delta_y = \frac{1}{3}\left(\frac{\sigma_y}{E}\right)\frac{l^2}{r}. \tag{2.14}$$

σ_y is roughly 400 MPa (Smithells) so

$$\delta_y \approx \frac{1}{3}\left(\frac{400 \text{ MPa}}{212 \times 10^3 \text{ MPa}}\right)\frac{150^2 \text{ mm}^2}{3.17 \text{ mm}} \approx 5 \text{ mm}. \tag{2.15}$$

This means that, until the end of the drill had deflected by about 5 mm, the instability was entirely elastic. Thereafter, a plastic hinge developed where the drill emerged from the workpiece, allowing the run-away to take place.

Design of Prop Shafts

The critical whirling speed is an important factor in the design of many rotating components. A good example is the prop shaft of a truck or automobile. This is usually supported at each end by a universal joint. As shown in Fig. 2.6, the shaft tends to bow out between the end mountings, driven by the centrifugal force of the mass of the shaft itself. Appendix 1B gives the critical speed for this situation as

$$f_{cr} = 1.571\left(\frac{EI}{Ml^3}\right)^{1/2}. \tag{2.16}$$

Now having a solid round bar is a very inefficient way of producing bending stiffness and in fact the optimum way of distributing the material of the shaft is in the

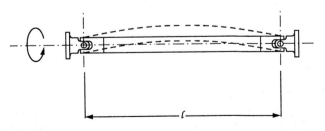

FIG 2.6   Whirling of a prop shaft.

form of a thin-walled tube. Then

$$I = \pi r^3 t,$$ (2.17)

$$M = 2\pi r t l \rho,$$ (2.18)

$$\frac{I}{M} = \frac{r^2}{2l\rho},$$ (2.19)

giving

$$f_{cr} = 1.571 \left(\frac{r^2}{2l^4} \times \frac{E}{\rho} \right)^{1/2}.$$ (2.20)

This reduces to

$$r = \sqrt{2}l^2 \left(\frac{f_{cr}}{1.571} \right)^2 \left(\frac{\rho}{E} \right)^{1/2}.$$ (2.21)

l and f_{cr} are fixed by the design, so the most compact shaft is obtained by selecting a material to minimise $(\rho/E)^{1/2}$. Values of minimum allowable tube diameter d ($=2r$) are given in Table 2.1 for $f_{cr} = 4000$ rpm (66.7 s^{-1}) and $l = 2$ m.

Although steel is the orthodox material for prop shafts, it is clear from Table 2.1 that a shaft made from CFRP is much more compact. However, GFRP is not an attractive proposition, as it is even bulkier than steel. Since prop shafts have to transmit power, there is an additional requirement that the prop shaft must support a specified torsional moment Γ without making the cross-section of the tube deform plastically. As shown in Fig. 2.7 the value of Γ which will just cause the section to yield is given by

$$\Gamma \approx (2\pi r t)(\sigma_y/2)r$$ (2.22)

TABLE 2.1

Material	ρ (Mg m^{-3})	E (GPa)	$(\rho/E)^{1/2}$	d (mm)
Steel	7.8	212	0.19	92
Aluminium	2.7	69	0.20	95
GFRP	2.0	30	0.26	124
CFRP	1.5	120	0.11	54

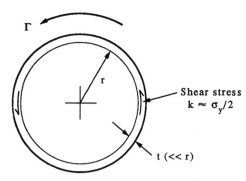

FIG 2.7   Torsional strength of the prop shaft.

TABLE 2.2

Material	E (GPa)	ρ (Mg m^{-3})	σ_y (MPa)	$\left(\dfrac{E^{1/2}\rho^{1/2}}{\sigma_y}\right)$
Mild steel	212	7.8	220	0.19
High-strength steels	212	7.8	400–1000	0.10–0.041
Aluminium alloy	69	2.7	300	0.046
GFRP	30	2.0	200	0.039
CFRP	120	1.5	700	0.019

from which

$$t \approx \frac{\Gamma}{\pi r^2 \sigma_y}. \tag{2.23}$$

Combining Eqns. (2.18), (2.21) and (2.23) we get

$$M \approx \sqrt{2}\left(\frac{\Gamma}{l}\right)\left(\frac{1.571}{f_{cr}}\right)^2\left(\frac{E^{1/2}\rho^{1/2}}{\sigma_y}\right). \tag{2.24}$$

The lightest prop shaft is thus obtained by minimising $(E^{1/2}\rho^{1/2}/\sigma_y)$. Typical values are given in Table 2.2.

Table 2.2 shows that CFRP gives the lightest shaft. The nearest competitors are GFRP, aluminium alloy and the top range of high-strength steels, but these are all about twice the weight (and, as Table 2.1 shows, twice the size) of a CFRP shaft. This is why CFRP is increasingly being used as a replacement for steel in the prop shafts of trucks, where the additional expense is justified.

A Note on Vibration Frequencies

We now return to the drill and chuck which introduced us to the phenomenon of whirling. The equation for the critical whirling speed for this situation was

$$f_{cr} = \frac{1}{2\pi}\left(\frac{3EI}{Ml^3}\right)^{1/2} \tag{2.25}$$

and the force required to deflect the end of the beam by amount δ was

$$F = \frac{3EI\delta}{l^3}. \tag{2.26}$$

The stiffness of the beam is given by

$$S = \frac{F}{\delta} = \frac{3EI}{l^3}. \tag{2.27}$$

Thus Eqn. (2.25) may be rewritten as

$$f_{cr} = \frac{1}{2\pi}\left(\frac{S}{M}\right)^{1/2}. \tag{2.28}$$

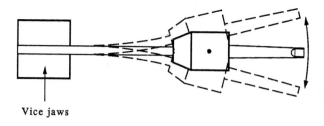

Vice jaws

FIG 2.8   Transverse mode 1 vibrations of the drill and chuck. Note that these components
are *not* rotating about the axis of the drill.

This is the standard result for the natural frequency of transverse vibration (mode 1) of the system. In other words, if we had gripped the drill and chuck in a vice as shown in Fig. 2.8 and "twanged" the end, the chuck would have vibrated from side to side with a fundamental frequency equal to f_{cr}. However, the physics of whirling is quite different from that of transverse vibration, and it is only a mathematical coincidence that the equations are the same for both.

Appendix 1B lists expressions for f_{cr} for six different situations. Obviously, these results can be used both for finding critical whirling speeds and mode 1 transverse vibration frequencies.

When components or structures are made to vibrate at their natural frequencies, failures can occur, for example by fatigue. If the amplitude of vibration is large enough, gross structural damage can result. This phenomenon is at its most dramatic in earthquakes. An earthquake usually has a characteristic frequency, or set of frequencies, which depend on the event that gave rise to the tremor and the geometry and properties of the geological strata through which the disturbance travels. It is quite common, for example, for some buildings to suffer total collapse in an earthquake when apparently similar buildings next door suffer only minor damage. This is because each building has its own natural frequency which depends on its mass and stiffness: the buildings which collapse have a natural frequency which couples into one of the frequencies of the tremor and they are shaken to pieces as a result. Because of this there is great interest in trying to predict the characteristics of earthquakes in the various seismically active areas in the world and designing new buildings so they do not couple into the expected frequencies of the tremors.

Reference

C. J. Smithells, *Metals Reference Book*, 6th edition, Butterworth, 1984.

CHAPTER 3

LOOSE WHEELS ON A RAILWAY LOCOMOTIVE

Background

Figure 3.1 is a photograph of a steam-powered railway locomotive of the Standard Class 7 type built by British Railways between 1951 and 1954. Railways in the United Kingdom were nationalised in 1948 and the new administration decided to bring out a set of twelve new designs of steam locomotive, called the Standard types, to replace the older locomotive types inherited from the pre-nationalisation railway companies. The Class 7 were the prestige locomotives of the new British Railways and, although they were officially classed as mixed-traffic engines, they were often used on the top express passenger trains.

The locomotives used the "Pacific" wheel arrangement, which is denoted by "4-6-2" according to the standard classification of steam locomotive wheel arrangements. It consists of a four-wheeled carrying bogie at the front of the locomotive, followed by six coupled wheels driven by the connecting and coupling rods, and a two-wheeled carrying truck at the rear. The front bogie has two axles, with a wheel pressed on to each end of both axles, giving the total of four bogie wheels. There are three coupled axles with a wheel pressed on to each end, giving the total of six coupled wheels. Finally, the trailing truck consists of one axle with a wheel pressed on to each end.

The class consisted of fifty-five locomotives, numbered from 70 000 to 70 054. Shortly after the class had been introduced, no. 70 014 "Iron Duke" failed in service. It was found that all six coupled wheels had shifted on their axles and the front coupling rods had bent as a result. The failure caused some concern, because "Iron Duke" had only been finished one month earlier. Over the next year a further six engines failed, culminating in a particularly embarrassing and widely publicised incident involving no. 70 004 "William Shakespeare" whilst hauling the "Golden Arrow" Pullman boat-train express. All twenty-five Class 7 engines then in service were immediately withdrawn from service for investigation and modification.

It was surprising that no. 70 005 "John Milton" had been tested to the limits of performance on both the main line and the Rugby stationary testing plant without any problems, and many other engines had experienced severe main-line service with complete success. The evidence pointed to a variable degree of fit between the axles and the pressed-on coupled wheels, with wheels on some locomotives not

21

FIG 3.1   A working model of the British Railways Standard Class 7 steam locomotive. The full-size engines were built at Crewe works from 1951 to 1954 and were withdrawn from service between 1965 and 1968. The model runs on a gauge of $3\frac{1}{2}$ inches. It is coal-fired just like the prototype and can easily pull ten adult passengers. Two of the full-size engines are preserved in a fully operational condition: "Oliver Cromwell", and the first of the class, no. 70 000 "Britannia".

adequately held in place. In this Case Study we attempt a historical reconstruction to try to establish what the causes of the failures might have been.

Wheel-Axle Assembly

Figure 3.2 is a photograph of one of the coupled wheels and Fig. 3.3 is a drawing of the wheel-axle assembly. Figure 3.4 shows a simplified version of the wheel that we will be using later on for modelling the elastic behaviour of the assembly. The steel axle is hollow, both to save weight and to allow coolant to flow down the centre during the quenching stage of the heat treatment process.

The diameter of the axle at the wheel seat is slightly less than it is at the bearing seat. As a result the inner race of the roller bearing can be passed over the wheel seat but is made a press fit on the bearing seat. The coupled wheel is a press fit on the wheel seat. It is also keyed in place, which prevents it rotating on the axle; but it relies on the press fit to stop it coming off the axle or developing a wobble.

The wheel itself is a single cast-iron casting consisting of a hub joined to a rim by twenty spokes. The hub is oval in shape to give room for the crank-pin hole. A separate flanged steel tyre is fixed over the rim of the wheel. The wheel casting contains a large balance weight which is intended to produce a static counterbalance

FIG 3.2   Close-up showing two of the coupled wheels.

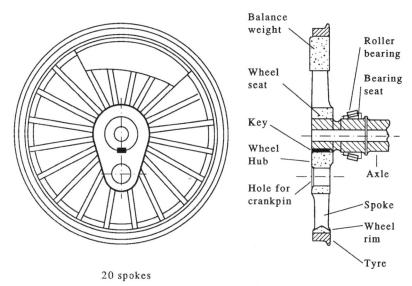

20 spokes

FIG 3.3   Drawing of the wheel-axle assembly. Both axle and tyre are made from medium-carbon steel. The wheel itself is cast iron.

to the weight of the coupling and connecting rods which hang on the crank pin.

The normal procedure when "wheeling" an engine is as follows. First, the centre of the hub is bored out to a diameter slightly less than that of the wheel seat on the axle. The surfaces are lubricated and the wheel is pressed on to the axle using

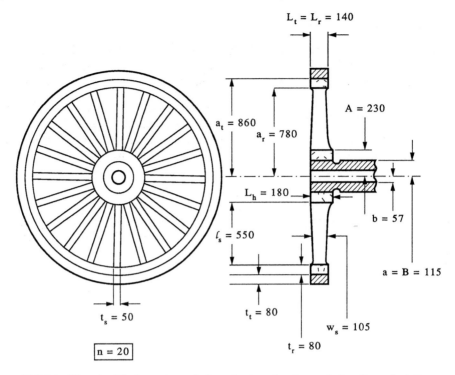

FIG 3.4   The simplified version of the wheel and axle used for the calculations.
Dimensions in mm.

a pressing force usually equal to 0.4 tonnef per mm of axle diameter (see the contemporary edition of the ASME Design Handbook). The key is inserted at the same time to ensure proper alignment between the two keyways. The wheel and axle are then rotated in a wheel lathe and the outside of the wheel rim is turned true.

The steel tyre is swung in a large lathe and the inside is bored out to a diameter that is less than the diameter of the wheel rim by about $1.3 \times 10^{-3} \times$ (rim diameter). The tyre is then heated until it has expanded enough that it can be dropped over the rim. The tyre is left to cool slowly in air to room temperature: as it cools it contracts on to the rim and puts a considerable pressure on the wheel. This compresses the hub even further on to the axle and gives a big increase in the interference fit between wheel and axle.

On the Class 7 a slightly different wheeling procedure was used. The wheel was first pressed on to a dummy axle which had been machined to give only a small interference between the hub and the wheel seat. The tyre was then fitted and the tyred wheel pressed off the dummy axle. Finally, the wheel was pressed on to the real axle. It is not obvious why this was done, but the reason might have been to stop cutting oil and swarf finding their way into the roller bearings during the turning operations.

Wheel-Axle Fit

Figure 3.5 is a schematic drawing of the press fit between the hub and the axle. The radial interference between the two components is e. In order to get the axle to end up inside the hub, we have to go through the following conceptual sequence.

(a) Apply a radial pressure p to the outside of the axle. The axle will shrink slightly by elastic deflection. The radial shrinkage of the wheel seat is Δa.

(b) Apply the same radial pressure p to the inside of the hub. The hub will expand slightly by elastic deflection. The radial expansion of the bore is ΔB.

(c) The value of p needed is determined by the requirement that

$$e = \Delta a + \Delta B. \tag{3.1}$$

In practice, as Fig. 3.6 shows, the pressure on both components is generated automatically when the hub is pressed on to the axle by the axial pressing force F. Both hub and axle are chamfered as shown to allow the axle to enter

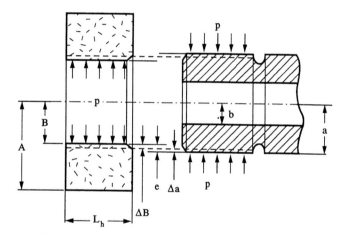

FIG 3.5   Schematic of the press fit between hub and axle.

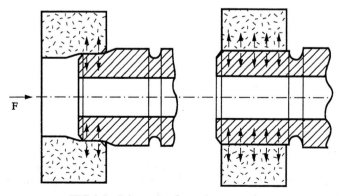

FIG 3.6   Schematic of pressing operation.

the hub at the start of the pressing operation. Once entry has occurred, "bow waves" of radial elastic deformation move along the wheel seat and the hub bore as shown in Fig. 3.6 until the wheel has been pressed on fully.

The pressing force is related to the radial pressure by the friction law, with

$$F = \mu R = \mu 2\pi a L_h p. \tag{3.2}$$

μ is the coefficient of friction and R is the normal reaction force between hub and axle. We know already that $a = 115$ mm, $L_h = 180$ mm and $F \approx 0.4$ tonnef per mm of axle diameter. The ASME Handbook gives $\mu \approx 0.15$ for wheel-pressing operations. Thus

$$F \approx 0.4 \text{ tonnef mm}^{-1} \times 2 \times 115 \text{ mm} \approx 92 \text{ tonnef} \tag{3.3}$$

and

$$p = \frac{F}{\mu 2\pi a L_h} \approx \frac{92 \times 10^3 \times 9.81 \text{ N}}{0.15 \times 2\pi \times 115 \text{ mm} \times 180 \text{ mm}}$$

$$= 46 \text{ N mm}^{-2} = 46 \text{ MPa}. \tag{3.4}$$

Since we now know p we are in a position to calculate Δa and ΔB using the equations for the elastic behaviour of thick pressure vessels that are listed in Appendix 1A.

For the axle,

$$\Delta a = \left(\frac{p}{E}\right) a \left\{\left(\frac{a^2 + b^2}{a^2 - b^2}\right) - v\right\}. \tag{3.5}$$

For medium-carbon steel $E = 212$ GPa and $v = 0.29$ (data from Smithells). In addition $a = 115$ mm and $b = 57$ mm. Putting these values into Eqn. (3.5) gives

$$\Delta a \approx 34 \text{ μm}. \tag{3.6}$$

For the hub,

$$\Delta B = \left(\frac{p}{E}\right) B \left\{\left(\frac{A^2 + B^2}{A^2 - B^2}\right) + v\right\}. \tag{3.7}$$

For cast iron $E = 152$ GPa and $v = 0.27$ (data from Smithells). $A = 230$ mm and $B = 115$ mm so we get

$$\Delta B \approx 67 \text{ μm}. \tag{3.8}$$

Thus

$$e = \Delta a + \Delta B \approx 101 \text{ μm}. \tag{3.9}$$

This gives an interference on *diameter* of about 202 μm which, at 8 thousandths of an inch, is typical of the fits given on locomotive works drawings of the period.

Effect of Spokes and Rim

Of course, when we applied Eqn. (3.7) to the hub, we assumed that the spokes and rim had no significant stiffening effect on the hub. This assumption needs to be checked before we can go any further. If the hub were totally free of external constraint we would be able to write

$$\Delta A = \left(\frac{p}{E}\right)\left(\frac{2AB^2}{A^2 - B^2}\right) \tag{3.10}$$

using the appropriate result in Appendix 1A. Given that $p = 46$ MPa, $E = 152$ GPa, $A = 230$ mm and $B = 115$ mm we then get

$$\Delta A = 46 \ \mu\text{m} \tag{3.11}$$

for the radial expansion of the outside of the hub. The question then is, what force is generated if this outwards movement is applied to the inner end of each spoke, and is this force significant or not?

Figure 3.7 shows schematically what happens. The axial force that the hub applies to the end of each spoke is f_s. This force compresses the spoke elastically, making it shorter by u_s. The rim is stretched elastically by the radial force from the spokes and its radius grows by u_r. We then have

$$\Delta A = u_s + u_r. \tag{3.12}$$

Now the compressive stress in the spoke is given by

$$\sigma_s = \frac{f_s}{t_s w_s}. \tag{3.13}$$

The strain in the spoke is

$$\varepsilon_s = \frac{u_s}{l_s} = \frac{\sigma_s}{E}. \tag{3.14}$$

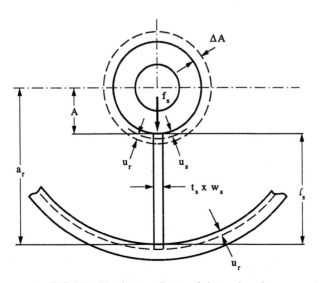

FIG 3.7   Elastic compliance of rim and spokes.

These equations can be combined to give

$$u_s = \frac{l_s f_s}{t_s w_s E}. \tag{3.15}$$

We can estimate u_r by assuming that the spokes apply an equivalent hydrostatic pressure to the inside of the rim, and treating the rim as a piece of thin-walled tube of length L_r. If the number of spokes is n, the equivalent pressure is

$$p_r = \frac{n f_s}{2\pi a_r L_r}. \tag{3.16}$$

The tensile hoop stress in the rim is given by

$$\sigma_r = \frac{p_r a_r}{t_r} = \frac{n f_s}{2\pi a_r L_r} \times \frac{a_r}{t_r}. \tag{3.17}$$

The strain in the rim is given by

$$\varepsilon_r = \frac{2\pi(a_r + u_r) - 2\pi a_r}{2\pi a_r} = \frac{u_r}{a_r}. \tag{3.18}$$

These equations combine to give

$$u_r = \varepsilon_r a_r = \frac{\sigma_r a_r}{E}$$

$$= \frac{n f_s}{2\pi a_r L_r} \times \frac{a_r}{t_r} \times \frac{a_r}{E} = \frac{n f_s a_r}{2\pi L_r t_r E}. \tag{3.19}$$

Thus

$$\Delta A = \frac{l_s f_s}{t_s w_s E} + \frac{n f_s a_r}{2\pi L_r t_r E}. \tag{3.20}$$

The reaction from the force in each spoke applies an equivalent hydrostatic pressure to the outside of the hub, given by

$$p_h = \frac{n f_s}{2\pi A L_h}. \tag{3.21}$$

Finally, the radial "pressure compliance" of the spoke–rim system is equal to

$$\frac{\Delta A}{p_h} = \frac{\dfrac{l_s f_s}{t_s w_s E} + \dfrac{n f_s a_r}{2\pi L_r t_r E}}{\dfrac{n f_s}{2\pi A L_h}}$$

$$= \frac{2\pi A L_h l_s}{t_s w_s E n} + \frac{A L_h a_r}{L_r t_r E}. \tag{3.22}$$

We use parameter values $A = 230$ mm, $L_h = 180$ mm, $l_s = 550$ mm, $t_s = 50$ mm, $w_s = 105$ mm, $E = 152$ GPa, $n = 20$, $a_r = 780$ mm, $L_r = 140$ mm and $t_r = 80$ mm to get

$$\frac{\Delta A}{p_h} = 28 \ \mu m \ MPa^{-1}. \tag{3.23}$$

We can see from Eqn. (3.23) that when $\Delta A = 46 \ \mu m$, as it is in our case, then $p_h = 1.6 \ MPa$. This is small compared to the pressure of 46 MPa which acts on the inside of the hub, so the effect of the spokes and rim can safely be neglected. Physically, this is because the spoke-rim structure has a much greater elastic compliance than the hub.

Effect of Hollow Axle

One of the explanations suggested at the time of the incidents was that they were caused by the fact that the axles were hollow. Obviously, a hollow axle has a larger elastic compliance than a solid one. Thus a given interference $2e$ will result in a lower contact pressure p and a less effective grip with a hollow axle. The circumstantial evidence for this, however, is not strong. In fact, hollow axles had been used long before, apparently with complete success, for example on the Great Western Railway (Stars and Kings) and the London, Midland and Scottish (Stanier Pacifics and Class 8F). In this section we use elastic analysis to see whether the hollow axles could have had a role in the failures.

For a solid axle $b = 0$ and Eqn. (3.5) becomes

$$\Delta a = \left(\frac{p}{E}\right) a(1 - v). \tag{3.24}$$

Setting $p = 46 \ MPa$ as before we get $\Delta a = 18 \ \mu m$. This is less than the radial deflection for the hollow axle by $34 - 18 = 16 \ \mu m$. So the diameter of a solid axle must be made 32 μm less than that of a hollow axle if the contact pressure is to be kept at 46 MPa.

Now the machining tolerance on both the bore of the hub and the diameter of the wheel seat is probably no better than $\pm 15 \ \mu m$, so the variation in fit could be as much as $\pm 30 \ \mu m$. This is comparable to the difference of 32 μm between the hollow and solid axles. The hollow-axle scenario does not therefore seem to be very likely. In spite of this, when the Class 7 engines were all re-wheeled after the failures, the bores of all the axles were plugged for the length of the wheel seat. Our analysis suggests that this was not necessary. We must wonder whether the design staff ever did the calculations, or whether they did and were just "playing safe".

Effect of Tyre

As we have said already, in the usual method of assembly the wheel is pressed on to the axle first, and the tyre is shrunk on afterwards, increasing the effective interference between the wheel and the axle. But by how much does the tyring operation actually increase the fit? This problem can be solved using a method rather like the one that we have already used in determining the behaviour of the spokes and rim. We will not give details of the calculations here, but they are set out in the Solutions to Examples 3.1 and 3.2. The sums show that adding the tyre decreases the internal radius of the hub by 70 µm. This gives a total interference on diameter of

$$2e = 2(70 \text{ µm} + 101 \text{ µm}) = 342 \text{ µm}. \tag{3.25}$$

This is a good deal more than the original value of 202 µm. As we can see from Eqns. (3.1), (3.5) and (3.7) this will push the contact pressure up by a factor of (342/202) to a value of 78 MPa.

Effect of Dummy Axle

As we know already, the assembly method used on the Class 7 engines was unusual because the wheels were pressed on to a dummy axle, then tyred and finally transferred to the real axle. This variation must be a prime suspect for the failures.

The ASME Handbook gives the results of experiments in which steel wheels were repeatedly pressed on and off their axles. The interference of the fit was measured before each pressing. The axles had diameters of 150 to 200 mm, quite close to the diameter of the Class 7 axles. The results are given in Table 3.1.

What is clear from the table is that the first pressing reduced the interference by about 30%. A further five pressings led to an additional reduction of only 15%. Virtually all of the metal lost came from the bore of the wheel. Of course, our cast iron wheels may have behaved differently from steel ones. There is a lot of graphite in the microstructure of cast iron and a machined cast iron surface has less tendency to seize up on steel because the graphite acts as a boundary lubricant.

However, the data on steel wheels does tend to support the idea that the interference of fit could have been reduced substantially when the wheels were pressed off the dummy axles.

Other Factors

We still have to explain why only seven out of the original batch of twenty-five locomotives failed and also why, when a particular locomotive failed, all six coupled wheels appear to have shifted. These statistics tend to indicate that there might have been systematic variations from one engine to another, or one batch to another. Were there differences in the lubrication conditions? Did one of the fitters have a wrongly set micrometer? Were some of the wheel castings inadequately stress relieved? We will probably never be able to identify the cause with complete

TABLE 3.1

Pressing number	Interference 2e before pressing (µm)
1	330
2	230
7	200
21	180

certainty. Indeed, it is not unknown for plausible reasons to be put forward by officials in situations of this sort in order to cover up the real reason!

Design Implications

The case study has shown that securing components by press fitting is a deceptively simple operation. In fact, it depends on complex elastic interactions between the components and is strongly influenced by effects due to friction, lubrication and wear that are difficult to reproduce consistently. Because of this, designers have tried hard to find alternatives.

One method is shrink fitting. A good example of this can be found on the Ffestiniog narrow-gauge railway in North Wales. The piston-valve liners on the steam locomotive "Blanche" were fitted by cooling them in liquid nitrogen (which boils at $-196°C$) and then popping them into the bores of the valve chests. When the liners warmed back up to room temperature they expanded to be a tight fit in the bores. The advantages were that there was no need for a heavy hydraulic press, the assembly could be done without removing the cylinder block, there was no risk of damage to the fragile liners, the machining tolerances were not critical, and there was no risk of seizure.

Another method is to use anaerobic adhesives such as Loctite. These set when they are taken out of contact with oxygen and are therefore ideal for close-fitting components. It is, however, essential to have a slight *gap* between the components, of around 25 to 100 µm, so that the glue is not squeezed out of the joint. Any surplus left outside the joint will not set and can be wiped off. And of course the glue will not go off in the bottle as long as there is an air space above it! Setting times are typically about 15 minutes, so there is plenty of time to manipulate the joint. A disadvantage, of course, is that it is difficult to get the components apart again, although this is sometimes possible if they are heated above the decomposition temperature of the polymerised glue.

A third method, again used on the Ffestiniog Railway and developed at Leeds University, is called Oil Injection. Figure 3.8 shows how this is set up. The bore of the wheel and the wheel seat are both machined to a shallow matching taper, although the minimum diameter of the bore is a little less than that of the wheel seat. The wheel seat has a series of interconnecting circumferential and longitudinal grooves machined into its surface. These extend over most of the length of the wheel seat, but stop short of the ends. They communicate with a tapped hole drilled into the end of the axle.

During the assembly operation the wheel is first engaged with the wheel seat. A high-pressure oil pump is then connected to the tapped hole by means of a reinforced hose. The oil pressure is increased and at the same time the wheel is gradually pulled on to the axle using a pair of nuts and bolts and a spanner. The oil pressure is enough to compress the axle and dilate the wheel and very little effort is required to pull the wheel up the taper. Because a high-viscosity oil is used there is little leakage from the ends of the wheel seat.

The beauty of the method is, of course, that it is just as easy to reverse the procedure to remove the wheel. Another great advantage is that the wheel can be

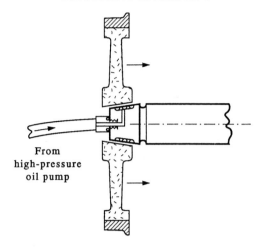

FIG 3.8   Schematic showing the oil injection process for fitting wheels to axles.

pressed on and off many times without in any way altering the fit between the wheel and the axle. Finally the method uses simple equipment and the operation can be performed without removing the axle from the locomotive. Naturally the oil grooves are shallow and are well rounded at the bottom—we certainly do not want to start any fatigue cracks from them!

References

ASME Handbook of Metals Engineering Design, 1st edition, American Society of Mechanical Engineers, 1953.

R. P. Bradley, *The Standard Locomotives of British Railways*, David and Charles, 1984.

O. S. Nock, *The British Railway Steam Locomotive 1925–1965*, Ian Allan, 1966.

J. E. Shigley, *Mechanical Engineering Design*, 3rd edition, McGraw-Hill, 1977.

C. J. Smithells, *Metals Reference Book*, 6th edition, Butterworth, 1984.

CHAPTER 4

BUCKLING OF A CHEMICAL REACTOR

Background

Figures 4.1 and 4.2 are schematic drawings of a chemical reactor made from steel. The reactor is part of a plastics-processing plant. The reactor itself consists of a circular tube of internal radius $r_1 = 800$ mm, wall thickness $t_1 = 12$ mm and length 1900 mm with a dished cap welded on at each end. It is surrounded over most of its length by a cylindrical jacket made from a circular tube of internal radius $r_2 = 862$ mm and wall thickness $t_2 = 9$ mm. The tube is welded to the reactor by means of two circular rings of width $w = 50$ mm. The distances between the two rings, L, is 1600 mm.

The chemical process is a batch operation, sequenced as follows. First, the chemicals to be reacted together are poured into the reactor through the filling nozzle at the top. They are then heated by circulating low-pressure oil at about 200°C through the jacket. The oil comes from a heating main which also supplies heat to other parts of the plant, as shown in Fig. 4.3. When the reactants have reached a temperature of about 160°C the rate of the chemical reaction becomes quite rapid, and since the reaction is exothermic, heat is generated inside the reactor. In order to avoid a thermal run-away the heat of reaction must be removed. This is done by opening and closing valves in the oil circuit (see Fig. 4.3) so that hot oil is no longer drawn from the main. Instead, oil is circulated in a closed loop, passing first through a steam-cooled heat exchanger, then through a pump and finally through the jacket before returning again to the heat exchanger. When the reaction is finished the products are drained out of the bottom of the reactor and another batch is processed by following the same cycle of events.

In this incident a fault occurred in the steam supply to the heat exchanger, allowing the circulating oil to heat up to an estimated temperature of 290°C. When the reactor was emptied and inspected it was found that a pronounced inward bulge had occurred in the wall (see the dotted outline in Figs. 4.1 and 4.2). The bulge was about 30 mm deep at its centre, extended about 600 mm along the reactor axis, and went about 300 mm around the circumference.

It was decided to carry out an analysis of the failure to see what needed to be done to return the reactor to service and to prevent the incident occurring again.

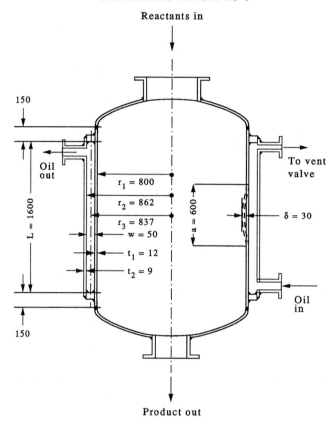

FIG 4.1   Vertical cross-section through the chemical reactor. The contents of the reactor are stirred using an electrically powered agitator which has been left out of the drawing for the sake of clarity. Dimensions in mm.

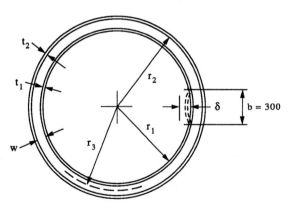

FIG 4.2   Horizontal cross-section through the middle of the reactor. Dimensions in mm.

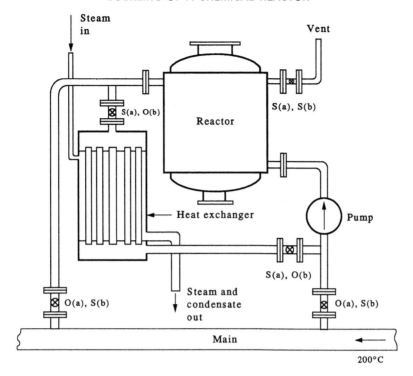

FIG 4.3   Schematic arrangement of oil flow (a) when heating reactor up, (b) when removing heat of reaction. O = valve open, S = valve shut.

Design Data

The first step in analysing the failure was to assemble the available design and materials property data. These are:

Reactor
   Austenitic stainless steel (316 type).
   Linear thermal expansion coefficient $\alpha = 15 \times 10^{-6}\,°C^{-1}$ (data from Smithells).
   Poisson's ratio $v = 0.28$ (data from Smithells).
   Working pressure = 0.06 MPa gauge.
   Design pressure = 0.10 MPa gauge.
   Hydraulic test pressure = 0.20 MPa gauge.

Jacket
   Carbon–manganese steel (BS 1501-151-430).
   $\alpha = 12 \times 10^{-6}\,°C^{-1}$ (data from Smithells).
   $v = 0.29$ (data from Smithells).
   Working pressure = 0.50 MPa gauge.
   Design pressure = 0.60 MPa gauge.
   Hydraulic test pressure = 1.20 MPa gauge.

Oil
   Synthetic heat-transfer fluid.

$\alpha = 23 \times 10^{-5}\,°C^{-1}$ (data from suppliers).
Operating limit $= 350°C$ in closed system.
Vapour pressure at $350°C = 1$ bar absolute.
Bulk modulus $K \approx 1.7 \times 10^3$ MPa (estimated from data for oils in Kaye and
Laby).

Failure Mechanism

The first thing that one notices from the design data is the big difference between
the coefficient of thermal expansion of the oil ($\alpha = 23 \times 10^{-5}\,°C^{-1}$) and the
coefficients of expansion of the steels ($15 \times 10^{-6}\,°C^{-1}$ and $12 \times 10^{-6}\,°C^{-1}$). The α
value for oil is seventeen times greater than the average α value of $13.5 \times 10^{-6}\,°C^{-1}$
for the steels. This discrepancy is not a fluke. As Fig. 4.4 shows, liquids generally
expand much faster than solids, the only exceptions being polymers, alkali metals
like sodium and potassium, and cold water. This is, of course, the principle behind
conventional mercury-in-glass or alcohol-in-glass thermometers. In these, the glass
expands much less than the liquid as the temperature is increased, and the liquid
column rises up the thermometer stem as a result. If one is silly enough to exceed

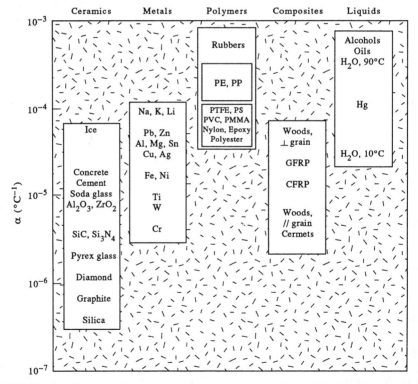

FIG 4.4   Coefficients of linear thermal expansion α for solids and liquids. Note that the
thermal expansion coefficients of liquids are normally listed in data books as coefficients
of volume expansion, 3α. Data from Ashby and Jones, Smithells, and Kaye and Laby.

the working range of the thermometer the liquid column will run out of space into which to expand, pressure will build up, and the glass will soon burst.

It is a reasonable assumption that differential thermal expansion of this type was responsible for causing the bulge in the reactor wall. Because the oil was circulating in a closed, pressure-tight loop when the temperature rise took place, and there was no space into which the oil could have expanded, the pressure built up until it reached the critical pressure for buckling of the vessel wall.

Additional Data Requirements

The equations for the elastic buckling of a cylindrical vessel under external pressure are given in Appendix 1B. However, before we can use these equations we must obviously show that the reactor wall was still in the elastic regime when buckling started. And we must also show that the jacket was strong enough to take the critical buckling pressure (the pressure in the jacket can never be more than the pressure at which the jacket deforms plastically). In order to carry out these checks we need to know σ_y for both the reactor and the jacket at the failure temperature of around 300°C.

Even if we can show that neither the reactor nor the jacket deformed plastically, we must still calculate the *elastic* compliance of the system: if the jacket "gives" under the pressure this will obviously absorb some of the thermal expansion of the oil and reduce the build-up of pressure. In order to do this we need values for E at 300°C as well.

Test data were not available for E or σ_y and we therefore have no choice but to refer to the literature. This is not a problem for E because this is a structure-insensitive property. However, σ_y is a structure-sensitive property which depends strongly on how the metal was processed and fabricated. Best estimates of E and σ_y can be obtained as follows.

Values of E and σ_y

Metals Handbook gives a value for E of 193×10^3 MPa for 316 stainless steel at room temperature. As a rule-of-thumb, E halves for most metals between absolute zero and the melting point, so we should reduce E to about 175×10^3 MPa at 300°C. Smithells gives $E = 212 \times 10^3$ MPa for mild steel at room temperature, which we reduce to 192×10^3 MPa at 300°C.

Metals Handbook gives a graph of σ_y at different temperatures for annealed 316 plate, which shows that σ_y should be about 130 MPa at 300°C. The reactor wall was produced by rolling up a piece of flat plate, which will have work-hardened the steel. 316 work-hardens rapidly, by about 75 MPa for each 1% plastic strain (see Honeycombe). Knowing the rate of work-hardening we can estimate the average yield stress in the reactor wall in the following way.

Figure 4.5 shows that the outer surfaces strain plastically by amount $t/2r$, whereas the neutral axis remains strain-free. The surface strain in the wall is given by 12 mm/(2 × 806 mm) (see Fig. 4.1), or 0.0074. The average strain through the tube wall is therefore about (0.0074 + 0)/2, or 0.0037 (0.37%). This should increase

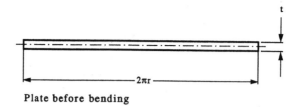

Plate before bending

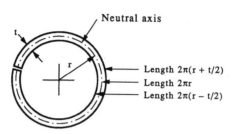

Plate after bending

FIG 4.5   Fabricating the reactor tube by rolling it up from flat plate increases the length of the outer surface by $2\pi t/2$ and decreases the length of the inner surface by the same amount. The outer surface strains plastically by $(2\pi t/2)/2\pi r = t/2r$ in tension. The inner surface strains by $(t/2r)$ in compression. The neutral axis is not strained.

σ_y by around 0.37×75 MPa $= 28$ MPa, giving a flow stress of $130 + 28 \approx 158$ MPa. This is very much a lower estimate since plate is rarely supplied in the annealed state, and has usually been work-hardened by finish-rolling and straightening.

The BS 1501 steel of the jacket has a specified minimum σ_y in plate form of 250 MPa. Data from Smithells show that closely similar steels have a yield stress of about 220 MPa at 300°C. Again, this is a minimum figure and as-supplied plate is usually significantly over specification.

To summarise, we use E and σ_y values as follows.

TABLE 4.1

	Reactor	Jacket
E (MPa)	175×10^3	192×10^3
σ_y (MPa)	158 minimum	220 minimum

Buckling Calculations

Using the information on external-pressure buckling in Appendix 1B we can model the part of the reactor between the circular rings as a circular tube of dimensions $r_1 = 800$ mm, $t_1 = 12$ mm and $L = 1600$ mm held circular at its ends. In order to find out where we are on the Tube Buckling Map we need to determine

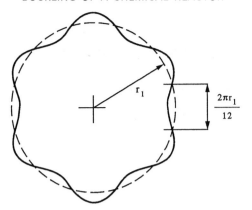

FIG 4.6   Fully developed six-lobed external pressure buckling.

the ratios (L/r_1) and (r_1/t_1). These are given by

$$\frac{L}{r_1} = \frac{1600 \text{ mm}}{800 \text{ mm}} = 2.00 \tag{4.1}$$

and

$$\frac{r_1}{t_1} = \frac{800 \text{ mm}}{12 \text{ mm}} = 67 \tag{4.2}$$

which intersect on the map at a point on the borderline between $n = 5$ and $n = 6$. In other words, a fully developed buckling failure should have either five or six lobes. If, for example, the failure were of the six-lobed type, the length of the inward-facing bulge should be about $\frac{1}{12}$ of the tube circumference (see Fig. 4.6) which works out to be $(2\pi \times 800 \text{ mm})/12 = 419 \text{ mm}$. This is similar to the circumferential length of about 300 mm measured on the bulge, and the failure is therefore consistent with a buckling mode involving about six lobes. The failure did not develop into a full multi-lobed failure presumably because the bulge created extra space in the jacket which would have relieved the pressure.

The relevant buckling equation (see Appendix 1B) is

$$p_{cr}(\text{theoretical}) = \frac{0.86E}{(1 - v^2)^{3/4}} \left(\frac{t_1}{L}\right)\left(\frac{t_1}{r_1}\right)^{3/2}. \tag{4.3}$$

Using the parameter values given above we find that the critical pressure for buckling is given by

$$p_{cr}(\text{theoretical}) = 2.2 \text{ MPa gauge}. \tag{4.4}$$

Equation (4.3) assumes that the tube is perfectly circular to begin with. This is not true in practice, of course. The ASME Code considers that in real tubes p_{cr} may be as much as 20% less than that calculated using Eqn. (4.3). Certainly, since the vessel was rolled-up from flat plate it probably departed quite significantly from a true cylinder. A working value of critical collapse pressure is therefore more likely to be given approximately by

$$p_{cr} = 1.8 \text{ MPa gauge}. \tag{4.5}$$

In order to check that the reactor wall is still elastic at the onset of buckling we calculate the compressive hoop stress generated in the wall by the critical pressure to see whether it is less than the yield stress. The hoop stress is given by

$$\sigma = \frac{p_{cr}r_1}{t_1} = 1.8 \text{ MPa} \times \frac{800 \text{ mm}}{12 \text{ mm}} = 120 \text{ MPa}. \tag{4.6}$$

This is well below the value of 158 MPa minimum for the yield stress of the reactor so our use of the elastic buckling model is reasonable.

As another check the hydraulic test pressure of 1.20 MPa gauge for the jacket is about 33% less than p_{cr} so the reactor should not have buckled during hydrostatic testing.

The tensile hoop stress in the jacket at p_{cr} is given by

$$\sigma = \frac{p_{cr}r_2}{t_2} = 1.8 \text{ MPa} \times \frac{862 \text{ mm}}{9 \text{ mm}} = 172 \text{ MPa} \tag{4.7}$$

which is well below the value of 220 MPa minimum for the yield stress. The jacket should therefore be capable of withstanding more than enough oil pressure to make the vessel collapse inwards.

Expansion Calculations

The dimensions of the heat exchanger and pipework in the closed-loop system are not known and as a first approximation (to be re-examined in due course) we have to assume that the jacket itself is the pressure-tight system. The jacket has dimensions $2\pi r_3 \times w \times L$ (see Fig. 4.1). On warming up from 200°C to 290°C each of these dimensions will increase owing to the thermal expansion of the steel. The increase in the capacity of the jacket caused by this expansion is given by

$$\Delta V_1 = 2\pi r_3 w L \times 3\alpha\Delta T. \tag{4.8}$$

Setting $\alpha = 13.5 \times 10^{-6}\,°\text{C}^{-1}$ and $\Delta T = 90°\text{C}$, and using the dimensions given in Fig. 4.1, we find that

$$\Delta V_1 = 1.53 \times 10^6 \text{ mm}^3 = 1.53 \text{ litres}. \tag{4.9}$$

This expansion is plotted schematically in Fig. 4.7. The increase in the volume of the oil in the jacket is given by Eqn. (4.8) with $\alpha = 23 \times 10^{-5}\,°\text{C}^{-1}$, i.e.

$$\Delta V_2 = 26.1 \times 10^6 \text{ mm}^3 = 26.1 \text{ litres}. \tag{4.10}$$

This is shown schematically in Fig. 4.7. At this point we need to remember that although the oil was originally at low pressure when the jacket was sealed the pressure of the trapped oil ultimately rose to about 1.8 MPa. The increase in pressure would have compressed the oil and reduced its volume slightly. The change in volume due to this compression can be calculated using the bulk modulus formula

$$\Delta V_3 = -\frac{Vp}{K}. \tag{4.11}$$

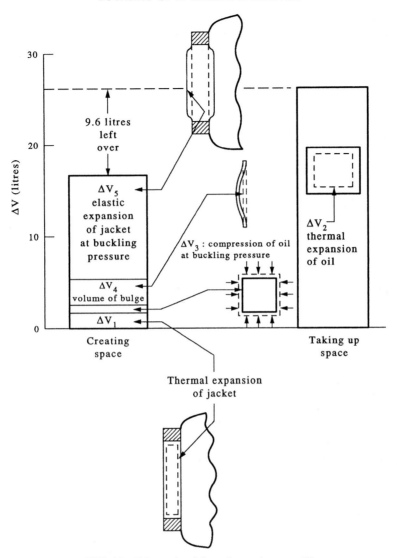

FIG 4.7   Schematic of the volume changes ΔV.

Given that $V = 2\pi r_3 wL$, $p = 1.8$ MPa and $K = 1.7 \times 10^3$ MPa, we find that

$$\Delta V_3 = 0.5 \times 10^6 \text{ mm}^3 = 0.5 \text{ litre} \tag{4.12}$$

as shown in Fig. 4.7.

Accommodation of Excess Oil

How is the excess volume of oil accommodated in practice? Obviously, some of the oil goes into filling the extra space created by the bulge. This is given approximately by

$$\Delta V_4 = a \times b \times \frac{\delta}{2}. \tag{4.13}$$

Inserting parameter values from Figs. 4.1 and 4.2 we obtain

$$\Delta V_4 = 2.7 \times 10^6 \text{ mm}^3 = 2.7 \text{ litres}. \tag{4.14}$$

Space is also created by the elastic deformations of the jacket and external pipework. As Fig. 4.7 shows, the pressure in the jacket will cause the reactor tube to decrease in diameter and the jacket tube to increase in diameter. We have already established that the reactor tube is still elastic at the onset of buckling and has a compressive hoop stress of 120 MPa at that point. The fractional decrease in the reactor circumference is therefore given by the Young's modulus formula, with

$$\varepsilon = \frac{\Delta(2\pi r)}{2\pi r} = \frac{\sigma}{E}. \tag{4.15}$$

The decrease in tube radius is then given by

$$\Delta r = \left(\frac{\sigma}{E}\right)r_1. \tag{4.16}$$

Setting $\sigma = 120$ MPa, $E = 175 \times 10^3$ MPa and $r_1 = 800$ mm, we obtain

$$\Delta r = 0.6 \text{ mm}. \tag{4.17}$$

Using a similar approach we find that the increase in the jacket radius is given by

$$\Delta r = \left(\frac{\sigma}{E}\right)r_2. \tag{4.18}$$

Setting $\sigma = 172$ MPa, $E = 192 \times 10^3$ MPa and $r_2 = 862$ mm we get

$$\Delta r = 0.8 \text{ mm}. \tag{4.19}$$

The combination of these changes in tube radius is to increase the jacket width, w, by $0.6 + 0.8 = 1.4$ mm. This produces extra space of amount

$$\Delta V_5 = 2\pi r_3 L \times 1.4 \text{ mm} = 11.8 \times 10^6 \text{ mm}^3 = 11.8 \text{ litres}. \tag{4.20}$$

Finally, as Fig. 4.7 shows, we are left with a volume of oil that is not accounted for which is given by

$$\Delta V_6 = \Delta V_2 - \Delta V_5 - \Delta V_4 - \Delta V_3 - \Delta V_1$$

$$= 9.6 \times 10^6 \text{ mm}^3 = 9.6 \text{ litres}. \tag{4.21}$$

This figure should be treated with some caution since it is strongly affected by the estimated size of the temperature excursion. For example, if the calculations are reworked using a value for ΔT of 70°C instead of the 90°C used above, the volume decreases to 4.1 litres.

Effect of External Pipework

We now return to the question of the heat exchanger and pipework external to the jacket in order to see what happens when we relax our assumption that the jacket alone can be treated as the closed system. In view of the lack of information (typical of many failure investigations) we have to make some reasonable guesses to act as a starting point. These are:

Piping—30 m of steel tube 70 mm bore by 5 mm wall thickness.
Heat exchanger—200 copper tubes each 2 m long having 20 mm bore and 2 mm wall thickness.

Calculations can be done for this geometry in exactly the way described above for the jacket. For the copper tubes, we have taken values for E and α of 130×10^3 MPa and $17 \times 10^{-6}\,°C^{-1}$ respectively. The results are:

TABLE 4.2

	Steel tube	Copper tube
Volume of bore (litres)	115	126
Net increase due to thermal expansion (litres)	6.7	7.2
Hoop stress at p_{cr} (MPa)	13	9
Increase in tube radius at p_{cr} (mm)	2.3×10^{-3}	0.7×10^{-3}
Increase in volume due to increase in tube radius (litres)	0.02	0.02
Compression of oil at p_{cr} (litres)	0.3	
Volume not accounted for (litres)	13.5	

So the effect of introducing the external circuit into the calculations is to *increase* the volume of oil not accounted for from 9 to 23 litres.

Where does this volume go? Almost certainly at such a high pressure there will be leakage of oil from the system, past valves and gasket seals for example. However, the point of the calculations has been to show that, even allowing for leakage, it is likely that there was enough expansion in the system to achieve the critical pressure for the buckling of the reactor wall.

Conclusions

A reasonable scenario for the bulging is that thermal expansion of oil in the closed recirculating system led to a build-up in pressure that was eventually enough to cause elastic buckling of the reactor under external hydrostatic pressure. This is what we had guessed early on, but a long and rather tedious set of calculations was needed to confirm that this sequence of events was actually probable in practice.

Because both the oil and the jacket are stiff elastic structures the pressure would have fallen off rapidly as the instability grew. The failure is thus a good example of a *displacement-controlled* event. Catastrophic collapse did not occur, and indeed could not have done, and the failure was a "safe" event.

Obviously, the incident would not have occurred if the pressure had not built up in the first place. A requirement for returning the system to service is therefore to fit a pressure-relief valve to the circuit. For safety reasons this is best arranged so that any oil blown through the valve is discharged back into the main.

Provided the bulge does not interfere with internal components such as the stirrer, it is probably safe to return the vessel to service as long as the pressure relief valve is set to a low pressure. For service at higher pressure, it would be necessary to jack the bulge back into position and carry out a repeat hydraulic test. Because of residual stresses, bulges can sometimes form again in service and it would obviously be wise to inspect the inside of the vessel at regular intervals.

References

M. F. Ashby and D. R. H. Jones, *Engineering Materials 2*, Pergamon, 1986.

ASME Boiler and Pressure Vessel Code (American Society of Mechanical Engineers). Section 8, Division 2: *Appendix J, Basis for Establishing External Pressure Charts*, 1986.

British Standards Institution, BS 1501: 1980: "Steels for Fired and Unfired Pressure Vessels: Plates": Part 1: "Specification for Carbon and Carbon–Manganese Steels".

R. W. K. Honeycombe, *Steels: Microstructure and Properties*, Arnold, 1981.

G. W. C. Kaye and T. H. Laby, *Tables of Physical and Chemical Constants*, 14th edition, Longmans, 1973.

Metals Handbook, 9th edition, Vol. 3: *Properties and Selection; Stainless Steels, Tool Materials and Special-Purpose Metals*, American Society for Metals, 1980.

C. J. Smithells, *Metals Reference Book*, 6th edition, Butterworth, 1984.

CHAPTER 5

PRESSURISATION DAMAGE TO A STORAGE TANK

Background

Figure 5.1 is a simplified drawing of a tank used for storing cooking oil at a food-processing factory. The tank is made from cold-rolled austenitic stainless steel of the 316 type. The shell of the tank is a cylinder 3000 mm in diameter and 6500 mm high. It consists of three sections welded together. The top section, 3000 mm high, is rolled up from 3.2-mm plate. The next section is 1500 mm high and is rolled from 4.75-mm plate. The lowest section is 2000 mm high and comes from 6.5-mm plate. A dished end cap is welded to the bottom of the shell and a conical lid is welded to the top. Finally, the tank is divided into an upper part and a lower part by a partition welded in at a location 4000 mm down from the top. As Fig. 5.2 shows, the partition is a spherical cap with a radius of 3000 mm.

The tank is used in the following way. Normally the valve in the transfer loop is kept closed and oil is drawn off from the bottom tank. The oil level can be followed using the glass level gauge, and when the oil is getting low more can be run in from the upper tank by opening the transfer loop. When the oil in the upper tank runs low, a further supply is added from a road tanker through the filling pipe. During the filling operation the level of the oil in the upper tank is followed using the level gauge to make sure that oil does not overflow through the vent pipe.

The standard method of getting the oil from the road tanker into the upper tank is shown in Fig. 5.3. Compressed air is fed into the top of the road tank. This forces oil out of the bottom of the road tank and up through a hose into the top of the upper tank. The air supply is controlled by the tanker driver using a throttle valve. With a fairly thin oil a pressure of about 1 bar gauge is enough, but a higher pressure has to be used if a thick oil is being off-loaded. The maximum pressure that can be delivered to the road tank is 2 bar gauge. This method of loading has the advantage that the oil only comes into contact with the insides of the tank and the hose whereas it might get contaminated if it were delivered through a pump.

A delivery of oil had just been finished when there was a loud "thump" from the storage tank. An immediate inspection showed that the top of the tank had distorted. As we can see from Fig. 5.4, the centre of the cone had lifted, pulling in the top of the shell. The edge of the cone had become corrugated all the way round and the top of the shell bulged in three places. When the tank had been drained and the inside inspected the reason for the noise became clear. The partition had "flipped"

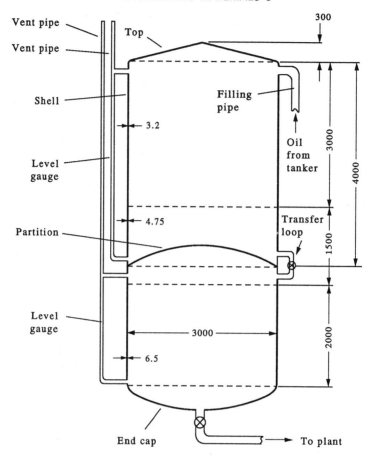

FIG 5.1   Simplified drawing of the storage tank. Dimensions in mm. ⊗ indicates a valve.

from being convex upwards to concave, as shown by the dashed outline in Fig. 5.2.

Failure Scenario

The damage to the top of the tank suggests that the upper tank had been exposed to internal pressure. This would also explain why the partition had flipped—a classic example of the buckling of a spherical cap under external pressure. Presumably the road tanker must have discharged its entire load so that, at the end of the filling operation, the compressed air in the road tank would have rushed into the upper tank, causing the damage. Certainly, the bore of the vent pipe was much smaller than that of the filling hose so that air would have entered the tank much faster than it could have escaped.

When the partition flipped the oil in the upper tank would have moved down to fill the extra space generated between the original and final surfaces of the partition (see Fig. 5.2). This would have created an additional air space at the top of the

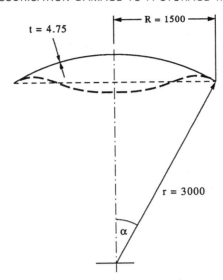

$t = 4.75$

$R = 1500$

$r = 3000$

α

FIG 5.2   Details of the partition. Dimensions in mm.

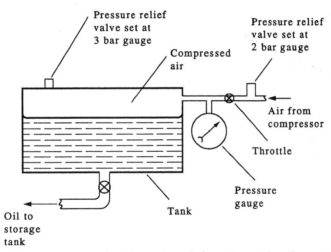

Pressure relief
valve set at
3 bar gauge

Compressed
air

Pressure relief
valve set at
2 bar gauge

Air from
compressor

Throttle

Pressure
gauge

Tank

Oil to
storage
tank

FIG 5.3   Arrangement used for discharging oil from the road tanker. ⊗ indicates a
valve.

tank. However, this extra space is small compared to the volume of air in the road
tank, so the pressure of the compressed air would not have decreased very much.
The failure of the partition is therefore a good example of a *load-controlled* event:
the load on the partition did not drop off very much as the deformation propagated
so catastrophic failure occurred.

Of course the partition can only flip over if there is a big enough air space in the
lower tank for it to move into without meeting much resistance. We therefore have
to postulate that the lower tank was not full of oil when the delivery was being
made and that the transfer loop was kept closed.

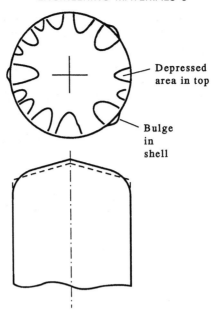

FIG 5.4   Distortion of the tank top after pressurisation. The depressions were as much as 60 mm deep and the bulges were up to 50 mm high.

Buckling Calculations

In order to be sure that our explanation for the failure of the partition is correct we need to show that the pressure in the road tank would have been enough to cause the spherical cap to buckle. The relevant equations are given in Appendix 1B. The ideal buckling pressure is given by

$$p_{cr}(\text{theoretical}) = \frac{2E}{\{3(1 - v^2)\}^{1/2}} \left(\frac{t}{r}\right)^2. \tag{5.1}$$

For 316 stainless, $E = 193 \times 10^3$ MPa (Metals Handbook) and $v = 0.28$ (Smithells). $r/t = 3000 \text{ mm}/4.75 \text{ mm} = 632$, so

$$p_{cr}(\text{theoretical}) = \frac{2 \times 193 \times 10^3 \text{ MPa}}{\{3(1 - 0.28^2)\}^{1/2}} \left(\frac{1}{632}\right)^2$$

$$= 0.58 \text{ MPa or } 5.8 \text{ bar}. \tag{5.2}$$

The actual collapse pressure is much less than this ideal value of course, and is given by

$$p_{cr}(\text{actual}) \approx (0.2 \text{ to } 0.6) \times 0.58 \text{ MPa}$$

$$\approx 0.12 \text{ to } 0.35 \text{ MPa or } 1.2 \text{ to } 3.5 \text{ bar}. \tag{5.3}$$

Equation (5.1) is valid for $5 \leq \lambda \leq 50$, where

$$\lambda = \{12(1 - v^2)\}^{1/4} \left(\frac{r}{t}\right)^{1/2} \alpha. \tag{5.4}$$

From Fig. 5.2 we can see that $\sin \alpha = 1500 \text{ mm}/3000 \text{ mm} = 0.5$. So $\alpha = 30° = 0.524$ radian. Setting $v = 0.28$, $r/t = 632$ and $\alpha = 0.524$ we get $\lambda = 24$, which is well within the range of validity of Eqn. (5.1).

What pressure is available at the top of the partition? As well as the air pressure (which is probably about 1 bar, but could be as high as 2 bar) we also have the hydrostatic head of the oil in the upper tank. The density of the oil is about $0.8 \times 10^3 \text{ kg m}^{-3}$ so the pressure head is

$$p = h\rho g = 4 \text{ m} \times 0.8 \times 10^3 \text{ kg m}^{-3} \times 9.81 \text{ m s}^{-2}$$

$$= 3.1 \times 10^4 \text{ Pa} = 0.31 \text{ bar}. \tag{5.5}$$

Thus the total pressure available is between about 1.3 and 2.3 bar gauge. This pressure range is comparable to the range given by Eqn. (5.3) which confirms our failure model.

Stresses in the Partition

The buckling analysis assumes that the cap is elastic at the critical pressure. The stress state is one of equi-biaxial compression, with the stress given by

$$\sigma = \frac{p}{2}\left(\frac{r}{t}\right). \tag{5.6}$$

Since the maximum pressure is about 2.3 bar, or 0.23 MPa, the maximum value that the stress can have is

$$\sigma(\text{max}) \approx \frac{0.23 \text{ MPa} \times 632}{2} = 73 \text{ MPa}. \tag{5.7}$$

As shown in Appendix 1C, the Von Mises yield criterion predicts that yielding should occur when $\sigma = \sigma_y$ in an equi-biaxial stress system. Smithells gives data for the yield stress of 316 stainless, with values ranging from 170 MPa for softened material to 700 MPa for heavily work-hardened material. Since our steel is cold rolled it is likely to have a yield stress that is well up in this range, and therefore the cap is well within the elastic limit at the failure pressure.

Stresses in the Shell

Is the shell of the tank able to resist the failure pressure? The hoop stress is given by

$$\sigma = \frac{pr}{t} \tag{5.8}$$

so the maximum value of the hoop stress is

$$\sigma(\text{max}) \approx \frac{0.23 \text{ MPa} \times 1500 \text{ mm}}{3.2 \text{ mm}} = 108 \text{ MPa}. \tag{5.9}$$

Appendix 1C shows that, in the case of a pressurised thin tube fitted with end caps, Von Mises' criterion indicates that yield will take place when the hoop stress is equal to $1.16 \times \sigma_y$. Taking the minimum value for the yield stress of 170 MPa, we need a hoop stress of 1.16×170 MPa = 197 MPa to get yield. This is much more than the actual hoop stress, so the shell will only deform elastically.

Stresses in the Conical Top

The dimensions of the tank top are shown in Fig. 5.5. As shown in Appendix 1A the stress in the membrane is biaxial, with

$$\sigma_1 = \left(\frac{pR}{t}\right)\frac{1}{(\cos \alpha)} \tag{5.10}$$

and

$$\sigma_2 = \left(\frac{pR}{2t}\right)\frac{1}{(\cos \alpha)}. \tag{5.11}$$

As Fig. 5.5 shows, $\tan \alpha = 1500$ mm/300 mm = 5. Thus $\alpha = 78.7°$ and $\cos \alpha = 0.196$. $R/t = 1500$ mm/3.2 mm = 469. The maximum pressure at the top of the tank is 2 bar, or 0.2 MPa. Inserting these values into Eqns. (5.10) and (5.11) we get maximum stress values of $\sigma_1 = 479$ MPa and $\sigma_2 = 239$ MPa.

As shown in Appendix 1C, the Von Mises equivalent stress in the conical top is given by

$$\sigma_e \text{ (MPa)} = [\tfrac{1}{2}\{(479 - 239)^2 + (239 - 0)^2 + (0 - 479)^2\}]^{1/2}$$

$$= 415 \text{ MPa}. \tag{5.12}$$

This is a high stress, comparable to the yield stress of moderately cold-worked 316 stainless, and it is conceivable that the top could yield during the pressure excursion.

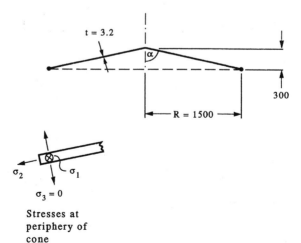

Stresses at
periphery of
cone

FIG 5.5   Details of the tank top. Dimensions in mm.

However, it is more important to see that the stress needed to restrain the edge of the cone is also high, at 239 MPa. The top end of the shell cannot stand this concentrated pull and it buckles, behaving just like a short length of tube under external pressure. Since the edge of the cone is no longer fully restrained by the shell, it pulls inwards, goes into circumferential compression, and develops regularly spaced corrugations as a way of relieving the compression stress.

Designing Pressure Vessels

The tank was subjected to a pressure that was only half that needed to make the shell yield, and yet severe distortion took place at the junction between the shell and the conical top. A cone is obviously far from an optimal shape for the end of a pressure vessel, although it is perfectly satisfactory to use it for capping the tank provided that it is not intended to take any pressure. In vessels that are intended to take internal pressure, however, the design of the end caps is very important.

Figure 5.6 shows the types of caps most commonly used for the ends of pressure vessels. The obvious shape, of course, is a hemisphere: the axial force required to hold the hemisphere on to the end of the shell is simply provided by a longitudinal force in the shell wall. Naturally, one is tempted to make the wall thickness of the cap the same as that of the shell, because this makes it easy to do the weld between the two components. It also makes the equi-biaxial stress in the cap the same as

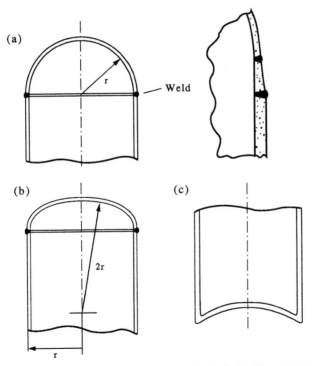

FIG 5.6   End caps for pressure vessels: (a) hemispherical, (b) ellipsoidal, (c) concave.

the longitudinal stress in the shell. However, this is not ideal because the hoop stress in the shell is then twice the stress in the cap. This means that there is a circumferential stress discontinuity where the shell is joined to the hemisphere and this causes local elastic distortions. It also means that the cap is understressed by a factor of two, which wastes metal and weight.

Of course, one way around these problems is to make the cap half as thick as the shell, making the equi-biaxial stress in the cap the same as the hoop stress in the shell. However, the stress in the cap is then twice the *longitudinal* stress in the shell, so we now have a *longitudinal* stress discontinuity instead. A way to mitigate this problem is to have a transition section at the junction, with the wall tapering in thickness as shown in Fig. 5.6(a). But this is expensive, and involves a second weld. One simply cannot win!

These problems can be solved fairly well by using an end cap of the ellipsoidal type, shown in Fig. 5.6(b). These are usually made so that they have the same thickness as the shell. However, the radius of the end is made about twice that of the shell so that the equi-biaxial stress in the end of the cap is the same as the hoop stress in the shell. The transition between the end of the cap and the shell is designed to minimise the stress discontinuities. Nonetheless, it is not unknown for ellipsoidal end caps to distort under pressure just like the top of our tank if they are made too flat!

An interesting alternative is the concave end shown in Fig. 5.6(c). This is a spherical cap, just like our failed partition, but hopefully with a much bigger ratio of t/r. This is useful where the vessel has to stand on its end without additional supports. Of course, this type of end cap must be surrounded by a very strong ring support, in just the same way that a masonry bridge needs very strong abutments to take the end thrust from the loaded arch.

Some Familiar Examples

Although we may not be aware of it, we are surrounded by pressure vessels in the most unexpected places. We can see some examples in Fig. 5.7.

The sparkling mineral water bottle would have held water saturated with carbon dioxide under pressure. The top end cap is a fair approximation to a hemisphere but the bottom has an interesting shape which allows the bottle to be stood on its end whilst at the same time resisting the internal pressure. The shape of the bottom makes full use of our ability to mould complex shapes in thermoplastic materials— such a shape would scarcely be possible in a welded metal pressure vessel.

The aerosol can is a classic example of a concave bottom. No one would buy aerosols if they could not be stood upright on a dressing table or workbench! The propellant in aerosols is a mixture of propane and butane. As we can see from the tables in Appendix 1H, propane, and to a lesser extent butane, have high vapour pressures at room temperature. They are also highly flammable. In fact, the design of aerosol cans is a considerable technical achievement, for they have not only to be strong and safe but also extremely cheap to make. As an example of the ingenuity in their design, note the way in which the bottom end of the shell is rolled over several times to make the strong ring abutment required for the end cap.

FIG 5.7   Some familiar pressure vessels.

The fire extinguisher, which contains BCF at high pressure (very unfashionable when we are worried about the ozone layer) is a nice example of a hemispherical top coupled with a concave bottom. The drinks can also has a concave bottom but actually has a *flat* top. Before you open a can observe that the top is actually slightly convex under the pressure of the carbon dioxide in the drink. But this is not a problem because the steel that the top is made from is heavily work-hardened and has a high yield stress; and also because the ring around the edge of the top is rolled over for strength and will not buckle as the top of our storage tank did.

A Final Note

To end this chapter we return to the phenomenon of external-pressure buckling. A classic example of load-controlled buckling can be found in the design of pipelines. A substantial proportion of the world's supply of oil and natural gas is extracted offshore and has to be brought onshore in pipelines run along the sea bed. The pressure exerted on the outside of the pipe by the hydrostatic head of the water above is very high, and if the pressure of the oil or gas inside the pipe is significantly less there will be a net pressure tending to make the pipe buckle inwards. Even if the pipe is designed to resist the theoretical buckling pressure with a suitable safety factor buckling can still be initiated by accidental damage to the pipe. For example, if a heavy object like an anchor is dropped on to the pipe the local dent that this will make will greatly lower the critical bulging pressure of the damaged area.

As shown in Fig. 5.8, once a buckling failure starts in this way it will not stop. The hydrostatic head will keep the load on all the time and the failure will run in a catastrophic way, spreading along the pipe in both directions away from the site of

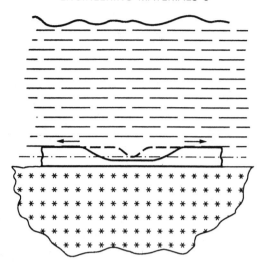

FIG 5.8   Running buckling failure of an underwater pipeline starting from an accidental dent.

initiation and at a speed limited only by the velocity of the pressure wave. Many miles of pipeline can be destroyed by such an event. The cure is to design on the basis of the critical pressure for a *running* failure. The desired strength is then achieved by making the pipe wall thick enough, although extensive experimental and theoretical work is required to establish exactly what the wall thickness needs to be. The same problem can be encountered in the design of other tubular components, such as the legs of offshore platforms or well liners.

References

Metals Handbook, 9th edition, Vol. 3: *Properties and Selection; Stainless Steels, Tool Materials and Special-Purpose Metals*, American Society for Metals, 1980.
C. J. Smithells, *Metals Reference Book*, 6th edition, Butterworth, 1984.

CHAPTER 6

STRESSES IN BOLTS

Background

Bolts and other threaded fasteners like studs and screws are used in an enormous range of engineering applications. In many situations the safe operation of a component depends critically on the integrity of the fasteners holding it together and a failure would have disastrous consequences. We can find some good examples in the engine of the ordinary automobile. A typical application involving the use of studs or bolts is to secure the bearing cap around a rotating shaft (see Fig. 6.1). This arrangement is found in the big-ends that make the mechanical connection between the connecting rods and the crankshaft, in the main bearings that secure the crankshaft to the engine block and in the camshaft bearings that secure the camshaft to the top of the cylinder head. In all these applications the bearing cap is subjected to a large number of loading cycles and the prime cause of failure is likely to be the fatigue fracture of the threaded fasteners. In fact, in the engineering industry as a whole, it is estimated that 75–85% of all fastener failures are caused by fatigue.

Threaded parts are especially prone to fatigue because of the stress concentrations that are caused by the presence of the threads themselves. These are most acute where there is a sudden change in the stress state in the fastener, usually where the thread terminates in the parallel shank and where it enters a tapped hole or a nut. There is also a stress concentration where the shank joins the head of a bolt. The stress concentration factors for this last situation are given in Appendix 1A. The SCF depends strongly on the radius of the fillet where the shank meets the head, but for a typical bolt the SCF is generally around 2.

In spite of this nasty environment it is very unusual for an engine to smash itself up because of a fatigue failure. The reason why this is the case is because of a clever application of elementary elastic stress theory. Looking at Fig. 6.1 we can see that, if the stud or bolt is almost slack to begin with, then the load F in each fastener will cycle from zero to P about a mean load F_m of $0.5P$. This situation can be plotted on the Goodman diagram, as shown in Fig. 6.2. Here, ΔF is the load range that would make the fastener fail by fatigue after N_f cycles and F_{TS} is the breaking load of the bolt. ΔF_0 is the load range that would make the fastener fail after N_f cycles in a test with $F_m = 0$. The example that we have plotted in Fig. 6.2 is for $P = 0.5F_{TS}$ which gives $F_m = 0.25F_{TS}$. The Goodman equation then gives

$$\Delta F = \Delta F_0\left(1 - \frac{F_m}{F_{TS}}\right), \tag{6.1}$$

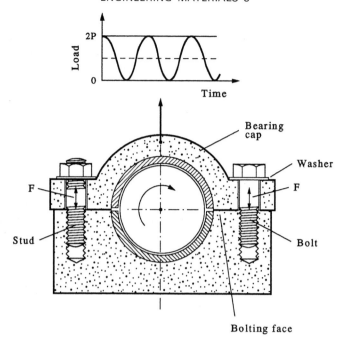

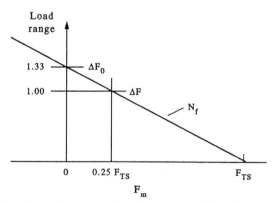

FIG 6.1   Schematic of a typical bearing housing showing how studs and bolts are used
to hold down the bearing cap.

FIG 6.2   The Goodman diagram for a bearing-cap bolt subjected to a load range $\Delta F = P$.

from which

$$\Delta F_0 = \frac{\Delta F}{0.75} = 1.33 \, \Delta F. \tag{6.2}$$

In practice, of course, bolts are "torqued up" before use in order to give them a tensile preload, F_i. In less critical applications it is usual to apply a specified torque to the head using a torque wrench. This is not satisfactory for critical applications because the tension generated by applying a torque depends on the friction between the mating threads and between the head and the washer. Normal variations in the coefficient of friction can lead to a scatter of $\pm 25\%$ in F_i. The procedure for assemblies like bearing caps is instead to apply a small torque to make the components fit snugly together and then to turn the head through a specified angle, typically 90°. This puts a definite strain into the bolt and produces a much more reproducible preload. As we shall show in the next section a correct preload is the vital ingredient in preventing the fatigue failure of the fasteners.

Effect of Preload

When the bolt is tightened up it is put into tension and the part of the bearing cap that sits underneath the head (the "member") is put into balancing compression. The bolt behaves like a very stiff spring in tension and gets slightly longer; the member behaves as a stiff spring in compression and gets slightly shorter. The situation can be modelled as shown in Fig. 6.3, with the elasticity of the bolt modelled by a tension spring between the top and bottom parts of the bolt and the elasticity of the member modelled by a compression spring between the top and bottom parts of the member. The spring constant of the bolt is given by

$$k_b = \frac{f_b}{u_b}, \tag{6.3}$$

where f_b is the tensile force in the bolt and u_b is the amount by which the bolt is extended when it is loaded from zero to f_b. In the same way the spring constant of the member is

$$k_m = \frac{f_m}{u_m}, \tag{6.4}$$

where f_m is the compressive force in the member and u_m is the amount by which the member is compressed when it is loaded from zero to f_m.

When $P = 0$ force equilibrium gives

$$F_i = f_b = f_m, \tag{6.5}$$

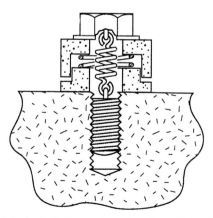

FIG 6.3   Elastic model of a bolted assembly when the bolt is subjected to a preload F_i.

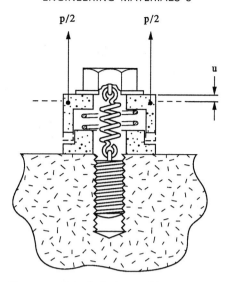

FIG 6.4   Response of the bolted assembly to an external load P.

and

$$F_i = u_b k_b = u_m k_m. \tag{6.6}$$

When the maximum load P is applied to the member it moves upwards by amount u, as shown in Fig. 6.4. The bolt is extended by u and the member relaxes back by u. Force equilibrium then gives

$$P + k_m(u_m - u) = k_b(u_b + u). \tag{6.7}$$

Thus

$$P = u(k_b + k_m), \tag{6.8}$$

and

$$F_p - F_i = k_b u, \tag{6.9}$$

where F_p is the tension in the bolt at the peak load P. Finally,

$$F_p = F_i + \left(\frac{k_b}{k_b + k_m}\right) P. \tag{6.10}$$

This result is valid only if $P < F_p$. When $P = F_p$ the member ceases to be in load-bearing contact with the bolting face. Equation (6.10) has two limiting cases:

(a) $k_b \ll k_m$. In practice, this can be obtained by putting a spring washer under the head, as shown in Fig. 6.5. Equation (6.10) then gives

$$F_p = F_i. \tag{6.11}$$

This means that there is no variation at all in the tension in the bolt even though the applied load is cycling between 0 and P. The result only holds if $F_i > P$. If $F_i = P$ the applied load will pull the member away from the bolting face and $F_p = P$.

(b) $k_b \gg k_m$. In practice, this can be obtained by putting a springy gasket between

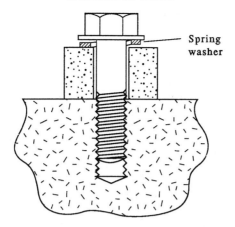

FIG 6.5   An example of $k_b \ll k_m$.

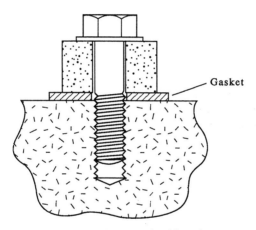

FIG 6.6   An example of $k_b \gg k_m$.

the member and the bolting face, as shown in Fig. 6.6. Equation (6.10) then gives

$$F_p = F_i + P. \tag{6.12}$$

This means that the bolt will see the full stress cycle. The member will only lift off the bolting face if $P = F_p = F_i + P$, i.e. if $F_i = 0$.

   These limiting cases are seldom met with in practice. In order to get an idea of the stress cycle likely to be experienced by a bearing cap bolt we can model the assembly as shown in Fig. 6.7. The end of the member through which the bolt passes is approximated to a cylinder with an outside diameter of $6b$ and an inside diameter of $2b$. The bolt is assumed to be a close fit in the hole. In order to find the values of k_b and k_m we write equations as follows.

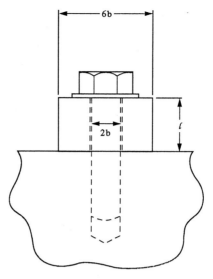

FIG 6.7   Elastic model of one end of the bearing cap.

$$E_b = \frac{f_b}{\pi b^2} \times \frac{l}{u_b}, \tag{6.13}$$

$$k_b = \frac{f_b}{u_b} = \frac{\pi b^2 E_b}{l}, \tag{6.14}$$

$$E_m = \frac{f_m}{\pi\{(3b)^2 - b^2\}} \times \frac{l}{u_m}, \tag{6.15}$$

$$k_m = \frac{f_m}{u_m} = \frac{8\pi b^2 E_m}{l}. \tag{6.16}$$

Both the bolt and the bearing cap are normally made from steel so $E_b = E_m$ and $k_m = 8k_b$. Equation (6.10) then gives

$$F_p = F_i + \frac{P}{9}. \tag{6.17}$$

This is an important result, because it tells us that the fatigue cycle experienced by the bearing cap bolts is only one-ninth that applied to the cap itself.

The member will lift off the bolting face when

$$P = F_p = F_i + \frac{P}{9}, \tag{6.18}$$

i.e. when $F_i = (8/9)P$. It is therefore important to specify the preload so that it is greater than $(8/9)P$ by a suitable safety margin.

Fatigue Strength Requirements

In Eqn. (6.2) we looked at the case where the load P on the member was cycled from 0 to $0.5F_{TS}$, but the bolt was not preloaded. In order for the bolt to survive N_f cycles it had to have a fatigue strength ΔF_0 of $1.33P$ as measured in a conventional push–pull fatigue test. But what is the strength requirement if the bolt, instead of being slack, is preloaded to $0.5F_{TS}$? Equation (6.17) then tells us that the load in the bolt will cycle between $0.5F_{TS}$ and $0.5F_{TS} + (\frac{1}{9}) \times 0.5F_{TS} = 0.56F_{TS}$. The mean stress is obviously $0.53F_{TS}$. The Goodman equation then gives

$$\Delta F_p = \Delta F_0\left(1 - \frac{0.53F_{TS}}{F_{TS}}\right) \approx 0.5\Delta F_0. \tag{6.19}$$

The required fatigue strength is then

$$\Delta F_0 = \frac{\Delta F_p}{0.5} = 2\Delta F_p = \frac{2}{9}P = 0.22P. \tag{6.20}$$

This is smaller by a factor of $1.33/0.22 = 6$ than the strength needed when there is no preload, which shows just how effective preloading is. The moral is that if you are investigating the fatigue failure of a threaded fastener the first things you should find out are the specification for tightening it up and whether the specified procedure was actually followed. If you are in charge of an engineering installation, or even replacing the big-end shells on your vintage car, do not even start on the job without a good torque wrench and the correct set of torque specifications.

Keeping Threads Tight

Threads that are subjected to either a varying load or to mechanical vibration tend to work loose. This can do two things. The first and obvious consequence is that the assemblies fall apart, which is not a good thing. And if a nut or a bolt falls off into machinery, like a gearbox, the effect is usually rather devastating. The second consequence is that the preload will decrease and the fastener may fail by fatigue before it falls off. In fact the preload is instrumental in keeping the threads tight. This is due to two factors. The first, of course, is that the preload decreases the variation in the load applied to the thread and therefore decreases the driving force for loosening. The second is that the preload forces the fastener into firm contact with the mating threads and with the top of the member. The friction between the contacting surfaces is high, and this opposes any tendency for the fastener to rotate. A final benefit of a high preload is that it greatly reduces the risk that a bolt or stud will fail under impact loading.

Residual Stresses in Bolts

There are two ways in which the thread on a fastener can be formed. One is to cut the thread, using a die or a screwcutting tool. The other is thread-rolling, where the thread form is generated by plastic deformation and no metal is removed

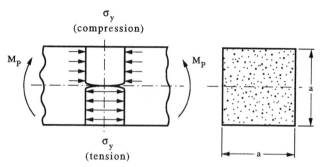

FIG 6.8   Plastic stress distribution in a fully plastic beam.

from the blank. Thread-rolling has two advantages over screwcutting. Because of the plastic flow around the root of the thread the "grain" of the metal tends to follow the thread profile which helps to prevent fatigue cracks propagating into the bolt from the thread root. The forming process also leaves the thread root in a state of residual elastic compression. As the Goodman diagram in Fig. 6.2 shows, a compressive mean stress will improve the fatigue resistance of the thread root. Because of these advantages thread-rolled fasteners are essential for high-performance applications. But how is the residual compression generated?

The geometry of the stress field at the root of a thread is quite complex, but we can understand the principles of what is happening perfectly well by analysing a much simpler situation. Figure 6.8 shows a bar of metal that is subjected to a bending moment which is just sufficient to make the whole of the cross-section go plastic. For simplicity we assume that the cross-section is a solid square with the neutral axis parallel to two opposite edges of the square. From Appendix 1C we see that the plastic moment is

$$M_\mathrm{p} = \frac{\sigma_y a^3}{4}. \tag{6.21}$$

When the plastic moment is removed the elastic "springback" leaves a residual elastic stress distribution in the bar. The trick for finding this distribution is to cancel out the plastic moment by applying an equal *elastic* moment of opposite sign. The required elastic moment is given by

$$M_\mathrm{el} + M_\mathrm{p} = 0. \tag{6.22}$$

From Appendix 1B

$$M_\mathrm{el} = \frac{\sigma_\mathrm{el} I}{z}, \tag{6.23}$$

where

$$I = \frac{a^4}{12}. \tag{6.24}$$

Thus

$$M_\mathrm{el} = \frac{\sigma_\mathrm{el} a^4}{12z} = -\frac{\sigma_y a^3}{4}, \tag{6.25}$$

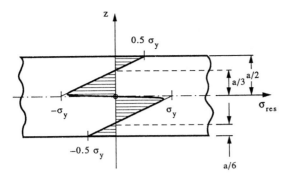

FIG 6.9   Residual elastic stress distribution in a fully plastic beam after springback.

which gives

$$\sigma_{el} = -3\sigma_y\left(\frac{z}{a}\right). \tag{6.26}$$

The residual stress distribution is found by summing the plastic and elastic stresses to give

$$\sigma_{res} = \sigma_y + \sigma_{el} = \sigma_y\left\{1 - 3\left(\frac{z}{a}\right)\right\}. \tag{6.27}$$

This function is plotted in Fig. 6.9. Because of the elastic springback the top part of the cross-section, which was previously yielding in compression, has gone into tension; and the bottom part of the section has gone into compression. There are balancing regions of both tension and compression in the central part of the cross-section. The residual stress at the top surface is $+0.5\sigma_y$ and the residual stress at the bottom surface is $-0.5\sigma_y$. The zone of compression extends in from the bottom surface by $a/6$.

The material at the root of the thread behaves in the same way when it unloads from the thread-rolling operation. The stress state at the thread root is compressive, with a zone of balancing tensile stress further in from the root. The process achieves an optimum distribution of stress because the region where fatigue cracks are most likely to initiate (the stress concentration at the thread root) is also the region of maximum compressive residual stress. Of course, when the bolt is preloaded the stress at the root will increase and the material may go into tension. But the stress state will be much less dangerous than if there were no residual stress to start with.

Reference

J. E. Shigley, *Mechanical Engineering Design*, 3rd edition, McGraw-Hill Kogakusha, 1977.

B. Plastic deformation

CHAPTER 7

BUCKLING OF A MODEL STEAM BOILER

Background

Figure 7.1 shows a cross-section of a model steam boiler. The boiler is of the vertical fire-tube type and comes from a model of a railway locomotive designed for use on tramway systems. The boiler has a working pressure of 0.69 MPa and is coal fired. It produces enough power to haul a load of several people on a model railway track of 5-inch gauge.

The components of the boiler are made from pure copper. The shell is a length of thin-walled cylindrical tube 130 mm in diameter. The firebox tube is 110 mm in diameter and has a wall thickness of 2.34 mm. The unsupported length of the firebox, measured between the foundation ring and the firebox tubeplate, is 120 mm.

The parts are all joined using silver solder, which melts at about 610°C. This is above the annealing temperature of copper so, although the tubes and plates are supplied in a work-hardened state, they are all fully softened by the soldering operations.

Before using the new boiler it was tested hydraulically to a pressure of 1.38 MPa, twice the working pressure. The test was successful and the model was put into service. However, when a repeat test was carried out two years later, two bulges were found in the firebox wall, as shown in Fig. 7.1.

The failure was obviously caused by the buckling of the firebox tube under the external hydrostatic pressure applied during the test. However, as we will show in this case study, the buckling initiated as a plastic event and is very different from the failures analysed in Chapters 4 and 5 which initiated as elastic events.

Material Properties

Smithells gives the following data for pure annealed copper.

Poisson's ratio $v = 0.34$.
Young's modulus $E = 117 \times 10^3$ MPa.
Limit of proportionality = 15 MPa.
0.2% proof stress = 48 MPa.

Figure 7.2 is a graph of stress against strain with the proportional limit and the proof stress plotted on it. Now it is usual to represent the strain-hardening of a metal by the equation

$$\sigma = a\varepsilon^m \tag{7.1}$$

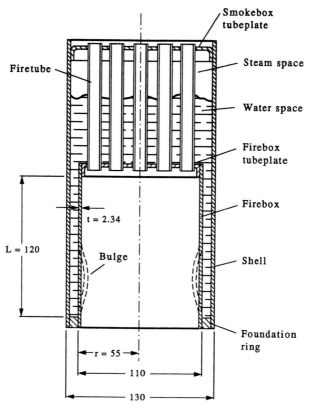

FIG 7.1   Cross-section through the model steam boiler. Dimensions in mm.

where a and m are constants. As a first try, to be verified later, we assume that $m = 0.5$. We also assume that Eqn. (7.1) is valid from the origin of the stress–strain graph, i.e. we assume that the annealed copper is so soft that it does not have a yield point. Then

$$a = \frac{\sigma}{\varepsilon^m} = \frac{48 \text{ MPa}}{(2.41 \times 10^{-3})^{0.5}} = 0.98 \times 10^3 \text{ MPa}. \tag{7.2}$$

The stress–strain equation is then

$$\sigma = 0.98 \times 10^3 \text{ MPa} \times \varepsilon^{0.5}. \tag{7.3}$$

This is plotted in Fig. 7.2. In fact the limit of proportionality is quite close to the curve, so our neglect of the yield point seems to be a reasonable approximation.

Stresses in the Firebox Wall

As we can see from Fig. 7.1, the firetubes function as stays which hold the smokebox and firebox tubeplates together against the action of the pressure inside the boiler. There is therefore no net downward force on the top of the firebox

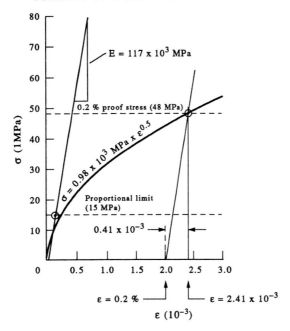

FIG 7.2 Stress–strain data for pure annealed copper. The elastic strain at the 0.2% proof stress is given by $\varepsilon = \sigma/E = 48$ MPa$/(117 \times 10^3$ MPa$) = 0.41 \times 10^{-3}$.

and the longitudinal stress in the firebox wall is zero. The principal stresses in the firebox wall are then

$$\sigma_1 = \frac{pr}{t}, \qquad \sigma_2 = 0, \qquad \sigma_3 = 0. \tag{7.4}$$

This corresponds to a state of uniaxial compression. The firebox then yields when the hoop stress is equal to the uniaxial yield stress, i.e. when

$$\sigma = \frac{pr}{t} = \sigma_y. \tag{7.5}$$

At the test pressure the hoop stress is given by

$$\sigma = 1.38 \text{ MPa} \times \frac{55 \text{ mm}}{2.34 \text{ mm}} = 32.4 \text{ MPa}. \tag{7.6}$$

This is well beyond the proportional limit and an elastic analysis of the failure is obviously not appropriate.

Plastic Buckling Analysis

Shanley showed in 1947 that the equations for the elastic buckling of struts could be applied to situations where the buckling began as a plastic hinge as long as

Young's modulus was replaced by the tangent modulus, given by

$$E_t = \frac{d\sigma}{d\varepsilon}. \tag{7.7}$$

The ASME Code adopts this approach for the plastic buckling of tubes under external pressure and uses the elastic equations with E replaced by E_t. In order to analyse the firebox failure we simply modify the tube-buckling equations given in Appendix 1B to give

$$p_{cr}(\text{theoretical}) = \frac{E_t}{4(1 - v^2)} \left(\frac{t}{r}\right)^3 \tag{7.8}$$

for long tubes, and

$$p_{cr}(\text{theoretical}) = \frac{0.86E_t}{(1 - v^2)^{3/4}} \left(\frac{t}{L}\right)\left(\frac{t}{r}\right)^{3/2} \tag{7.9}$$

for tubes where the end constraints are significant. To find out which of these equations is appropriate to the failure we have to plot the ratios of L/r and r/t on the Tube Buckling Map (see Appendix 1B). The ratios are

$$\frac{L}{r} = \frac{120 \text{ mm}}{55 \text{ mm}} = 2.18 \tag{7.10}$$

and

$$\frac{r}{t} = \frac{55 \text{ mm}}{2.34 \text{ mm}} = 23.5 \tag{7.11}$$

which intersect in the $n = 4$ field on the map. The relevant equation is then Eqn. (7.9), and a fully developed buckling failure should have four lobes.

The next task is obviously to find the appropriate value for E_t. This is done by writing equations as follows.

$$\sigma = a\varepsilon^{0.5} = \frac{pr}{t} \tag{7.12}$$

which gives

$$\varepsilon^{-0.5} = \frac{at}{pr}. \tag{7.13}$$

Now

$$\frac{d\sigma}{d\varepsilon} = 0.5a\varepsilon^{-0.5} = \frac{0.5a^2t}{pr} \tag{7.14}$$

and

$$E_t = \left(\frac{d\sigma}{d\varepsilon}\right)_{\text{critical}} = \frac{0.5a^2t}{p_{cr}(\text{theoretical})r}. \tag{7.15}$$

Thus

$$p_{cr}(\text{theoretical}) = \frac{0.86}{(1 - v^2)^{3/4}} \times \frac{0.5a^2t}{p_{cr}(\text{theoretical})r} \left(\frac{t}{L}\right)\left(\frac{t}{r}\right)^{3/2} \tag{7.16}$$

which gives

$$p_{cr}(\text{theoretical}) = \frac{0.66a}{(1 - v^2)^{3/8}} \left(\frac{t}{L}\right)^{1/2} \left(\frac{t}{r}\right)^{5/4}. \tag{7.17}$$

Since $a \approx 0.98 \times 10^3$ MPa (see Eqn. (7.3)) we get

$$p_{cr}(\text{theoretical}) = 1.83 \text{ MPa} \tag{7.18}$$

using the values for v, t, L and r given above. Finally,

$$p_{cr}(\text{actual minimum}) \approx 0.8 \times 1.83 \text{ MPa} \approx 1.46 \text{ MPa}. \tag{7.19}$$

This is very close to the actual collapse pressure of 1.38 MPa, with the theory over-predicting the answer by only 6%.

Verifying the Analysis

The theoretical predictions are obviously sensitive to the values chosen for a and m in Eqn. (7.1) and it would be comforting to have an independent verification that the values we have used are reasonable. Experimental results for the collapse pressures of annealed copper tubes are sparse, although Markham gives some data for long tubes, where Eqn. (7.8) is appropriate. We should be able to verify our method by comparing Markham's data with Eqn. (7.8), provided we incorporate the expression for E_t given in Eqn. (7.15). Then

$$p_{cr}(\text{theoretical}) = \frac{1}{4(1 - v^2)} \times \frac{0.5a^2t}{p_{cr}(\text{theoretical})r} \left(\frac{t}{r}\right)^3 \tag{7.20}$$

giving

$$p_{cr}(\text{theoretical}) = \frac{a}{2.83(1 - v^2)^{1/2}} \left(\frac{t}{r}\right)^2 \tag{7.21}$$

and

$$p_{cr}(\text{actual minimum}) \approx \frac{0.8a}{2.83(1 - v^2)^{1/2}} \left(\frac{t}{r}\right)^2. \tag{7.22}$$

Figure 7.3 is a plot of $p_{cr}(\text{actual minimum})$ against t/r taken from Eqn. (7.22) with $a = 0.98 \times 10^3$ MPa. Remember that Eqn. (7.22) assumes that $m = 0.5$. Markham's experimental results are also plotted for comparison. Theory and experiment agree reasonably well, with the theory over-predicting the collapse pressure by up to 18%. Our method of calculating E_t thus seems to be fairly accurate.

Design of Model Boilers

The parts of model boilers that are in tension under operating conditions are usually designed so that the working stress is about one-eighth of the tensile strength of annealed copper at the working temperature. So there is an ample margin of safety against failure. This is important, because if a shell were to fail the contents of the boiler would discharge rapidly into the environment and could cause serious damage or injury. However, parts that are subjected to external pressure—like

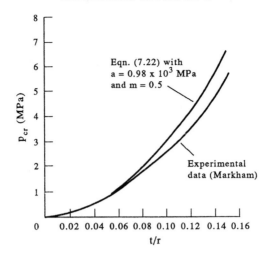

FIG 7.3   Collapse pressures of long copper tubes in the annealed condition.

firetubes, or the tubular firebox in our vertical boiler—are not consciously designed against buckling.

This is probably because the designs of model boilers were evolved before the present design codes had been formulated, and also because the people involved in model engineering did not generally have the resources or expertise to design against buckling in a predictive and quantitative way. Instead, designs were evolved on the basis of experience, and the dimensions arrived at for tubes were known to be adequate to resist buckling under hydraulic test.

Since hydraulic testing is generally done at twice the working pressure there will usually be a factor of safety of at least 2 against buckling failure at the working pressure. This is adequate because a buckling failure is contained and is most unlikely to lead to a rupture in the strongly bonded and highly ductile assembly that is a silver-soldered copper boiler. However, silver solder and copper are expensive, and with the average model boiler costing about £500 ($750) one wants to avoid failures even under hydraulic test.

The vast majority of steam-driven models built by model engineers are railway locomotives, steam rollers and road traction engines. These almost always use boilers of the locomotive type, an example of which we show in Fig. 7.4. The firebox is made from flat plates which are tied to the outside of the boiler by stays. These take the pressure loading in tension. As a result the firebox is not liable to collapse by buckling. The only parts that could buckle are the firetubes.

As we see from Fig. 7.4 most boilers for railway locomotives use two sizes of firetube. The larger tubes have to accommodate tubes for superheating the steam. Table 7.1 lists dimensions and test pressures for the flue tubes of a range of published boiler designs together with the calculated collapse pressures. Details of the calculations are given in the solution to Example 7.1.

In most cases there is an ample factor of safety at the test pressure. There is also a large variation in the factor of safety, indicating that the designs are far from optimal. In fact the specified tube dimensions are probably dictated more by the

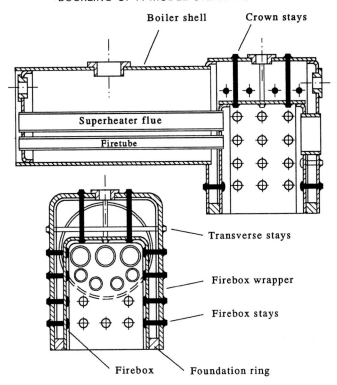

Boiler shell Crown stays

Superheater flue

Firetube

Transverse stays

Firebox wrapper

Firebox stays

Firebox Foundation ring

FIG 7.4 Drawings of a typical locomotive-type boiler for a model steam railway engine.

TABLE 7.1 *Flue Tubes for Model Boilers*

Type of model	Test pressure (MPa)	L (mm)	r (mm)	t (mm)	Actual minimum p_{cr} (MPa)	Factor of safety under test
0-6-0 locomotive	1.56	203	5.5	0.91	8.09	5.2
4-6-2 locomotive	1.10	259	5.5	0.71	4.91	4.5
			9.0	0.91	2.63	2.4
2-6-2 locomotive	1.24	318	6.3	0.91	6.16	5.0
			12.7	1.22	2.73	2.2
4-6-2 locomotive	1.24	559	8.0	0.91	3.82	3.1
			12.7	1.63	4.86	3.9
4-4-2 locomotive	2.08	419	6.3	0.91	6.16	3.0
			12.7	1.22	2.73	1.3
Traction engine	1.10	203	4.8	0.71	6.46	5.9
Portable engine	1.24	280	5.5	0.56	3.06	2.5
Vertical boiler	1.38	89	17.5	1.22	2.28	1.7
Vertical boiler	1.38	108	12.7	0.91	1.84	1.3

TABLE 7.2   *Tubular Fireboxes*

Type of model	Test pressure (MPa)	L (mm)	r (mm)	t (mm)	Actual minimum p_{cr} (MPa)	Factor of safety under test
0-6-0 locomotive	1.56	175	48	3.25	2.55	1.6
Vertical boiler	0.82	70	38	2.34	3.05	3.7
Centreflue boiler	1.10	165	19	0.91	0.90	0.8
Scotch boiler	0.82	89	25	1.63	2.42	3.0
Marine boiler	1.38	114	32	2.03	2.30	1.7
0-4-0 locomotive	1.10	133	66	3.18	1.87	1.7
Vertical boiler	1.38	76	51	2.34	2.02	1.5

standard sizes of tube available than by any design considerations. Interestingly, the wall thickness decreases slowly in service because the ash particles that are drawn through the firetubes with the flue gases are abrasive. Thus we need a definite factor of safety at the test pressure if the boiler is to have a long life.

Table 7.1 shows two cases where the safety factor is only 1.3. In the case of the 4-4-2 locomotive superheater flue this is caused by an unusually high working pressure. In the case of one of the vertical boilers the ratio of r/t is large. The designers of these two boilers were probably not aware that the tubes were close to collapse at the test pressures!

Because boilers with tubular fireboxes are much easier to make than locomotive-type boilers they have become more popular in recent years. Table 7.2 gives sizes and pressures for some published designs of tubular fireboxes. Details of the calculations are given in the solution to Example 7.2.

There is one boiler where the safety factor is less than 1. The firebox of this unit was fitted with cross water tubes which would have helped to resist buckling, and this is presumably why failures were not reported. In all other cases the safety factor is at least 1.5. This might seem adequate, but because the actual collapse pressure depends strongly on the geometrical perfection of the tube it would be wise to have a safety factor of at least 2. As shown in Eqn. (7.17) the predicted collapse pressure is proportional to $t^{1.75}$ so only a small increase in wall thickness would be needed to increase the safety factor from 1.5 to 2.

References

ASME Boiler and Pressure Vessel Code (American Society of Mechanical Engineers). Section 8, Division 2: *Appendix J, Basis for Establishing External Pressure Charts*, 1986.

M. Evans, *Manual of Model Steam Locomotive Construction*, Model Aeronautical Press, 1967.

H. Greenly, *Model Steam Locomotives*, Cassell, 1954.

K. N. Harris, *Model Boilers and Boilermaking*, Model and Allied Publications, 1972.

B. G. Markham, "Collapsing Tubes", *Model Engineer*, **159**, 586 (1987).

Pressure Vessel and Piping Design: Collected Papers 1927–1959, American Society of Mechanical Engineers, 1960.

F. R. Shanley, "Inelastic Column Theory", *J. Aero. Sci.*, **14**, 261 (1947).

C. J. Smithells, *Metals Reference Book*, 6th edition, Butterworth, 1984.

CHAPTER 8

FAILURE OF A CHAIN WRENCH

Background

In this case study we investigate why a hand tool called a chain wrench fractured when it was being used to unscrew a length of threaded pipe from a screwed fitting. The wrench was roughly 1.5 m long, was made from steel, and weighed about 25 kg. When it fell apart it injured one of the operators so it was important to identify the likely cause of the accident.

Figure 8.1 shows a schematic of the chain wrench being used to unscrew the pipe. One end of the chain is anchored permanently inside the head of the wrench. From there the chain is taken over the outside of the pipe and the trailing end is secured back inside the head as shown. The handle of the wrench is pressed downwards by the operator. This puts the chain into tension near the fixed anchoring point and it is this tension which applies a moment to the pipe. The tension in the chain is reacted by a compressive force between the pipe and the wrench head, which is serrated so that it will not slide around the outside of the pipe.

Figures 8.2 and 8.3 are scale drawings showing details of the wrench head and the chain. The diameter of the pipe is about 160 mm. As we can see from Fig. 8.2, for this diameter of pipe there is an offset of about 75 mm between the line of action of the leading end of the chain and the fulcrum between the head and the pipe. The trailing section of the chain is offset by about 39 mm from the fulcrum.

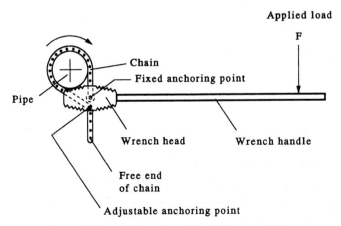

FIG 8.1   Schematic drawing of the chain wrench in use.

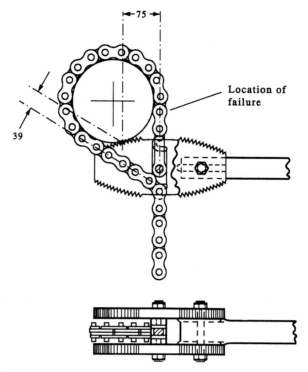

FIG 8.2   Close-ups of the wrench head. To scale. Dimensions in mm.

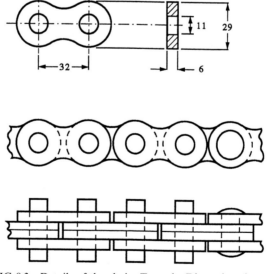

FIG 8.3   Details of the chain. To scale. Dimensions in mm.

As Fig. 8.3 shows, the chain is made up from separate links which are joined either by rivets or short pieces of rod. We can see from Fig. 8.2 that the leading section is assembled with rivets and the trailing section with rods. The purpose of having the rods is that they can be used to secure the trailing section of the chain back inside the head. There is a pair of hook-shaped depressions on either side of the head which are designed to receive the protruding ends of one of the rods. Thus the anchoring point can be adjusted to the nearest suitable link. The leading end of the chain is connected to the fixed anchoring point by a single forged link. The link is able to swing about the anchoring point in order to adjust itself to the direction of pull of the chain.

Details of Failure

In the accident the chain broke near the fixed anchoring point at the location shown in Fig. 8.2. Details of the fracture are shown in Fig. 8.4. A pair of links had broken in two at the rivet holes, allowing the rivet to pull out. The fracture surfaces appeared to be fibrous in nature. As shown in Appendix 1C this is consistent with a ductile failure caused by a tensile overload. There was not much reduction in area but this is what one might expect from a steel which had been hardened and tempered. The rivet had deformed in double shear, although it was still intact.

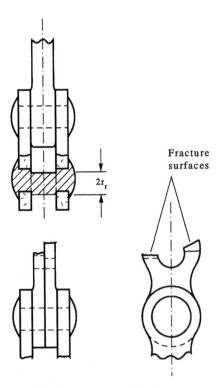

FIG 8.4   Details of the fracture.

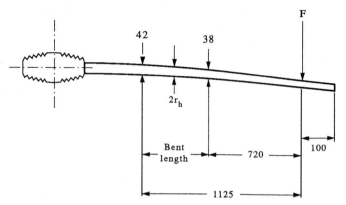

FIG 8.5   The bend in the wrench handle. Dimensions in mm.

Not only had the chain broken but the handle had become bent, as shown in
Fig. 8.5. The bend started about 1225 mm from the free end of the handle, at a
location where the handle was 42 mm in diameter, and ended about 820 mm from
the end, where the diameter was 38 mm. Obviously, the load which had been applied
to the handle had been big enough to create a fully plastic bending moment over
the length of the bent section.

There were two scenarios for the incident. The first was that the wrench had
failed because there was a defect in the chain. The second was that the wrench had
been overloaded. The investigation was complicated by the fact that any tests done
on the parts had to be non-destructive. In practice this meant that the investigation
was restricted to a theoretical analysis of the forces involved and a non-destructive
evaluation of the mechanical properties of the parts using hardness testing.

Force Equations

The forces acting on the wrench are shown in Fig. 8.6. We have assumed that in
normal use the best that the operator can do is to press down on the handle at a

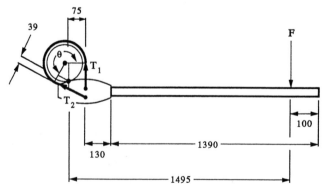

FIG 8.6   Forces acting on the wrench. Dimensions in mm.

distance of about 100 mm from the free end. Since the average human hand is about 100 mm wide the operator would not be able to exert a greater leverage without a hand falling off the end of the wrench.

The moment that the applied force exerts about the fulcrum is balanced by the moments that are exerted about the fulcrum by the tensions in the chain. We call the tension in the leading end of the chain T_1 and the tension in the trailing end T_2. Because of friction between the chain and the pipe, T_2 is less than T_1. The two tensions are related by the standard mechanics equation

$$T_1 = T_2 e^{\mu\theta}. \tag{8.1}$$

μ is the coefficient of friction and θ is the angle of wrap in radians (see Fig. 8.6). Smithells quotes a value of about 0.5 for the friction of hard steel on steel under clean, dry conditions. Measurements made on Fig. 8.2 show that θ is about 245° or 4.3 radian. With these values we get

$$T_2 = 0.116 T_1. \tag{8.2}$$

Taking moments about the fulcrum we have

$$F \times 1495 + T_2 \times 39 = T_1 \times 75. \tag{8.3}$$

Combining this with Eqn. (8.2) gives

$$T_1 = 21F. \tag{8.4}$$

Bending the Handle

We can estimate F by finding out how big the applied force needs to be to generate a fully plastic moment in the handle. From Appendix 1C we have

$$M_p = \frac{4\sigma_y r_h^3}{3} \tag{8.5}$$

where r_h is the radius of the handle.

In order to estimate the yield strength we measured the Vickers hardness on the outside of the handle. The average hardness was 185 ± 2. Using the correlation between hardness and tensile strength given in Appendix 1C we get

$$\sigma_{TS} \text{ (MPa)} \approx 3.2 \times HV \approx 3.2 \times 185 \approx 590 \text{ MPa}. \tag{8.6}$$

The specification for the handle called for a forging of plain carbon–manganese steel containing 0.25 to 0.35% carbon and 0.60 to 0.80% manganese. Smithells quotes a yield stress of about 435 MPa for a steel having a tensile strength of 590 MPa with this composition.

Using this estimate of the yield stress we can calculate the plastic moment at two positions: the start of the bend, where r_h is 21 mm and the end of the bend, where r_h is 19 mm. The moment cannot be found accurately in the bent section: because of work-hardening the yield stress will be greater than 435 MPa because the handle has experienced a definite amount of plastic flow.

The plastic moment at the start of the bend is

$$M_p = \frac{4 \times 435 \text{ N mm}^{-2} \times 21^3 \text{ mm}^3}{3}$$

$$= 5.37 \times 10^6 \text{ N mm.} \qquad (8.7)$$

At the end of the bend it is

$$M_p = \frac{4 \times 435 \text{ N mm}^{-2} \times 19^3 \text{ mm}^3}{3}$$

$$= 3.98 \times 10^6 \text{ N mm.} \qquad (8.8)$$

Taking moments about the start of the bend we get

$$M_p = F \times 1125 \text{ mm} \qquad (8.9)$$

which gives

$$F = \frac{M_p}{1125 \text{ mm}} = \frac{5.37 \times 10^6 \text{ N mm}}{1125 \text{ mm}} = 4.77 \times 10^3 \text{ N.} \qquad (8.10)$$

Taking moments about the end of the bend we get

$$M_p = F \times 720 \text{ mm} \qquad (8.11)$$

which gives

$$F = \frac{M_p}{720 \text{ mm}} = \frac{3.98 \times 10^6 \text{ N mm}}{720 \text{ mm}} = 5.53 \times 10^3 \text{ N.} \qquad (8.12)$$

The mean of these two forces is

$$F = \frac{(4.77 + 5.53) \times 10^3 \text{ N}}{2} = 5.15 \times 10^3 \text{ N} = 525 \text{ kgf.} \qquad (8.13)$$

Finally Eqn. (8.4) shows that this load will generate a tension in the leading end of the chain given by

$$T_1 = 21 \times 5.15 \times 10^3 \text{ N} = 1.08 \times 10^5 \text{ N.} \qquad (8.14)$$

Shearing the Rivet

The load needed to make the rivet fail in double shear is given by

$$T_1 = k_u 2\pi r_r^2 \qquad (8.15)$$

where k_u is the shear failure strength and r_r is the radius of the rivet shank. Since we were not able to measure the mechanical properties of the rivet directly the shear failure strength had to be found from the hardness using the empirical correlations given in Appendix 1C. Unfortunately it was not possible to measure the Vickers hardness of the cylindrical body of the rivet because the very flat pyramid of the

Vickers indenter fouled the chain links. But it was possible to get a Rockwell C-scale indenter onto the body of the rivet because this indenter is more compact. The average Rockwell hardness was 30 ± 3. The corresponding value of Vickers hardness is given by

$$HV \approx \frac{HRC}{0.1} = \frac{30 \pm 3}{0.1} = 300 \pm 30. \qquad (8.16)$$

Note that we have used a figure of 0.1 in the correlation rather than the figure of 0.09 listed in Appendix 1C. The correlation term does vary over the full range of hardness values and the detailed conversion tables given by Smithells indicate that 0.1 is a more accurate figure in the region of 30 HRC. The tensile strength of the rivet shank is given approximately by

$$\sigma_{TS} \approx 3.2 \times (300 \pm 30) \text{ MPa} = 960 \pm 96 \text{ MPa}. \qquad (8.17)$$

Finally the shear failure strength is

$$k_u \approx \frac{(960 \pm 96) \text{ MPa}}{1.6} = 600 \pm 60 \text{ MPa}. \qquad (8.18)$$

Equation (8.15) then gives us

$$T_1 = (600 \pm 60) \text{ N mm}^{-2} \times 2\pi(5.5)^2 \text{ mm}^2$$

$$= (1.14 \pm 0.11) \times 10^5 \text{ N}. \qquad (8.19)$$

This value is in very good agreement with the chain tension of 1.08×10^5 N that is generated when the handle is bent.

Fracturing the Link

Figure 8.3 gives the dimensions of a typical link. If we use these with the details of the tensile failure shown in Fig. 8.4 we can estimate the area of the tensile fracture surface on each link. This is

$$(29 - 11) \text{ mm} \times 6 \text{ mm} = 108 \text{ mm}^2. \qquad (8.20)$$

The total area of the fracture surface is twice this value (because a pair of links broke) and is

$$2 \times 108 \text{ mm}^2 = 216 \text{ mm}^2. \qquad (8.21)$$

We can estimate the load needed to break the links by multiplying this area by the tensile strength. We are not able to measure the tensile strength of the links directly but we can estimate it by using the correlation between tensile strength and hardness given in Appendix 1C.

In order to do this the Vickers hardness was measured on the surface of both broken links, giving an average of 250. The tensile strength is thus

$$\sigma_{TS} \approx 3.2 \times 250 \text{ MPa} = 800 \text{ MPa}. \qquad (8.22)$$

Finally we get a breaking load of

$$T_1 \approx 800 \text{ N mm}^{-2} \times 216 \text{ mm}^2 = 1.73 \times 10^5 \text{ N}. \tag{8.23}$$

It is interesting to see that this is about 50% more than the chain tension that we calculated on the basis of handle bending or rivet shear.

Diagnosis

One's first reaction to this 50% discrepancy is that there must have been a defect in the links. But we have to bear in mind that tensile strength is measured using uniaxial tensile specimens whereas the metal around the holes in the links was in anything but uniaxial tension. On this basis we might well expect the links to have failed at a lower average stress than the tensile strength and this alone might explain the discrepancy.

Of course it is possible that the failure was not entirely tensile but involved an element of shear as well. Figure 8.7 shows a failure mechanism which takes place entirely by shear. In this case the fracture area is the same as it was in the tensile failure mode, but the failure stress is now the shear failure stress, given by

$$k_u \approx \frac{800 \text{ MPa}}{1.6} = 500 \text{ MPa}. \tag{8.24}$$

The breaking load in this case is

$$T_1 \approx 500 \text{ N mm}^{-2} \times 216 \text{ mm}^2 = 1.08 \times 10^5 \text{ N} \tag{8.25}$$

which is in close agreement with the chain tension estimated from both handle bending and rivet shear.

There is one way in which the links could conceivably have been defective and that is if they had not been heat treated properly. Unfortunately no information was available on the composition of the links but a typical steel for such parts has about 0.36% carbon and 1.2% manganese with a tensile strength of 780 to 930 MPa after hardening and tempering (see Smithells). This is comparable to our estimated value of 800 MPa.

FIG 8.7   Failure of a link in double shear.

Appendix 1G gives data for the hardness of martensite as a function of the carbon content. From this information we would expect a steel which contained 0.36% carbon to have a surface hardness in the as-quenched condition of about 650 Vickers. The hardness of 250 Vickers that we measured on the links is much less than this value so we have no evidence to suggest that the steel had not been tempered adequately after quenching.

Although there is little evidence that there was a prior defect in the chain there is ample evidence that the wrench was overloaded. As we showed earlier we would have to apply a force of about 525 kgf to the end of the handle to bend it. This is 5.8 times the body weight of an average person which seems to give a generous factor of safety. In order to bend the handle we would have to double its length and stand three people on the end. However, it is by no means unknown for people to slip a length of pipe over the handle of a wrench or spanner to increase the leverage. The author has done this himself, but as this case study shows it is not a practice to be encouraged!

Reference

C. J. Smithells, *Metals Reference Book*, 6th edition, Butterworth, 1984.

CHAPTER 9

SHEAR OF A TRUCK STEERING SHAFT

Background

In this case study we examine a broken steering shaft from a truck that had been involved in a road accident. There were two possibilities:

(a) the failure had occurred before the accident and the loss of steering had caused the accident;
(b) the accident had been caused by something else and the shaft had broken purely as a result of the accident.

Figures 9.1 and 9.2 show scale drawings of the steering shaft. The shaft is supported by two needle roller bearings. Between the bearings is a toothed sector which is driven by a worm connected to the steering wheel. The steering arm, which transmits movement to the track rods, is attached to the end of the shaft by a splined connection.

As Fig. 9.3 shows, the steering shaft had been subjected to a torsional overload. The splined section of the shaft had been twisted permanently so that the ends of the splines were offset by 1.3 mm. The shaft had broken where it met the toothed sector. As we can see from Appendix 1C, this is a classic shear type failure. Presumably the torque had been generated as a result of a large force applied to the end of the steering arm, as shown in Fig. 9.2.

Photographs of the Failure

Figure 9.4 shows the twisted splines. Figure 9.5 shows the matching fracture surfaces. Most of the fractured area is flat and smooth, but there is a region near the centre which is a good deal rougher. The flat area is the shear part of the failure. The rough area is where the two parts of the shaft finally separated by tensile failure. Figure 9.6 is a view taken in the scanning electron microscope of the shear part of the fracture surface. It shows the classic features of a shear failure including the smearing patterns made by the ductile metal sliding over itself in the direction of shear. Figure 9.7 is a scanning electron micrograph taken from the tensile failure: it shows the classic features of fibrous failure reviewed in Appendix 1C. There were no signs of any defects on the fracture surface which could have weakened the cross-section and led to premature failure.

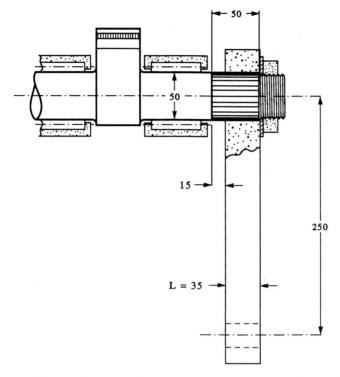

FIG 9.1   Side view of the steering shaft. Dimensions in mm.

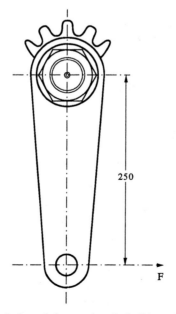

FIG 9.2   End view of the steering shaft. Dimension in mm.

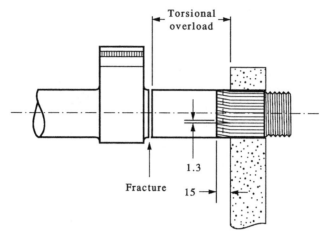

FIG 9.3   Side view of the shaft showing the plane of the fracture and the region of torsional overload. Dimensions in mm.

FIG 9.4   Photograph of the splined section of the steering shaft showing the twisted splines.

Finding the Shear Failure Stress

In order to show that the torsional failure could have been caused by the collision we need to estimate the torsional moment needed to shear the shaft. But before we can do this we need to have data for the shear failure stress k_u. This is best done by making hardness measurements on a cross-section cut from the shaft and then using the correlation between hardness and shear failure strength given in Appendix 1C to find k_u.

In order to do this a thin slice was cut out of the shaft using a high-speed abrasive cutting disc. Plenty of coolant was poured over the shaft during the cutting operation to stop it overheating and to avoid "running the temper" of the structure.

FIG 9.5   Photograph of the matching fracture surfaces.

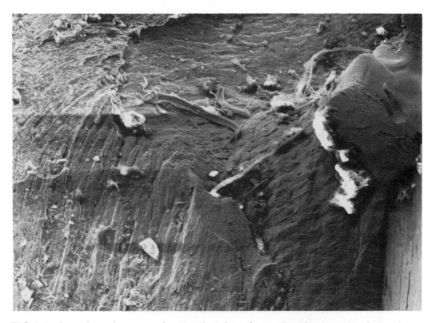

FIG 9.6   Scanning electron micrograph taken from the shear part of the failure.
Magnification: ×170.

One side of the slice was ground and polished to give a good surface on which to
do the hardness tests and also to have a look at the structure.

Figure 9.8 is a macrograph of the polished cross-section after it had been etched
in 2% nital (a solution of 2 ml of concentrated nitric acid in 100 ml of methyl
alcohol). The most noticeable feature is that there is a dark etching skin on the
outside of the shaft, about 1.5 mm thick. This is a case-hardened layer, needed to
give a hard bearing surface for the needle rollers in the bearings.

The Vickers hardness was measured on the polished surface as a function of
the radial distance r from the centre of the shaft. The results are given in

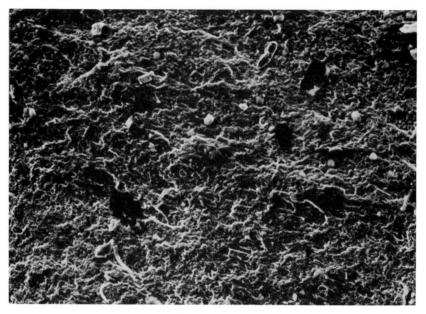

FIG 9.7   Scanning electron micrograph taken from the fibrous part of the failure.
Magnification: ×325.

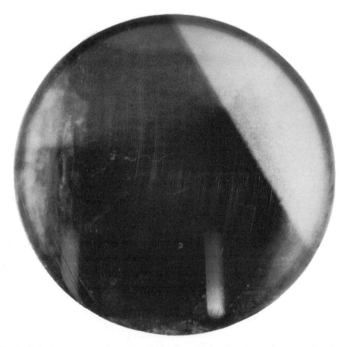

FIG 9.8   Polished cross-section through the shaft showing the case-hardened layer
1.5 mm thick.

TABLE 9.1

r (mm)	HV	σ_{TS} (MPa)	k_u (MPa)
0	350	1120	700
10	360	1152	720
17.5	375	1200	750
22.5	400	1280	800
Case	880	2816	1760

Table 9.1. Note that each result is the average of eight separate readings taken from sites that were spaced uniformly around a concentric circle. This was done in order to get reliable data and to even out any variation in properties at a particular radial distance.

Using the empirical equations given in Appendix 1C we can estimate the tensile strengths from

$$\sigma_{TS} \text{ (MPa)} \approx 3.2 \times HV \tag{9.1}$$

and the shear failure stresses from

$$k_u \approx \frac{\sigma_{TS}}{1.6} \tag{9.2}$$

which is how we get the values listed in Table 9.1. The shear failure stresses are also plotted in Fig. 9.9. Note that there is a big variation from the inside to the outside of the shaft.

Finding the Torsional Moment

Because of the dependence of k_u on r we cannot use the equations given in Appendix 1C to calculate the torsional moment because they assume uniform properties. Referring to Fig. 9.10, we see that the torque needed to shear a narrow concentric strip is

$$d\Gamma = 2\pi k_u r^2 \, dr. \tag{9.3}$$

As Fig. 9.9 shows, k_u can be related to r by the empirical expression

$$k_u \text{ (MPa)} = 700 + 8.78 \times 10^{-3}(r/mm)^3 \tag{9.4}$$

over the core of the shaft (the region from the centre of the shaft to the inner edge of the case). The torque needed to shear the core is then given by

$$\Gamma = 2\pi \int (700 + 8.78 \times 10^{-3}r^3)r^2 \, dr \text{ N mm}$$

$$= 2\pi \left[\frac{700r^3}{3} + \frac{8.78 \times 10^{-3}r^6}{6} \right]_0^{23.5} \text{ N mm}$$

$$= 2\pi(3.03 \times 10^6 + 0.25 \times 10^6) \text{ N mm}$$

$$= 2.06 \times 10^7 \text{ N mm}. \tag{9.5}$$

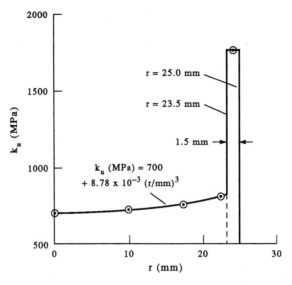

FIG 9.9   Plot of shear strength against radial distance from the centre of the shaft.

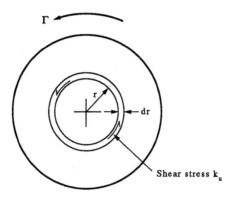

FIG 9.10   Calculating the torsional moment.

The standard formula given in Appendix 1C for the torque needed to shear a tube can be used for the case. We then have

$$\Gamma = \frac{2\pi k_u (r_1^3 - r_2^3)}{3}$$

$$= \frac{2\pi \times 1760 \times (25^3 - 23.5^3) \, \text{N mm}}{3}$$

$$= 0.98 \times 10^7 \, \text{N mm.} \tag{9.6}$$

The total torque is then

$$\Gamma = (2.06 + 0.98) \times 10^7 \, \text{N mm}$$

$$= 3.04 \times 10^7 \, \text{N mm.} \tag{9.7}$$

It is interesting to notice that, even though the case is only 1.5 mm thick, it is responsible for nearly one-third of the total torsional strength. This is partly because the case is more than twice as strong as the core and partly because the torsional strength depends on the cube of the radius.

Calculating *F*

Having found the torsional moment we are now in a position to find the value of the force *F* that must have been applied to the end of the steering arm. Taking moments about the axis of the steering shaft we get

$$F \times 250 \text{ mm} = 3.04 \times 10^7 \text{ N mm} \tag{9.8}$$

which leads to

$$F = 1.22 \times 10^5 \text{ N} = 12.4 \text{ tonnef.} \tag{9.9}$$

It it easy to show that forces of this size can be generated in a collision. If we assume a vehicle weight of 20 tonnes and a deceleration of, say, $4g$ then the force required to stop the truck can be found from Newton's second law, with

$$F = ma = 20 \times 10^3 \text{ kg} \times 40 \text{ m s}^{-2}$$
$$= 8 \times 10^5 \text{ N.} \tag{9.10}$$

This is nearly seven times the value of *F* required to shear the shaft so the steering arm only needs to take one-seventh of the stopping force to make the failure possible.

Specification of the Shaft

Before concluding our analysis of the failure we need to check that the shaft complied with the specifications. These called for the shaft to be machined from a nickel–chromium–molybdenum steel with the following composition.

TABLE 9.2

Element	Weight %	Element	Weight %
Carbon	0.17	Phosphorus	0.035 max
Silicon	0.25	Chromium	1.6
Manganese	0.5	Molybdenum	0.3
Sulphur	0.035 max	Nickel	1.55

The mechanical properties of the core in the quenched condition were specified as follows.

Yield stress = 736 MPa minimum.
Tensile strength = 1079 to 1324 MPa.
Elongation = 8% minimum.
Impact energy = 59 J cm^{-2} minimum.

The case was required to have a Rockwell C-scale hardness of 59 to 63.

The data in Table 9.1 show that our estimates for the tensile strength of the core lie in the range 1120 to 1280 MPa, which is within the specified range.

The elongation can be found from the offset of the splines shown in Fig. 9.3. The maximum engineering shear strain that the splined section has suffered is given by

$$\gamma = \frac{1.3 \text{ mm}}{15 \text{ mm}} = 0.087 = 8.7\%. \tag{9.11}$$

As shown in Appendix 1C, the equivalent plastic strain in uniaxial tension is

$$\frac{\gamma}{1.732} = 5.0\%. \tag{9.12}$$

This is five-eighths of the minimum requirement, so it is not surprising that fracture did not occur in the parallel part of the shaft. The fact that fracture did occur at the transition between the shaft and the toothed sector reflects the fact that conditions at this point were far from uniaxial.

Using the hardness conversion tables given by Smithells we find that the specified values for the hardness of the case are equivalent to Vickers hardnesses of 680 to 780. As Table 9.1 shows, the actual Vickers hardness of the case is 880 HV. This is higher than the specification but there is no evidence that this influenced the failure.

Metallurgy of the Shaft

The way in which the hardness of the shaft depends on the distance from the centre is a good example of how the properties of steel sections can be modified by case hardening and quenching.

The case hardening would have involved heating the steel for several hours in a carburising atmosphere at about 900°C. At this temperature the steel is in the austenite phase and can dissolve as much as 1.3 weight % carbon. However, the carbon concentration would only have been 1.3% at the surface of the shaft: below the surface the concentration would have fallen off rapidly because of the concentration gradient needed to drive the diffusion of carbon into the steel. After the carburising treatment the shaft would have been quenched to transform the structure to martensite. The quenching medium would almost certainly have been oil, because quenching in water usually causes quench cracking. As we can see from the data in Appendix 1G the hardness of the case (880 HV) is typical of martensite containing more than about 0.8% carbon. Presumably the hardness readings in the case were taken at a position where the carbon concentration was at least 0.8%.

The structural changes that the quench induces in the core can be followed from the continuous cooling transformation diagram for the steel of the shaft. This is shown in Fig. 9.11. It tells us that if we quench our 50-mm shaft in oil the structure at the centre will be 50% martensite and 50% bainite. The as-quenched hardness should be 340 HV at the centre. As we can see from Table 9.1, the measured hardness at the centre is in fact 350 HV, in excellent agreement with the predictions of the diagram. However, this close agreement does not necessarily indicate that the shaft had not been tempered after quenching. As Parrish and Harper point out, it

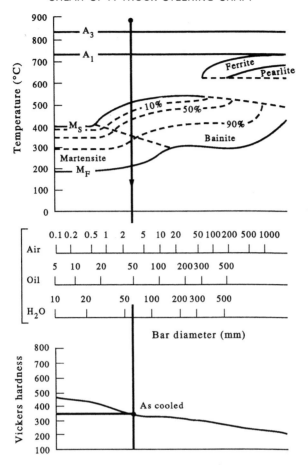

FIG 9.11   Continuous cooling transformation diagram for the steel of the shaft (Atkins).
Before the quench the samples were first heated at 900°C for 4 hours to simulate the
thermal history that the steel would experience during case carburising.

would have been normal practice to temper the case-hardened shaft in the range
150 to 180°C in order to increase the toughness. But only if the tempering had been
carried out above 300°C would there have been any decrease in the hardness of the
core.

Table 9.1 shows that the maximum hardness of the core is 400 HV. The diagram
shows that this hardness is typical of martensite containing 0.17% carbon: the layers
of the core near the case obviously cooled fast enough even in oil to give a fully
martensitic structure.

Design of Splined Connections

The steering shaft is an interesting example of how splined connections are an
outstanding solution to the problem of transmitting large torques. The dimensions
of the connection are shown in Fig. 9.12.

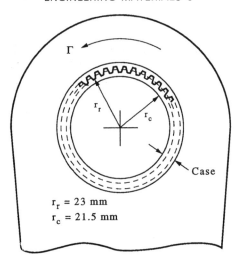

$r_r = 23$ mm
$r_c = 21.5$ mm

FIG 9.12   Dimensions of the splined connection. As shown in Fig. 9.1 the splines were engaged over a length L of 35 mm.

In order to calculate the torsional strength of the connection we need to consider two different failure modes. The first is the shear of the core just under the case-hardened layer. The shear takes place over a cylindrical surface of radius r_c and length L. The shearing torque is given by

$$\Gamma = k_u 2\pi L r_c^2 = 800 \times 2\pi \times 35 \times (21.5)^2 \text{ N mm}$$

$$= 8.1 \times 10^7 \text{ N mm.} \qquad (9.13)$$

The second is shear at the roots of the splines with

$$\Gamma \approx \frac{k_u 2\pi L r_r^2}{2} = \frac{1760 \times 2\pi \times 35 \times 23^2}{2} \text{ N mm}$$

$$= 10.2 \times 10^7 \text{ N mm.} \qquad (9.14)$$

Note that the torsional strength of the splines is actually greater than that of the core even though metal has been machined away to generate the splines. This is because the splines have been case hardened and are therefore much harder than the core. But if the splines are not case hardened the strength of the connection is almost halved.

It is interesting to see that the limiting strength of the joint (8.1×10^7 N mm) is 2.67 times greater than the torsional strength that we determined earlier for the shaft. This confirms what we have already seen about the strength of this type of joint. In fact splined connections are used in many applications, such as the sliding joints at the ends of prop shafts and the mountings of suspension arms.

References

M. F. Ashby and D. R. H. Jones, *Engineering Materials 2*, Pergamon, 1986.

M. Atkins, *Atlas of Continuous Cooling Transformation Diagrams for Engineering Steels*, British Steel Corporation.

G. Parrish and G. S. Harper, *Production Gas Carburising*, Pergamon, 1985.

J. E. Shigley, *Mechanical Engineering Design*, 3rd edition, McGraw-Hill, 1977.

C. J. Smithells, *Metals Reference Book*, 6th edition, Butterworth, 1984.

CHAPTER 10

FREEZE BURSTING OF WATER PIPES

Introduction

Water pipes in plumbing installations are often made from copper. Copper is invariably used for pipes supplying water for drinking and washing, is popular for recirculating heating and cooling systems and is also used in small heat exchangers. It has the following advantages for these applications.

(a) It resists corrosion well. This means that pipe walls can be thin (0.65 mm on standard 15-mm diameter small-bore pipe) with considerable savings in weight and cost. It also means that water for drinking and washing is not contaminated or discoloured by corrosion products.
(b) Because annealed copper is soft and ductile it can readily be extruded into thin-walled seamless tube.
(c) Copper water pipes are easily cut to length and bent to shape. They can also be used with compression fittings (see Example 26.1).
(d) Copper is easily wetted by lead–tin solders. The thin-walled tubes have a low thermal mass so they can be used with soldered capillary fittings.
(e) Copper has a very high thermal conductivity. This makes it ideal for heating modules in boilers, spirals in hot-water tanks, and heat exchangers.
(f) Although copper is an expensive metal the cost penalty is offset by
  (1) using thin sections,
  (2) modest labour costs at installation,
  (3) fuel savings in heat-transfer applications.

It is important to prevent water freezing in systems which use thin-walled copper pipes. At 0°C water has a specific volume of $1.000 \text{ m}^3 \text{ Mg}^{-1}$ whereas ice has a specific volume of $1.091 \text{ m}^3 \text{ Mg}^{-1}$. This means that if all the water freezes the volume required to contain the contents of the system must increase by 9.1%. Unless water is able to escape from the system during the freezing process pressure will build up inside the pipes, eventually bursting them open. As we will see in this case study the ability to resist bursting is governed by the shape of the stress–strain curve for copper and the initial yield stress of the pipe wall.

Burst Pipes in Heat Exchangers

Figure 10.1 shows a length of copper tube removed from the heat exchanger in an air conditioning plant. The tube had bulged over a length of 17 mm. A tensile

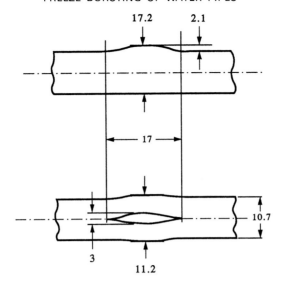

FIG 10.1   Drawings of a burst copper tube from an air conditioning heat exchanger.
Dimensions in mm.

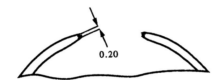

FIG 10.2   Geometry of the fracture. Dimension in mm.

fracture had initiated in the centre of the bulge and had spread longitudinally over the length of the bulge. Away from the bulge the tube had an external diameter of 10.7 mm and a wall thickness of 0.37 mm. As shown in Fig. 10.2, the wall had drawn down to a thickness of 0.20 mm at the edge of the fracture. The fracture surface was essentially at right angles to the tube wall and had the fibrous morphology typical of ductile tensile fracture. At the centre of the bulge the split had opened up by 3 mm. The top of the bulge was 2.1 mm above the rest of the tube. Two very similar bursts had occurred at other positions in the system. The heat exchanger had been mounted close to the cooling coil of the air conditioning plant. The water in the heat exchanger had accidentally been cut off, allowing the cooling coils to lower the temperature of the water in the tubes to 0°C.

Figure 10.3 shows a length of copper tube removed from the condenser in a refrigeration plant. The tube had bulged and split in a similar manner. Away from the bulge the tube had an external diameter of 16.0 mm and a wall thickness of 1.0 mm. The top of the bulge was 1.2 mm above the rest of the tube. Figure 10.4 is a schematic diagram of a typical refrigeration cycle (see Haywood). The function of the condenser is to remove the latent heat of condensation from the refrigerant during the condensation part of the evaporation–condensation cycle. In the present case this was achieved by using a shell-and-tube heat exchanger. The refrigerant was passed through the shell and cooling water was passed through the tubes. The

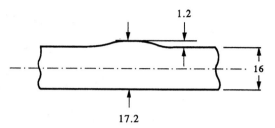

FIG 10.3   Drawing of a burst copper tube from the condenser of a refrigeration plant.
Dimensions in mm. An adjacent tube had a bulge 0.45 mm high but had not burst.

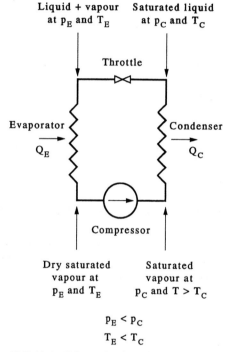

$$p_E < p_C$$
$$T_E < T_C$$

FIG 10.4   Schematic of a refrigeration cycle.

failure was caused by an accidental loss of pressure in the shell. As a result the
liquid refrigerant boiled off, taking latent heat from the heat exchanger and freezing
the water in the pipes. Data for some common refrigerants are given in Appendix
1H. At atmospheric pressure the liquids boil at $-25°C$ to $-30°C$, so severe freezing
was inevitable.

Analysing the Failure

Stable Plastic Deformation

We first look at what happens if we have an annealed copper tube containing
freezing water at a pressure p. The geometry of the tube is shown in Fig. 10.5.

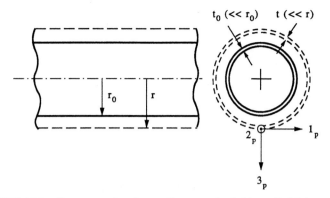

FIG 10.5   Geometry of an internally pressurised thin-walled tube.

The only way of getting the pressure to build up is for the ends of the tube to be blocked by ice, and this means that the tube must behave as if it has end caps. The stresses for this situation are given in Appendix 1A, and are

$$\sigma_1 = \sigma_1, \qquad \sigma_2 = \sigma_1/2, \qquad \sigma_3 = 0. \tag{10.1}$$

The true stress–true strain curve for annealed copper is shown in Fig. 10.6. As we will show later it has the form

$$\sigma \approx A\varepsilon^{0.5}. \tag{10.2}$$

A is about 500 MPa. For practical purposes copper yields when σ reaches the 0.2% proof stress, which is about 50 MPa (Smithells).

   On the face of it, we might think that the tube should yield when $\sigma_1 = \sigma$. But the stress states that relate to σ_1 and σ are different: the tube is in a biaxial stress field; the stress–strain curve is measured in uniaxial tests. The uniaxial stress

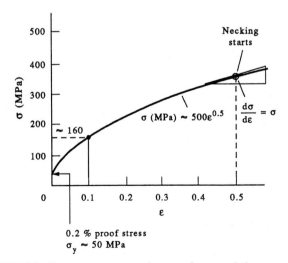

FIG 10.6   True stress–true strain curve for annealed copper.

that produces an equivalent effect to the biaxial stress state in the tube can be found by using the expression for the Von Mises equivalent stress given in Appendix 1C. This is

$$\sigma = [\tfrac{1}{2}\{(\sigma_1 - \sigma_2)^2 + (\sigma_2 - \sigma_3)^2 + (\sigma_3 - \sigma_1)^2\}]^{1/2}. \tag{10.3}$$

Inserting the stress components given in Eqn. (10.1) we get

$$\sigma = \left[\frac{1}{2}\left\{\frac{\sigma_1^2}{4} + \frac{\sigma_1^2}{4} + \sigma_1^2\right\}\right]^{1/2} = \frac{\sigma_1}{1.16}. \tag{10.4}$$

The tube will therefore yield when

$$\sigma_1 = 1.16 \times 50 \text{ MPa} = 58 \text{ MPa}. \tag{10.5}$$

If we want to increase the diameter of the tube by a definite amount, as shown in Fig. 10.5, we must increase the pressure to balance the effects of strain hardening. As an example, to get a true strain of 0.1 we must increase the yield, or flow, stress to about 160 MPa (see Fig. 10.6). The hoop stress must then be

$$\sigma_1 = 1.16 \times 160 \text{ MPa} = 186 \text{ MPa}. \tag{10.6}$$

On the face of it we might think that the tube achieves a strain of 0.1 on the stress–strain curve when the hoop strain is 0.1. This is not so because of the biaxial stress state in the tube. The plastic strains in the tube can be found by putting the stresses given in Eqn. (10.1) into the Levy–Mises equations for plastic flow. We will not go through the procedure here: the flow of an internally pressurised tube with end caps is analysed in Appendix 1C. But the answer is

$$\varepsilon_1 = \varepsilon_1, \qquad \varepsilon_2 = 0, \qquad \varepsilon_3 = -\varepsilon_1. \tag{10.7}$$

This is an interesting result: the axial strain is zero; as the diameter increases the thickness decreases to conserve volume. A frozen pipe may burst, but at least its length does not change!

The uniaxial strain that is equivalent to this strain field is found from the expression for the equivalent plastic strain given in Appendix 1C. This is

$$\varepsilon = [\tfrac{2}{9}\{(\varepsilon_1 - \varepsilon_2)^2 + (\varepsilon_2 - \varepsilon_3)^2 + (\varepsilon_3 - \varepsilon_1)^2\}]^{1/2}. \tag{10.8}$$

Inserting the strain components from Eqn. (10.7) we get

$$\varepsilon = [\tfrac{2}{9}\{\varepsilon_1^2 + \varepsilon_1^2 + 4\varepsilon_1^2\}]^{1/2} = 1.16\varepsilon_1. \tag{10.9}$$

The hoop strain needed to achieve a strain of 0.1 on the stress–strain curve is therefore

$$\varepsilon_1 = \frac{0.1}{1.16} = 0.086. \tag{10.10}$$

Unstable Plastic Deformation

Above a critical value of equivalent strain the tube stops deforming in a stable uniform way. A plastic instability intervenes which grows into a bulge and leads to the fracture of the tube. In order to test the tube for instability we proceed as follows. At the start of the "test" the tube has dimensions r and t and is subjected to an internal pressure p. In order to conserve volume during the deformation we must have

$$2\pi rt = 2\pi r_0 t_0, \tag{10.11}$$

giving

$$\frac{1}{t} = \frac{r}{r_0 t_0}. \tag{10.12}$$

Thus

$$\sigma_1 = \frac{pr}{t} = \frac{pr^2}{r_0 t_0} \tag{10.13}$$

and

$$\sigma = \frac{\sigma_1}{1.16} = \left(\frac{p}{1.16 r_0 t_0}\right) r^2. \tag{10.14}$$

The tube is then allowed to strain plastically at constant pressure. The stress increases at a rate given by

$$\left(\frac{d\sigma}{d\varepsilon}\right)_p = \left(\frac{p}{1.16 r_0 t_0}\right) 2r \frac{dr}{d\varepsilon}. \tag{10.15}$$

From Eqn. (10.9) we have

$$d\varepsilon = 1.16 \, d\varepsilon_1 = 1.16 \left(\frac{dr}{r}\right). \tag{10.16}$$

This gives

$$\frac{dr}{d\varepsilon} = \frac{r}{1.16} \tag{10.17}$$

and Eqn. (10.15) becomes

$$\left(\frac{d\sigma}{d\varepsilon}\right)_p = \left(\frac{p}{1.16 r_0 t_0}\right) \frac{2r^2}{1.16}. \tag{10.18}$$

Combining Eqns. (10.18) and (10.14) we then get

$$\left(\frac{d\sigma}{d\varepsilon}\right)_p = \frac{2\sigma}{1.16}. \tag{10.19}$$

From Eqn. (10.2) the slope of the stress–strain curve is

$$\frac{d\sigma}{d\varepsilon} \approx 0.5 A \varepsilon^{-0.5}. \tag{10.20}$$

The plastic instability will develop when

$$\left(\frac{d\sigma}{d\varepsilon}\right)_p \geq \frac{d\sigma}{d\varepsilon} \tag{10.21}$$

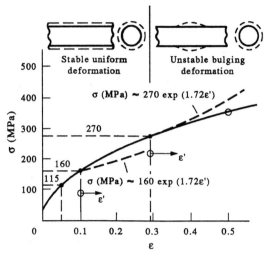

FIG 10.7   Deformation modes of copper tubes. When the flow stress exceeds about
270 MPa the pressurised tube is unstable and a bulge develops.

because the equivalent stress increases at a faster rate than the uniaxial flow stress
of the material. The strain at which the instability starts is given by

$$\frac{2\sigma}{1.16} \approx 0.5 A \varepsilon^{-0.5} = \frac{2A}{1.16} \varepsilon^{0.5} \tag{10.22}$$

from which

$$\varepsilon \approx 0.29. \tag{10.23}$$

Eqn. (10.19) can be integrated to give

$$\int \frac{d\sigma}{\sigma} = \frac{2}{1.16} \int d\varepsilon \tag{10.24}$$

and

$$\sigma = B \exp(1.72\varepsilon'). \tag{10.25}$$

B is the value of σ at the start of the instability "test", when $\varepsilon' = 0$. Equation (10.25)
is plotted in Fig. 10.7 for two stability "tests". When $\varepsilon = 0.1$ the tube is stable; when
$\varepsilon = 0.29$, Eqn. (10.25) is tangential to the stress–strain curve, showing the onset of
instability at an equivalent strain of 0.29.

Behaviour of Copper Tubes

Copper tubes generally come in two grades, "hard" and annealed. The hard grade
is work-hardened as a result of the extrusion process; the annealed grade is produced
by recrystallising the extruded tube at about 600°C. A typical Vickers hardness for
hard copper is 90 kg mm^{-2}, or 880 MPa. As shown in Appendix 1C,

$$\sigma_y \approx \frac{H}{3} = 293 \text{ MPa} \tag{10.26}$$

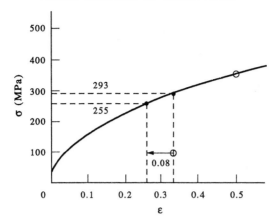

FIG 10.8   The flow stress of "hard" copper tube after a hardness test is about 293 MPa.
The stress–strain curve shows that this is equivalent to a flow stress before hardness testing
of 255 MPa.

where σ_y is the flow stress after the extra 8% plastic strain of the hardness test. Figure 10.8 shows that before hardness testing the pipe had a yield strength of 255 MPa. This corresponds to a strain of 0.26, which is close to the strain of 0.29 at which the tube should go unstable.

The way in which a hard copper tube behaves with increasing internal pressure is as follows. Initially the tube deforms elastically and its dimensions depart only slightly from t_0 and r_0. When the pressure reaches the value

$$p = \frac{1.16 \times 255 \text{ MPa } t_0}{r_0} \tag{10.27}$$

the tube will deform plastically. The deformation will be stable and uniform until the equivalent stress reaches 270 MPa. At this point the tube is unstable with respect to a small increase in radius and a bulge will form with no further increase in pressure.

The plastic strain needed to take the tube to the point of instability is $0.29 - 0.26 = 0.03$. This corresponds to an increase in the radius of the tube of only 3%. So hard tubes can burst under internal pressure even though the tube as a whole hardly swells at all. This is just what happened in the heat exchangers, which must have used hard tubes.

Annealed tubes are much more tolerant of frost damage. Consider the worst case, where the diameter of the pipe has to expand by 4.5% to absorb the volume change of 9.1% that occurs on freezing. The nominal hoop strain is 0.045, which gives an increment of uniaxial true strain

$$\varepsilon = 1.16 \ln(1.045) = 0.05. \tag{10.28}$$

As we can see from Fig. 10.7, this plastic strain work-hardens the tube to an equivalent stress of 115 MPa. The deformation is stable and uniform and the tube can safely absorb all the expansion caused by freezing. A second cycle of freezing increases the diameter by a further 4.5%, giving a cumulative equivalent strain of

0.10. After five cycles of freezing the accumulated strain is $5 \times 0.05 = 0.25$. Only towards the end of the sixth cycle of freezing does the tube go unstable, with a cumulative strain of 0.29.

Design Implications

Plastic bulging instabilities are not confined to copper tubes containing freezing water. Similar instabilities are possible whenever thin-walled tubes are used to contain fluids under pressure. Whalley gives an example of a hydraulic cylinder made from a carbon steel with a tensile strength of 485 MPa which bulged and split in the same way as the copper tubes. The diameter of the cylinder was 95 mm and the split was 180 mm long. The cylinder had deformed uniformly over the whole of its length with a radial strain of about 8%. Although the steel was in the normalised condition the yield strength was not quoted. Evidently the initial plastic strain of 8% had work-hardened the steel to the point at which the true stress–true strain curve was tangential to Eqn. (10.25).

The freeze-bursting of copper tubes is an example of a load-controlled event. The freezing rate is controlled simply by the rate at which the latent heat of freezing is removed from the system and this defines the rate at which the bulge expands under constant pressure. The rupture of the hydraulic cylinder was probably caused by an excessive dead load on the ram or an excessive delivery pressure from the pump.

The case study has shown that if tubes or pressure vessels are made from annealed material there will be ample warning of a pressure excursion: rupture will not take place until there has been a substantial increase in diameter. This is not the case when the material has been supplied in a cold-worked condition: bulging failures can then occur with little or no warning. When using materials with a flat true stress–true strain curve it is therefore important to make sure that the internal pressure can never exceed some fraction of the yield stress so that bulging cannot initiate.

An unlikely application of the phenomenon can be found in the musical instrument industry. Brass instruments like trumpets, french horns and trombones are made from lengths of thin-walled brass tube. In the finished instruments the tubes are not parallel, but are tapered to give a gradual expansion of the bore all the way from the mouthpiece to the bell. To complicate matters the bore is not a straight cone but has a non-linear variation of diameter with length. Pieces of brass tube are expanded to the correct profile by putting them inside an accurately tapered mould and pumping them up hydraulically. An obvious requirement is that the deformation is uniform and stable: but this can be ensured by using tubes that are sufficiently soft to start with.

A Note on the Stress–Strain Curve

Before finishing we need to explain the basis of the stress–strain curve shown in Fig. 10.6. The curve has been plotted again in Fig. 10.9 to show the data points used to construct it. Metals Handbook gives data for the mechanical properties of cold rolled electrolytic tough pitch copper as a function of the reduction of

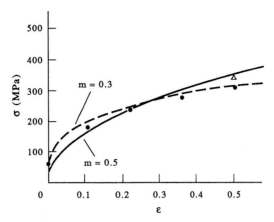

FIG 10.9   Experimental stress–strain data for copper taken from Table 10.1.

TABLE 10.1   *Data for Electrolytic Tough Pitch Copper*

0.2% Proof stress (true)	Reduction in thickness, $(t_0 - t)/t_0$	True strain, $\ln(t_0/t)$	Tensile strength (nominal)	Nominal strain to fracture, ε_f
65	0	0		
180	0.10	0.11		
240	0.20	0.22		
270	0.30	0.36		
310	0.40	0.51	340	0.04

thickness experienced during the rolling operation. This information has been used to obtain data for true stress and strain as shown in Table 10.1. Stresses in the table are given in MPa.

The data points were fitted to an equation of the form

$$\sigma = A\varepsilon^m. \tag{10.29}$$

The values of m given in the literature for annealed copper range from 0.30 (Hertzberg) to 0.54 (Dieter). As Fig. 10.9 shows, the curve with $m = 0.5$ is a reasonably good fit to the data provided the tensile strength is used at a strain of 0.51. Since the elongation to fracture is then only 4% this is a good approximation. The curve with $m = 0.3$ fits rather badly for strains below 0.2. There is likely to be some variation in the stress–strain curve between different batches of copper and a design analysis should, of course, use test data for the material actually used.

A decrease in the value of m will lower both the true strain at which necking starts and the strain at which bulging starts. The condition for necking is

$$\frac{d\sigma}{d\varepsilon} = \sigma, \tag{10.30}$$

from which

$$mA\varepsilon^{m-1} = A\varepsilon^m, \tag{10.31}$$

and $\varepsilon = m$. The condition for bulging is

$$\frac{2\sigma}{1.16} = mA\varepsilon^{m-1} = \frac{2A}{1.16}\varepsilon^m, \tag{10.32}$$

giving $\varepsilon = 1.16m/2 = 0.58m$. Thus, if $m = 0.3$, bulging starts at a strain of $0.58 \times 0.3 = 0.17$ and necking starts at a strain of 0.30. As a reminder, our analysis (which set $m = 0.5$) gave a buckling strain of 0.29 and a necking strain of 0.50.

References

G. E. Dieter, *Mechanical Metallurgy*, 2nd edition, McGraw-Hill Kogakusha, 1976.

R. W. Haywood, *Analysis of Engineering Cycles*, 2nd edition, Pergamon, 1975.

R. W. Hertzberg, *Deformation and Fracture Mechanics of Engineering Materials*, 3rd edition, Wiley, 1989.

Metals Handbook, 9th edition, Vol. 2: *Properties and Selection; Non Ferrous Alloys and Pure Metals*, American Society for Metals, 1980.

C. J. Smithells, *Metals Reference Book*, 6th edition, Butterworth, 1984.

B. G. Whalley, "The Analysis of Service Failures", *Metallurgist and Materials Technologist*, **15**, 137 (1983).

C. Creep

CHAPTER 11

CREEP RUPTURE IN A WATER-TUBE BOILER—1

Background

In Chapter 10 we looked at the situation where a thin-walled tube was subjected to an internal pressure p which was large enough to make the metal deform plastically. We saw that, if the stress–strain curve was sufficiently shallow, a small increase in the diameter of the tube at constant pressure made it go unstable. The instability grew into a bulge and the tube split open when the strain in the bulge reached the tensile ductility of the metal.

At high temperatures metals creep. They elongate with time at constant stress and fracture occurs when the accumulated strain ε reaches the creep ductility ε_f. A thin-walled tube containing a pressurised fluid at high temperature is therefore in an unstable situation to begin with, and the designer must ensure that:

(a) the creep strain that accumulates over the design life is acceptable;
(b) the accumulated strain is significantly less than the creep ductility;
(c) the design life is significantly less than the time to failure t_f.

If the service temperature is increased, then provided there is no major change in the microstructure (such as a polymorphic phase transformation) the creep rate will increase and the time to failure will decrease. This means that the maximum temperature of an internally pressurised tube must be carefully controlled in order to avoid the risk that the tube may bulge or even burst as a result of overheating.

In this case study we look at the creep rupture of a water tube in a large water-tube boiler. The failure caused considerable damage and disruption and it was considered important:

(a) to establish the temperature at which the failure occurred;
(b) to estimate the time it took for the tube to burst;
(c) to correlate these parameters with the operating records of the boiler.

Description of the Boiler

Figure 11.1 is a schematic of the water-tube boiler. For simplicity the figure shows only two banks of hot tubes ("risers") and two banks of cool tubes ("downcomers"). In the actual boiler there were twelve banks of risers and twelve banks of

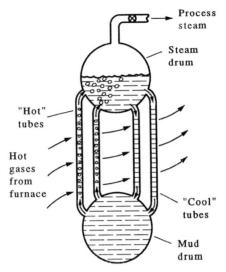

FIG 11.1   Schematic of the water-tube boiler.

downcomers. Each bank consisted of a parallel array of ninety tubes spaced regularly along the length of the drums. The total number of tubes in the boiler was 2160. The drums were 17 m long and 1.7 m in diameter. The average length of the water tubes was 10 m.

Technical data are as follows.

Boiler Operating Parameters
    Maximum evaporation rate = 500 tonnes/hour.
    Maximum heat flux through riser walls = $20 \, \text{kW m}^{-2}$.
    Gas temperature at inlet = 940°C.
    Gas temperature at mid-space = 650°C.
    Gas temperature at outlet = 490°C.
    Operating pressure = 4.8 MPa gauge.
    Normal operating temperature = 264°C.

Water-Tube Specifications
    Internal diameter = 80 mm.
    External diameter = 90 mm.
    Wall thickness = 5 mm.
    Operating hoop stress = 38 MPa.
    Material: carbon steel.
    Fabrication: single longitudinal diffusion weld.
    Composition (in weight %): ≤ 0.17 C, ≤ 0.35 Si, 0.40 to 0.80 Mn, ≤ 0.045 S.
    Mechanical properties at room temperature: $\sigma_y \geq 235$ MPa, $\sigma_{TS} = 360$ to 480 MPa,
        $\varepsilon_f \geq 25\%$.

Service History
    Approximately 1 year of operation at 85 to 95% of full load.

Description of Failure

Figure 11.2 is a diagram of the ruptured tube. The outside diameter of the tube had increased from the original value of 90 mm to a maximum of about 103 mm. A bulge had formed on one side of the tube and the wall at the centre of the bulge had thinned down to 3.2 mm. At this position a chisel-edged creep fracture had occurred, giving a fracture surface 0.35 to 0.80 mm wide. The length of the rupture was 370 mm.

The burst tube was in the sixth bank of riser tubes counting in from the inlet side of the boiler. Under normal operating conditions the gas temperature at this position would have been about 795°C. As a result of the rupture the contents of the boiler were discharged into the space occupied by the hot tube banks. Thirty tubes were cut out in order to gain access, and it was found that the force of the explosion had bent ninety of the neighbouring tubes. Another fifty tubes had enlarged diameters. The worst affected had swollen over a length of 1 m. The maximum diameters were 98 to 103 mm. The tube walls had thinned down from the original 5 mm to 4.6 and 4.3 mm respectively.

Apart from a thin layer of oxide scale, the tubes were clean on the outside. On the inside they were coated with a layer of hard-water deposit 0.25 mm thick.

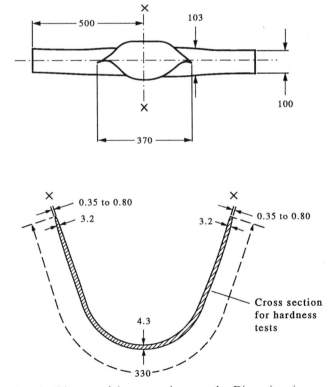

FIG 11.2   Diagram of the ruptured water tube. Dimensions in mm.

Failure Analysis—Causes of Overheating

Why should the tubes have overheated in the first place? The feature common to all water tubes is that they are meant to operate under conditions of the highest possible heat transfer. Heat is supplied to the outer surfaces of the tubes by the flue gases and removed from the inner surfaces by the water circulating around the boiler. Obviously, anything which interferes with the cooling action of the water will lead to overheating. Deposits of hard-water scale, layers of corrosion product and delaminations in the tube wall can provide disastrously effective thermal barriers. In many water-tube boilers the conditions for circulation may be marginal. Risers and downcomers may get mixed up, and there can be regions of the boiler where the water literally does not know whether it "is coming or going". Water-tube boilers are also prone to "steam blanketing" where a stable layer of steam forms between the water and the inner surface of the tube. This helps insulate the tube from the circulating water and allows it to warm up. The situation can become unstable: as the tube warms up the steam blanket will tend to grow and the whole tube can boil dry. Once this happens there is almost nothing to prevent the tube heating up to the temperature of the surrounding gases. Worse still, if a large number of tubes boil dry, then the furnace gases will not be cooled by the risers and the temperature of the gas around the dried-out tubes can in principle be as high as the inlet temperature of the flue gases (940°C in our boiler).

The temperature drop across the layer of hard-water deposit can be estimated from the heat conduction equation

$$\frac{dQ}{dt} = KA\frac{dT}{dx}. \tag{11.1}$$

The term $(1/A)(dQ/dt)$ is the heat flux through the tube wall, which we know is at most 20 kW m^{-2}. A rough estimate for the thermal conductivity K of the layer is $0.5 \text{ W m}^{-1}\,^{\circ}\text{C}^{-1}$ (Kaye and Laby). Since $dx = 0.25$ mm, $dT = 10^{\circ}\text{C}$. It is unlikely that this small thermal resistance would have allowed the tubes to heat up much. A more likely explanation for the temperature excursion is a major disruption in the circulation and boiling behaviour affecting a large number of the riser tubes.

Failure Analysis—Metallurgy

A sample of steel was removed from the tube just outside the damaged area. The sample was analysed chemically and specimens from it were tested in tension. The results gave the following information.

Composition (in weight %): 0.13 C, 0.18 Si, 0.65 Mn, 0.02 S, 0.02 P, 0.01 Cr, 0.01 Mo, 0.005 Ni.

Mechanical properties at room temperature: $\sigma_y = 240$ MPa, $\sigma_{TS} = 505$ MPa, $\varepsilon_f = 25\%$.

The longitudinal diffusion weld was barely visible even on a polished cross-section and consisted only of a very narrow band with a fine grain size. The split did not line up with the weld. The chemical composition and mechanical properties were consistent with the specifications for the tube. So there was no indication that the failure was caused by any defect in the material itself.

The Vickers hardness was measured on the cross-section shown in Fig. 11.2. Next to the fractured edge the hardness was 250 to 400 HV. Away from the fracture the hardness of the cross-section was 200 to 250 HV.

When the tube was made the structure would have been grains of ferrite (α) plus nodules of pearlite in the weight ratio 88 to 12. As soon as the steel is heated above the A_1 temperature (723°C) the nodules of pearlite are replaced by an equal weight of austenite (γ) grains. As the temperature rises above A_1 the α progressively transforms to γ until, at the A_3 temperature (which is 860°C for our 0.13% carbon steel) the structure is all γ.

As shown in Appendix 1G, if γ containing 0.13% C is quenched to martensite it will have a Vickers hardness of 400. This is the same as the maximum hardness next to the fractured edge, so the rupture must have happened above the A_3 temperature of 860°C. Obviously, when the water poured out of the boiler through the hole in the tube it must have quenched the γ to martensite. The large variation in hardness over the cross-section (200 to 400 HV) was caused by different cooling rates. Only at the fracture, where the metal is very thin (<1 mm) was the quench fast enough to turn the γ to martensite. Away from the fracture the metal was much thicker (up to 4.3 mm): the cooling rate would have been less and the γ would have transformed to the softer bainite instead.

Failure Analysis—Creep Fracture

Fracture Mechanisms

The way in which a creeping material fractures can often give us information about the conditions under which the failure took place. There are three basic mechanisms of creep failure (see Appendix 1D). At low stresses the most likely mechanism is *intergranular creep fracture*. In this mechanism voids or wedge-cracks nucleate at grain boundaries under the action of the applied tensile stress. The defects grow and eventually the remaining ligaments fail. The deformation is concentrated at the grain boundaries and the overall tensile ductility and reduction in area at break are usually small. At high stresses failure usually occurs by *transgranular creep fracture*. Voids nucleate and grow throughout the grains. Failure takes place by microvoid coalescence in a way that is similar to ductile failure at ordinary temperatures: the tensile ductility and reduction in area at break are usually quite large. At high temperatures and stresses, *dynamic recrystallisation* operates. Waves of recrystallisation sweep through the creeping material and continually remove the microstructural damage caused by the creep processes. As a result, voids do not nucleate and the metal necks down to a point (if the specimen was a round bar to begin with) or a chisel-edge (if the specimen was initially a flat plate). This mechanism is referred to as *rupture*. Note that the creep literature often uses the term "rupture" to mean failure by *any* creep mechanism, which is rather confusing.

As Fig. 11.2 shows, the reduction in area at the split is considerable. The width of the fracture surface is only 0.35 to 0.80 mm compared to the initial wall thickness of 5 mm. In other words the thickness of the plate has gone down by six to fourteen times. The fracture is decidedly "chisel-edged" and rupture seems the likely mechanism.

The regimes of stress and temperature for the three mechanisms of fracture can be displayed on a *fracture-mechanism map* (see Appendix 1D). Figures 11.3 and 11.4 show the maps for pure iron and for a low-alloy Cr–Mo steel containing 0.13%

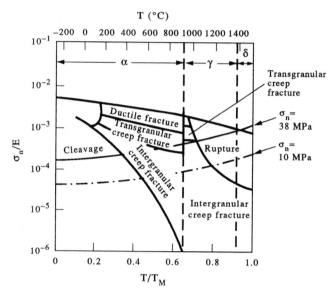

FIG 11.3   Fracture-mechanism map for pure iron.

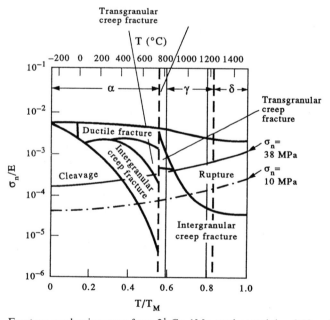

FIG 11.4   Fracture-mechanism map for a $2\frac{1}{4}$ Cr–1Mo steel containing 0.13 weight % C.

carbon (Fields *et al.*). The vertical axis plots σ_n/E, where σ_n is the nominal tensile stress in a uniaxial tensile creep test. Because E decreases with temperature, a line of constant tensile stress will rise as it goes from left to right across the diagram. The maps are complicated by the polymorphic phase transformations that take place in iron. Pure iron transforms from bcc to fcc at 914°C and transforms back to bcc at 1391°C. As we have seen, a 0.13% carbon steel transforms from (ferrite + pearlite) to austenite over the range 723 to 860°C. Each diagram is therefore like two separate adjoining maps, one for ferrite (plus pearlite if the metal contains carbon) and one for austenite. Because of this the diagrams actually contain five creep-mechanism fields.

Creep data at very high temperatures are sparse (no one designs structures for very short lives) and there is no map for a steel like the one used for the tubes. However, the creep behaviour of the tubes is bracketed by the behaviour of pure iron and low-alloy steel: pure iron contains no alloying elements and will creep faster than the tube steel; low-alloy steel is more highly alloyed than the tube steel and should creep more slowly. We have plotted the hoop stress of 38 MPa on both maps in order to identify the possible fracture mechanisms for the tube. As we increase the temperature the sequence of mechanisms is the same in both iron and low-alloy steel; it is: intergranular (ferritic)—transgranular (ferritic)—intergranular (austenitic)—rupture (austenitic). In pure iron rupture begins at about 1000°C; in the low-alloy steel it begins at about 900°C. This is an example of how creep properties are structure-sensitive: clearly we need to do experiments on samples of the tube steel to get reliable data for the onset of rupture.

Creep-Fracture Experiments

Fracture mechanisms are affected by the geometry of the test specimen. Creep data are almost always obtained from uniaxial tensile specimens. Until necking starts the only component of stress is the axial tension and there is no plastic constraint. However, the stress state in the creeping tube is one of plane strain. To see why this is so, we first look at the stress state in the tube wall. The tube geometry is shown in Fig. 11.5. The water tubes are open-ended, and at first sight it seems that the axial stress σ_2 should be zero. Figure 11.6 shows that this is not the case:

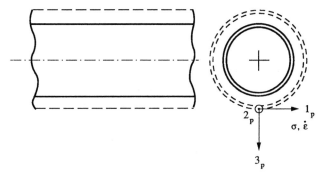

FIG 11.5   Geometry of the creeping tube.

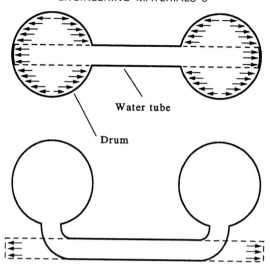

FIG 11.6   Why water tubes behave as if they had end caps.

the tubes behave as if they have end caps and the axial stress is half the hoop stress (see Appendix 1A). The stress state is therefore:

$$\sigma_1 = \sigma, \qquad \sigma_2 = \sigma/2, \qquad \sigma_3 = 0, \tag{11.2}$$

where σ is the hoop stress. The elongation rates along the three principal directions can be found by using the Levy–Mises equations for strain-rate (see Appendix 1D). Then

$$\dot{\varepsilon}_1 = C\left\{\sigma_1 - \frac{\sigma_2 + \sigma_3}{2}\right\} = C\left\{\sigma - \frac{\sigma}{4}\right\} = \tfrac{3}{4}C\sigma, \tag{11.3}$$

$$\dot{\varepsilon}_2 = C\left\{\sigma_2 - \frac{\sigma_1 + \sigma_3}{2}\right\} = C\left\{\frac{\sigma}{2} - \frac{\sigma}{2}\right\} = 0, \tag{11.4}$$

$$\dot{\varepsilon}_3 = C\left\{\sigma_3 - \frac{\sigma_1 + \sigma_2}{2}\right\} = C\left\{\frac{\sigma}{2} - \frac{\sigma}{4}\right\} = -\tfrac{3}{4}C\sigma. \tag{11.5}$$

In other words

$$\dot{\varepsilon}_1 = \dot{\varepsilon}, \qquad \dot{\varepsilon}_2 = 0, \qquad \dot{\varepsilon}_3 = -\dot{\varepsilon}. \tag{11.6}$$

This means that, as the circumference of the tube increases during creep, the wall thickness decreases so as to conserve volume. There is no change in the length of the tube (this is just as well, otherwise the creeping tubes could distort the whole boiler).

Figure 11.7 shows the specimen geometry used to approximate to plane strain. The gauge section was kept short relative to its width to help stop it contracting laterally during the test. Blanks for the specimens were cut from a nearby tube which had bulged by amounts varying between 5 and 13%. The blanks were flattened at red heat. They were then austenitised and air cooled to room temperature before machining. The tensile axis of each specimen was parallel to the hoop direction.

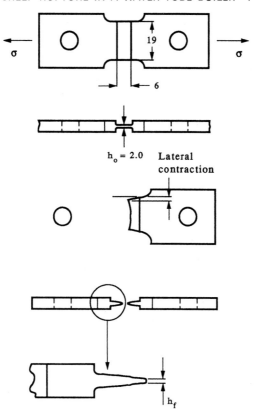

FIG 11.7   Geometry of the test specimens. Dimensions in mm.

Creep tests were carried out in a vacuum furnace to stop the steel oxidising. Each specimen was subjected to a constant axial load which was calculated to apply a nominal tensile stress of 38 MPa to the gauge section. Creep-fracture tests were carried out at various temperatures. After each test had been completed the specimen was removed from the furnace and measured to find the reduction in thickness, h_0/h_f.

A few tests were done using specimens with a gauge length of 19 mm instead of the usual 6 mm. The gauge sections in these specimens were less well constrained against lateral contraction and had a strain field that was essentially half way between plane strain and uniaxial tension. However, there was no significant difference between the results from the two types of specimen. In nearly all cases creep fracture started at the centre of the specimen and propagated out to the edges of the gauge section (see Fig. 11.7).

The results are shown in Fig. 11.8. As the temperature increases the reduction in thickness varies, reflecting the sequence of fracture mechanisms: the data show the relatively small ductility of the intergranular fields and the much larger ductility of the transgranular field. The rupture field starts at around 890 to 920°C, with reductions in thickness comparable to those seen in the split tube.

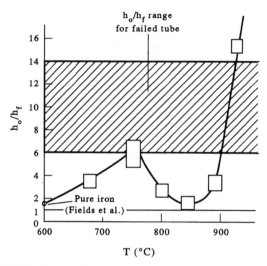

FIG 11.8   Results for reduction in thickness at fracture.

Failure Analysis—Temperature of Failure

To summarise, the creep-fracture data show that the tube burst at about 900°C. This is comfortably above the A_3 temperature of 860°C and explains why martensite formed at the edge of the fracture. The temperature of failure is also comfortably below the maximum inlet temperature of 940°C so the flue gases were capable of heating the tubes up to the failure temperature.

CHAPTER 12

CREEP RUPTURE IN A WATER-TUBE BOILER—2

Failure Analysis—Time to Failure

The creep tests described in Chapter 11 also gave values for the times to failure at the different testing temperatures. These are plotted in Fig. 12.1 in the form of $\ln t_f$ against T^{-1}. But why are these the appropriate variables? Looking at the creep curve in Fig. 12.2, we see that

$$\dot{\varepsilon} \approx C\left(\frac{\varepsilon_f}{t_f}\right). \tag{12.1}$$

C is a constant which depends on the mechanism of creep fracture: the maximum value of C is obviously 1.0, but it can be much smaller than this if there is a large strain during tertiary creep. Equation (12.1) then gives

$$t_f \approx C\varepsilon_f/\dot{\varepsilon} \tag{12.2}$$

and

$$\varepsilon_f \approx t_f\dot{\varepsilon}/C. \tag{12.3}$$

Creep is a thermally activated process, so

$$\dot{\varepsilon} = A\sigma^n e^{-(Q/RT)}. \tag{12.4}$$

Combining Eqns. (12.4) and (12.2) we get

$$t_f = \left(\frac{C\varepsilon_f}{A\sigma^n}\right)e^{(Q/RT)}. \tag{12.5}$$

Within a given mechanism field the pre-exponential terms are approximately constant, so

$$t_f \propto e^{(Q/RT)}. \tag{12.6}$$

This means that a plot of $\ln t_f$ against T^{-1} will be a straight line with a slope of Q/R.

The data in Fig. 12.1 show some interesting features. In the rupture field (above about 890°C) the failure time depends strongly on temperature: at 890°C t_f is about 40 minutes; at the maximum gas temperature of 940°C it is only 10 minutes. On the other hand, between 760°C and 860°C the times hardly change at all. This is because austenite is much more resistant to creep than ferrite: as the temperature

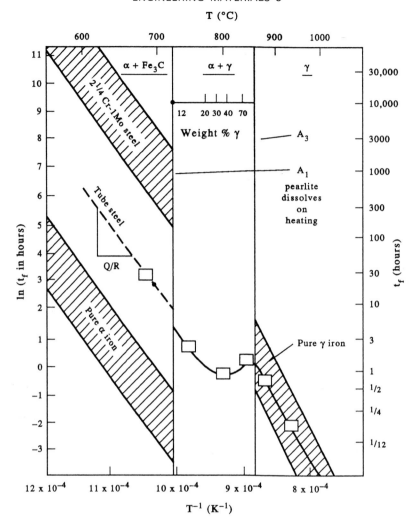

FIG 12.1   Failure times for the tube steel. The data for pure iron and the $2\frac{1}{4}$Cr–1Mo
steel are taken from the paper by Fields *et al.*

increases from A_1 to A_3 the ferrite progressively transforms to austenite, and this
compensates almost exactly for the increased thermal energy available for diffusion.
    The data also illustrate how sensitive creep rates are to changes in composition.
Below the A_1 temperature the tube steel resists creep fracture much better than
pure iron. This is mainly due to the pinning effect of both the dissolved carbon and
the pearlite nodules. On the other hand, the low-alloy Cr–Mo steel is much better
than the tube steel. Mo is a strong carbide former and the molybdenum carbides
are very good at pinning dislocations. Although there are not enough points to be
sure, there seems to be a hint of a sharp step at the A_1 temperature, which could
be caused by the transformation of the pearlite nodules to austenite. In the austenite
field the creep strength of the tube steel is in fact comparable to that of pure

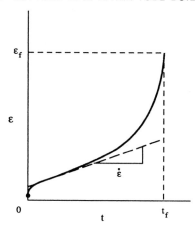

FIG 12.2   A typical creep curve for a constant-load test in tension.

iron: carbon dissolves easily in austenite and does not give much solid-solution strengthening.

Failure Analysis—Boiler Operating Conditions

We saw from Chapter 11 that the tube failed at about 900°C. At this temperature the test specimens failed in 30 minutes. The first stage in finding out what exactly caused the failure is therefore to inspect the plant records: were there changes in the operating parameters (e.g. firing rates, safety-valve discharges, boiler blow-downs) within 30 minutes of the failure? When carrying out this search it is as well to remember that the failure time of the tube is not necessarily the same as the failure time of the specimens. The reasons for this are as follows.

Changes in Hoop Load

In producing the final bulge the tube expanded from an initial outside diameter of 90 mm to 103 mm or so. The wall thickness went down from 5 mm to 4.3 mm as a result. This means that the bore of the tube increased from 80 mm to 94 mm. As shown in Fig. 12.3, the hoop load in the tube wall is given by

$$F_h = \frac{p2rl}{2},$$ (12.7)

where l is the length of the bulged section of tube. Since the diameter $2r$ increases by the ratio $94/80 = 1.18$, the hoop load increases by 1.18 times as well. At constant temperature the creep equation is

$$\dot{\varepsilon} = B\sigma^n,$$ (12.8)

so the strain rate will increase by $(1.18)^n$ times. But what value should we use for n? In order to find n we need first to establish the mechanism of creep during the

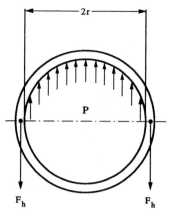

FIG 12.3 Cross-section of the tube showing the relationship between the internal
pressure and the hoop load.

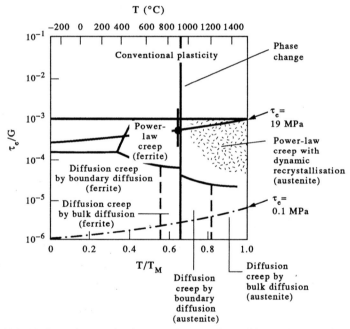

FIG 12.4 Deformation-mechanism map for pure iron with a grain size of 0.1 mm.

failure. This can be done by inspecting the deformation-mechanism map (see
Appendix 1D). Figures 12.4 and 12.5 show the maps for pure iron and a 1Cr Mo V
low-alloy steel (from Frost and Ashby). The vertical axis in the diagrams is the
equivalent shear stress τ_e normalised by the shear modulus G. Because the shear
modulus decreases with temperature a line of constant shear stress will rise as it
goes from left to right across the diagram. There is no map for the tube steel, but its
creep behaviour should be bracketed by the maps for pure iron and the low-alloy steel.

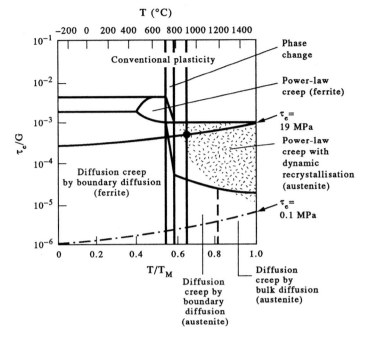

FIG 12.5   Deformation-mechanism map for a 1Cr Mo V steel with a grain size of 0.1 mm.

Appendix 1D shows that, for an internally pressurised tube with end caps,

$$\tau_e = \frac{\sigma}{2}, \tag{12.9}$$

where σ is the hoop stress. Since this is 38 MPa, the equivalent shear stress is 19 MPa. The line $\tau_e = 19$ MPa is plotted on each map as shown in Figs. 12.4 and 12.5. At 900°C the lines lie well inside the field for power-law creep. The value of n for the power-law creep of austenite is 4.5 (Fields and Ashby) and this is the value we use. The creep rate when the final bulge forms is therefore $(1.18)^{4.5} = 2.1$ times the creep rate in the test specimen after the same strain. The *average* creep rate during the expansion is about 1.6 times the creep rate in the specimen.

As we saw in Eqn. (12.2) the failure time varies inversely with the strain rate, so the failure time of the tube is about 1.6 times less than the failure time of the specimen. This reduces the time to failure at 900°C from 30 minutes to 20.

Differences in the Stress State

As we saw in Chapter 11, most of the creep tests were done under conditions approximating to plane strain, and the creep behaviour of the specimens and the tube should have been closely similar. In general, however, creep tests are done using uniaxial tensile specimens and a correction is needed when the test data are used to predict the rupture times of tubes. The uniaxial stress that produces an equivalent effect to the biaxial stress state in the tube can be found by using the expression for the equivalent stress given in Appendix 1D. This is

$$\sigma_e = [\tfrac{1}{2}\{(\sigma_1 - \sigma_2)^2 + (\sigma_2 - \sigma_3)^2 + (\sigma_3 - \sigma_1)^2\}]^{1/2}. \tag{12.10}$$

Inserting the stress components given in Eqn. (11.2) we get

$$\sigma_e = \left[\frac{1}{2}\left\{\frac{\sigma^2}{4} + \frac{\sigma^2}{4} + \sigma^2\right\}\right]^{1/2} = \left(\frac{\sqrt{3}}{2}\right)\sigma. \tag{12.11}$$

The uniaxial strain rate that is equivalent to the strain-rate field in the tube is found from the expression for the equivalent strain rate given in Appendix 1D. This is

$$\dot{\varepsilon}_e = [\tfrac{2}{9}\{(\dot{\varepsilon}_1 - \dot{\varepsilon}_2)^2 + (\dot{\varepsilon}_2 - \dot{\varepsilon}_3)^2 + (\dot{\varepsilon}_3 - \dot{\varepsilon}_1)^2\}]^{1/2}. \tag{12.12}$$

Inserting the strain-rate components from Eqn. (11.6) we get

$$\dot{\varepsilon}_e = [\tfrac{2}{9}\{\dot{\varepsilon}^2 + \dot{\varepsilon}^2 + 4\dot{\varepsilon}^2\}]^{1/2} = \left(\frac{2}{\sqrt{3}}\right)\dot{\varepsilon}. \tag{12.13}$$

The creep equation

$$\dot{\varepsilon}_e = B\sigma_e^n \tag{12.14}$$

then gives

$$\left(\frac{2}{\sqrt{3}}\right)\dot{\varepsilon} = B\left(\frac{\sqrt{3}}{2}\sigma\right)^n. \tag{12.15}$$

Finally

$$\dot{\varepsilon} = B\left(\frac{\sqrt{3}}{2}\right)^{n+1}\sigma^n. \tag{12.16}$$

Since $n = 4.5$, Eqn. (12.16) reduces to

$$\dot{\varepsilon} = 0.45B\sigma^n. \tag{12.17}$$

This result tells us that the tube creeps at about half the rate of the uniaxial specimen. The time to failure of the tube is therefore twice the time to failure of the specimen.

Some Related Failures

Boiler water tubes are not the only situation where an internally pressurised tube can get hot and fail by creep fracture. Other common examples are superheater elements in steam plant and reformer tubes in the chemical industry. As an idea of how important it is to avoid creep fractures of this type it is estimated that 10% of all power-plant breakdowns are caused by creep fractures of boiler water tubes. In general 30% of all tube failures in boilers and reformers are caused by creep. In this section we discuss some failures which have been reported in the literature and use them to show how the conditions of failure can be narrowed down by metallurgical detective work.

Failure of a Superheater Tube

Grogli describes a creep fracture which took place in a superheater tube which had been in service for about 7 years with an intended operating temperature of 375°C. The tube steel was to the same specification as the one discussed in our main case study. The fracture was of the "thick-lipped" type, with the width of the fracture surface comparable to the original thickness of the tube wall. This low creep ductility is typical of intergranular creep fracture (see Appendix 1D). Metallurgical sections taken close to the fracture showed a maze of fine intergranular cracks running into the wall from the surface. The plane of the cracks was approximately perpendicular to the direction of the hoop stress. This confirms that the failure mechanism was intergranular creep fracture. As we saw in Chapter 11, the hoop stress is the largest tensile stress in the wall: the axial stress is only half the hoop stress, and the through-thickness stress is essentially zero. We would expect the cracks to form at right angles to the maximum tensile stress, and this is just what happened.

As we can see from the fracture-mechanism maps (Figs 11.3 and 11.4) the sequence of fracture mechanisms that the tube encounters as it heats up depends on the value of the hoop stress. Unfortunately this is not given. If the hoop stress is below about 20 MPa the maps suggest that the sequence is intergranular (ferritic)—intergranular (austenitic)—rupture (austenitic). The fracture geometry should be thick-lipped up to about 900°C at which point it should change to a thin-lipped rupture. If the hoop stress is above about 30 MPa an additional (transgranular) field will appear between about 700 and 750°C: failure in this temperature range should then be quite ductile. To summarise, the thick-lipped failure could have occurred anywhere between the service temperature and 900°C, although the range 700 to 750°C is tentatively excluded if the hoop stress is high enough. But how do we find out where in this large range of temperature the failure actually took place?

To begin with the tube steel has a microstructure which consists of ferrite grains plus nodules of pearlite (see Fig. 12.6). If the steel is overheated to a temperature

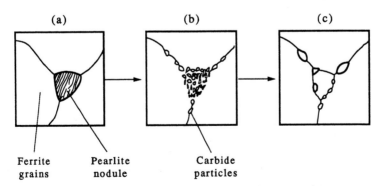

FIG 12.6   Schematic showing how pearlite spheroidises with time at a high temperature below A_1. (a) Original normalised structure, (b) significant redistribution of carbides, (c) complete redistribution of carbides and particle coarsening.

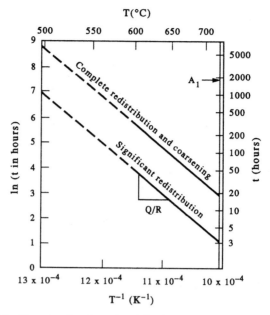

FIG 12.7   How the time for spheroidisation depends on temperature.

that is less than the A_1 temperature there will be no transformation to austenite. However, the pearlite is thermodynamically unstable because of the interfacial free energy of the carbide–ferrite boundaries. At high temperature diffusion can take place. This allows the structure to evolve in such a way that the total interfacial free energy is reduced. The process, called spheroidisation, is summarised in Fig. 12.6. The carbide plates gradually break down with time and the iron carbide reforms as roughly spherical particles. Eventually the plates vanish altogether. The final stage is the coarsening of the carbide particles themselves: the larger particles grow at the expense of the smaller ones and the distribution gradually shifts from a large number of small particles to a small number of large ones.

The time that the tube steel takes to spheroidise is plotted in Fig. 12.7. Because the process is controlled by diffusion it is thermally activated. This means that a plot of $\ln t$ against T^{-1} is a straight line with a slope Q/R. The value of Q found from the slope is 166 kJ mol^{-1}. This is close to the value of 174 kJ mol^{-1} for the diffusion of iron in ferrite grain boundaries (Frost and Ashby) which suggests that this is the rate-limiting diffusional process.

A metallurgical section taken next to the fracture showed in fact that the pearlite in the steel had spheroidised significantly. This tells us that the failure took place below 723°C. What we do not know, however, is *where* below 723°C the tube failed. The data in Fig. 12.7 can in principle be used to find the time that the tube spent at any selected temperature. However, the times given in the figure are an upper limit: experiments show (Chattopadhyay and Sellars) that the rate of spheroidisation can be increased enormously if the steel is deformed plastically at the same time.

Failure of a Boiler Tube

Thielsch describes the creep failure of a carbon-steel boiler tube with a diameter of 95 mm and a wall thickness of 6 mm. The split was 190 mm long and the wall had drawn down to a thickness of about 2 mm at the edge of the fracture. The tube had expanded noticeably on either side of the final bulge. The temperature of the failure was given as approximately 700°C.

The original wall thickness is three times the width of the fracture. This suggests that the failure could have taken place by a transgranular mechanism, and this possibility is reinforced by the quite large hoop strain that occurred before the final failure. Presumably the hoop stress was quite large, and the tube entered the transgranular field on the fracture-mechanism map. This would happen at about 700°C so the quoted failure temperature seems reasonable. The fracture mechanism can be verified by taking metallurgical sections next to the split: as shown in Appendix 1D these should show rounded voids forming in the grains as the precursor to microvoid coalescence.

Finding the temperature at which this sort of failure occurred can be difficult. If the pearlite next to the split has spheroidised, then T_f is below 723°C. Another trick is to take a specimen further along the tube from a section which has not expanded much (and which, therefore, has been at a lower temperature). If this has spheroidised but the steel next to the split has not then the fracture has almost certainly occurred above 723°C. If the failure takes place between the A_1 and A_3 temperatures the structure can be complicated. Depending on the cooling rate after the failure it can consist of mixtures of ferrite and bainite, or ferrite and martensite, with hardnesses to match. At this stage the eye of an experienced metallurgist is required, and even then it may be necessary to do heat treatments in the laboratory to simulate the structures seen next to the fracture.

References

S. Chattopadhyay and C. M. Sellars, "Kinetics of Pearlite Spheroidization During Static Annealing and During Hot Deformation", *Acta metall.*, **30**, 157 (1982).

R. J. Fields, T. Weerasooriya and M. F. Ashby, "Fracture Mechanisms in Pure Iron, Two Austenitic Steels, and One Ferritic Steel", *Metall. Trans. A*, **11A**, 333 (1980).

H. J. Frost and M. F. Ashby, *Deformation-Mechanism Maps*, Pergamon, 1982.

A. Grogli, "Creep Fractures on Tubes from Steam Generating Plants", in *Source Book in Failure Analysis*, American Society for Metals, 1974, p. 332.

G. W. C. Kaye and T. H. Laby, *Tables of Physical and Chemical Constants*, 14th edition, Longmans, 1973.

H. Thielsch, "Why High-Temperature Piping Fails", in *Source Book in Failure Analysis*, American Society for Metals, 1974, p. 60.

B. G. Whalley, "The Analysis of Service Failures", *Metallurgist and Materials Technologist*, **15**, 273 (1983).

D. Fast fracture

CHAPTER 13

FAST FRACTURE OF AN AIRCRAFT FUSELAGE

Background

Figure 13.1 is a drawing of the Comet aircraft. These were designed and built by de Havillands in the UK and were put into service with the British Overseas Airways Corporation (BOAC) in May 1952. The aircraft was powered by four de Havilland Ghost turbo-jet engines and had a cruising altitude of 35,000 feet. It was intended for long-distance high-speed flights on both passenger and freight routes with a maximum all-up weight of 49 tonnes. By comparison existing civil aircraft used turbo-prop engines and had an altitude ceiling of about 17,000 feet so the Comet was a very advanced concept for its time. Unfortunately, Comet G-ALYY crashed in an exceptionally severe tropical storm near Calcutta on 2 May 1953. The accident involved the structural failure of the airframe, but there was no evidence that there was any inherent weakness in the design or construction of the aircraft.

A much more disturbing failure occurred on 10 January 1954. Comet G-ALYP left Rome airport at 0931 hours bound for London. The crew kept in contact with the airport control tower after take-off and at 0950 they reported that they were over the Orbetello beacon. The BOAC flight plan indicated that the plane should have climbed to 26,000 feet at this stage in the ascent. A message from the Comet to another BOAC plane was broken off at 0951 in mid-sentence. At 1000 the control tower heard what they thought was an unmodulated transmission from the Comet. At the same time four eyewitnesses on Elba saw pieces of the plane fall into the sea to the south of the island. None of the twenty-nine passengers and six crew survived. It was estimated that the plane had reached 27,000 feet when the accident happened. The weather conditions were generally good and there was no evidence, as there had been in the Calcutta crash, that the weather had caused the accident.

On 8 April 1954 Comet G-ALYY left Rome at 1832 hours bound for Cairo. At 1857 the crew reported that they were alongside Naples and were approaching an altitude of 35,000 feet. At 1905 the crew radioed Cairo to give their estimated time of arrival. This was the last transmission from the aircraft and neither Rome nor Cairo was able to make contact again. Some light wreckage was recovered from the sea on the following day and it was presumed that the fourteen passengers and seven crew had all perished. The weather conditions were reasonable and there was no indication that the weather had caused the accident. There was now strong

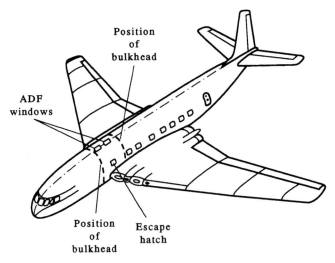

Position
of
bulkhead

ADF
windows

Position   Escape
of   hatch
bulkhead

FIG 13.1   The Comet aircraft.

circumstantial evidence that something was wrong with the aircraft itself and BOAC decided to ground the whole fleet of Comets.

The first priority was to recover as much of the wreckage of the two aircraft as possible. Unfortunately G-ALYY had gone down in water 1000 m deep and there was no realistic prospect of recovering any wreckage from such depths with the undersea technology of the time. G-ALYP had sunk in water 180 m deep and recovery was considered feasible if difficult. The search was carried out by two naval vessels which carried an eight-toothed lifting grab and an underwater inspection chamber. Interestingly the chamber was fitted with a closed-circuit television system: this was the first time that TV had been used in an underwater salvage operation. By August 1954 the recovery team had raised 70% of the structure, 80% of the engines and 50% of the equipment.

The wreckage of the fuselage was reconstructed and the failure was traced to a crack which had started at the corner of a window in the cabin roof. The window was one of a pair which were positioned one behind the other just aft of the leading wing spar. The windows housed the aerials for the automatic direction finding system (ADF). As Fig. 13.2 shows, the crack had started at the rear starboard corner of the rear ADF window and had then run backwards along the top of the cabin parallel to the axis of the fuselage. Circumferential cracks had developed from the main longitudinal crack and as a result whole areas of the skin had peeled away from the structure. A second longitudinal crack then formed at the forward port corner of the rear window and ran into the forward ADF window. Finally two more cracks nucleated at the forward corners of the forward window and developed into a pair of circumferential fractures. As one might expect the circumferential cracks all followed the line of a transverse frame member. Figure 13.3 is a close-up of the rear starboard corner of the rear window showing where the first crack probably originated. Fatigue markings were found on the fracture surface at this location: they had probably started at the edge of a countersunk hole which had been drilled through the skin to take a fastener.

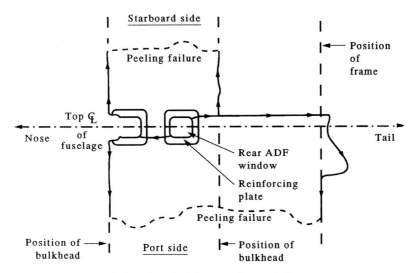

FIG 13.2 The failure on Comet G-ALYP.

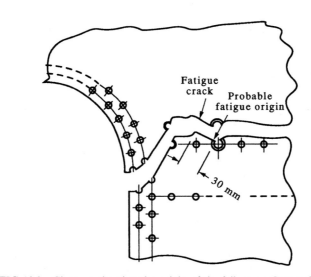

FIG 13.3 Close-up showing the origin of the failure on Comet G-ALYP.

Passengers cannot be carried at high altitude unless the cabin is pressurised. The Comet cabin was pressurised to 0.57 bar gauge, 50% more than the pressure in other civil aircraft at the time. During flight the fuselage functioned as a pressure vessel and was subjected to both an axial and a circumferential tensile stress as a result. With each flight the fuselage experienced a single cycle of pressure loading. Comets G-ALYP and G-ALYY had flown 1290 and 900 flights respectively. The inescapable conclusion was that this comparatively small number of cycles had generated a fatigue crack which was long enough to cause fast fracture.

Design Data

The skin of the fuselage was made from aluminium alloy DTD 546. This had the following composition in weight %: 3.5 to 4.8 Cu, ≤ 1.0 Fe, ≤ 1.5 Si, ≤ 0.6 Mg, ≤ 1.2 Mn, ≤ 0.3 Ti. This specification is now obsolete and has been replaced by alloys in the 2000 series. The minimum yield strength was 325 MPa and the minimum tensile strength was 418 MPa. The actual tensile strength was given as 450 MPa, 8% more than the minimum specified. If we assume that the yield strength was also 8% above the minimum then we get $\sigma_y \approx 350$ MPa. Al–Cu alloys have a relatively poor resistance to corrosion. In accordance with normal practice the sheet used for the skin was roll-clad with a thin layer of pure aluminium on both sides to protect the surface. The overall thickness of the sheet was 0.91 mm.

The fuselage was approximately 3.7 m in diameter and 33 m long. In principle it was possible to calculate the axial and hoop stresses generated in the skin by the cabin pressure p of 0.057 MPa. In practice this was difficult because of the stiffening effect of the transverse and longitudinal frame members. In modern aircraft design the stress distribution can be found to a fair degree of accuracy by using finite element analysis. This was not possible when the Comet was designed because the large high-speed computers which we now take for granted had not been developed. The stresses could have been measured experimentally with strain gauges but apparently this was not done at the design stage. However, de Havillands did calculate the stress in the skin near the corners of the windows using analytical methods: the average value was about 195 MPa measured over a distance of 50 to 75 mm.

The rules of the International Civil Aviation Organization (ICAO) contained two major requirements for pressurised cabins: there was to be no deformation of the structure when the cabin was tested to $1\frac{1}{3}p$; and the maximum stress in the structure at $2p$ was to be less than the tensile strength. de Havillands decided to go one better—they increased the "design" pressure from $2p$ to $2\frac{1}{2}p$ and decided to test a section of the fuselage to $2p$. Under this test pressure the maximum stress at the windows should have gone up to 390 MPa if the calculations were to be believed. This was still less than the minimum tensile strength of 418 MPa and de Havillands were sure that the cabin would stand up to the increased test pressure without excessive deformation. In the event the test section withstood two excursions to $2p$ without any problems, showing that there was an ample margin of safety on static loading.

The ICAO rules did not consider the possibility of fatigue caused by repeated cycles of pressure loading. But there was a growing awareness among aircraft designers in the UK that this was a potential problem. As a result de Havillands decided to determine the likely fatigue life of the fuselage. Between July and September 1953 the test section was subjected to a continuous fatigue test which involved cycling the pressure between zero and p. By 18,000 cycles the section had failed from a fatigue crack which had initiated at a small defect in the skin next to the corner of a window. The design life of the Comet was 10,000 flights so de Havillands considered that they had a reasonable margin of safety.

Tests after the Accidents

In order to verify that the failure had indeed been caused by fatigue it was decided to do a full-scale simulation test on a Comet. The aircraft chosen was G-ALYU, which had already made 1230 flights. The aim was to cycle the cabin between zero and p until it failed by fast fracture. This could not be done using air because the energy released when the cabin broke up would be equivalent to a 225-kg bomb. An explosion of this size would be dangerous and would destroy the experimental evidence. It was decided instead to pressurise the cabin using water. In order to balance the weight of the water in the cabin the whole of the fuselage was put into a large tank and immersed in water at atmospheric pressure. Finally the required pressure differential was generated by pumping a small volume of water into the fuselage.

The cabin failed after 1830 cycles in the tank, giving a total of $1230 + 1830 = 3060$ "flights". The crack started at an escape hatch which was positioned in the port side of the cabin just forward of the wing. As shown in Fig. 13.4, the fatigue crack initiated at the bottom rear corner of the hatch. Once the crack had reached the critical length it ran backwards along the axis of the fuselage. The crack was diverted in a circumferential direction at the site of a transverse bulkhead. A second crack then nucleated at the forward bottom corner of the hatch. This crack ran forward along the axis of the fuselage until it was diverted into a circumferential crack at a transverse bulkhead. The panel of skin detached by the cracks was pushed out of the fuselage by a few cm, relieving the pressure inside the cabin and arresting the failure process. If the cabin had been filled with air the panel would have blown right out and the whole fuselage would have fragmented.

The fuselage was repaired and strain gauges were stuck onto the skin immediately next to the edge of a window. The cabin was taken up to p and the stresses were found from the change in resistance of the strain gauges. The maximum stress appeared at the corner of the window and had a value of 297 MPa. This was 50% more than the maximum stress calculated by de Havillands and was only 3% below the minimum yield strength for DTD 546. Worse still, the bolt holes would have

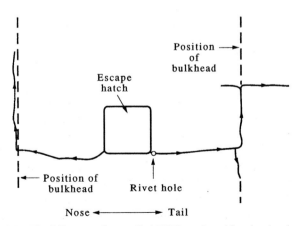

FIG 13.4   The failure on Comet G-ALYU produced by the simulation test.

acted as stress raisers. Appendix 1A gives the elastic stress-concentration factor (SCF) for the case of a hole in an infinite sheet. When the sheet is stressed in uniaxial tension the SCF is 3; when it is stressed in equibiaxial tension the SCF is 2. In the absence of plastic flow the stress at the edge of the bolt holes would therefore have been at least 2×297 MPa $= 594$ MPa. In reality, of course, the stress *was* relieved by plasticity and was truncated at $\approx \sigma_y$. Even if we use de Havilland's value of 195 MPa the elastic stress next to a hole is $\geq 2 \times 195$ MPa $= 390$ MPa. This is 40 MPa more than our estimate for the yield stress and the designers should have anticipated appreciable plasticity at the holes.

Failure Analysis—Fatigue

In retrospect it seems obvious that the Comet was bound to suffer from fatigue. With each cycle of loading the skin near the bolt holes would have yielded and cracks would have formed by low-cycle fatigue in less than 10,000 cycles. Even worse, cracks were sometimes found in the cabin components at the assembly stage: there was an official procedure for immobilising them which involved drilling a small hole at each end of the crack. Almost certainly the structure contained other cracks which were *not* detected during manufacture. But we should remember that the Coffin–Manson law for low-cycle fatigue

$$\Delta \varepsilon^{\mathrm{pl}} N_{\mathrm{f}}^{\mathrm{b}} = C \tag{13.1}$$

was not published widely until 1966. Paris's classic equation

$$\frac{\mathrm{d}a}{\mathrm{d}N} = A \, \Delta K^{\mathrm{m}} \tag{13.2}$$

appeared in an obscure scientific journal in 1961 after three leading journals had refused to publish it!

However, the accident investigators were able to point to two fallacies in de Havillands' figure for the fatigue life. The first was purely statistical. Data for low-cycle fatigue were already known to have a large amount of scatter. The probability that a single Comet would fail by fatigue could be found only if both the mean and the standard deviation of the test data were known. This information could only be obtained from a statistically significant number of tests. The second was that the test section had been pressurised to $2p$ before the fatigue test, whereas the aircraft themselves had been proof tested to only $1\frac{1}{3}p$. We saw in Chapter 6 what happens when a structure is loaded into the plastic range and is then unloaded again: the elastic springback generates a residual elastic stress. When the test section was taken to $2p$ the bolt holes would have yielded in tension. When the structure was let down to atmosphere again the elastic springback would have put the metal around the holes into compression. When, later on, the test section was taken up to p the metal around the holes would have experienced a much reduced tensile stress. To summarise, plastic shakedown would have smoothed out the stresses in the test section and the fatigue resistance would have been improved as a result.

Failure Analysis—Fast Fracture

The accident investigators did not calculate the length to which the fatigue crack must have grown to make the skin fail by fast fracture. This is not surprising: modern concepts of fracture mechanics such as stress intensity and fracture toughness only date from Irwin's classic work in 1957. In this section we apply the now routine methods of linear-elastic fracture mechanics (LEFM) to estimate how big the crack must have been to make the fuselage explode.

Analysis for Plane Strain

Most of the published data for fracture toughness are given in terms of the plane-strain fracture toughness K_{1c}. Plain-strain LEFM therefore provides a convenient starting point for our analysis.

The condition for fast fracture in plane strain is

$$K_{1c} = Y\sigma\sqrt{\pi a}, \tag{13.3}$$

where Y is a dimensionless factor which is governed by the geometry of the cracked component. The critical crack length is therefore

$$a_c = \frac{1}{\pi}\left(\frac{K_{1c}}{Y\sigma}\right)^2. \tag{13.4}$$

Figure 13.5 gives K_{1c} data for aluminium alloys in the 2000 series. The data are plotted against σ_y: they show the classic trend of decreasing toughness with increasing strength. For Comet skin $\sigma_y \approx 350$ MPa so $K_{1c} \approx 30$ MPa$\sqrt{\text{m}}$.

The next step is to estimate the hoop stress in the skin. In order to keep the analysis simple we neglect the load carried by the transverse stiffeners. This defines

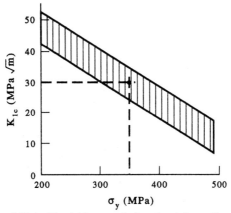

FIG 13.5   Variation of K_{1c} with yield strength for aluminium alloys in the 2000 series (data from Polmear).

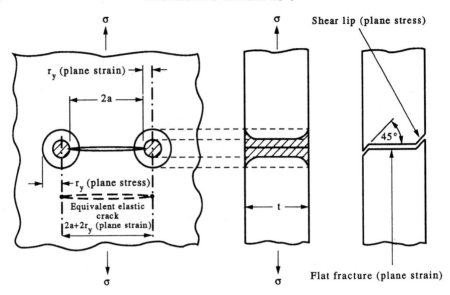

FIG 13.6   Fast fracture in plane strain.

an upper bound on the hoop stress which is given by

$$\sigma = \frac{pr}{t} = \frac{0.057 \text{ MPa} \times 1850 \text{ mm}}{0.91 \text{ mm}} = 116 \text{ MPa}. \qquad (13.5)$$

To simplify the analysis we choose a geometry which gives $Y = 1$. As shown in Appendix 1E an appropriate geometry is an infinite sheet containing a through-thickness crack of length $2a$. This is shown in Fig. 13.6. Equation (13.4) then gives

$$a_c = \frac{1}{\pi} \left(\frac{30 \text{ MPa} \sqrt{m}}{116 \text{ MPa}} \right)^2 = 21 \text{ mm}. \qquad (13.6)$$

The value of a given by Eqn. (13.6) is in fact the length of the *equivalent elastic crack*. As shown in Fig. 13.6 the length of the real crack is given by

$$a_c(\text{actual}) = a_c(\text{equivalent elastic}) - r_y(\text{plane strain}). \qquad (13.7)$$

How big is the radius of the plastic zone? Using the standard result given in Appendix 1E we get

$$r_y(\text{plane strain}) = \frac{1}{6\pi} \left(\frac{K_{1c}}{\sigma_y} \right)^2 = 0.40 \text{ mm}. \qquad (13.8)$$

Thus

$$a_c(\text{actual}) = 21 \text{ mm} - 0.40 \text{ mm} \approx 21 \text{ mm}, \qquad (13.9)$$

so the plastic zone correction is negligible.

Two requirements must be satisfied before we can use a plane-strain LEFM analysis. The first is that the crack-tip plastic zone must be small compared to both the length of the crack and the width of the plate. The standard requirement in plane strain is that

$$a \geq 50 r_y \text{(plane strain)}. \tag{13.10}$$

This gives $a \geq 50 \times 0.40 \text{ mm} = 20 \text{ mm}$. Since $a_c \approx 21 \text{ mm}$, the requirement is satisfied. The second requirement is that the skin is thick enough to get plane-strain conditions at the crack tip. The standard equation is

$$t \geq 2.5 \left(\frac{K_{1c}}{\sigma_y} \right)^2. \tag{13.11}$$

This gives

$$t \geq 2.5 \left(\frac{30 \text{ MPa} \sqrt{m}}{350 \text{ MPa}} \right)^2 = 18 \text{ mm}. \tag{13.12}$$

Unfortunately the skin is only 0.91 mm thick so a plane-strain analysis is not valid. As we show in the next section a *plane-stress* analysis must be used instead.

Analysis for Plane Stress

Figure 13.7 shows how fracture occurs in plane stress. Appendix 1E gives

$$r_y \text{(plane stress)} = \frac{1}{2\pi} \left(\frac{K_c}{\sigma_y} \right)^2. \tag{13.13}$$

If we compare Eqns. (13.13) and (13.8) (and assume to begin with that $K_c \approx K_{1c}$) we can see that the plastic zone in plane stress is three times as big as the plastic

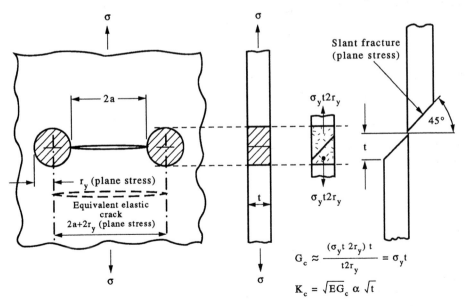

FIG 13.7   Fast fracture in plane stress.

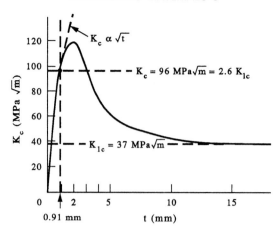

FIG 13.8   Variation of K_c with plate thickness for aluminium alloy 7075-T6 (data from Knott).

zone in plane strain. The plastic work done when the crack advances is much greater in plane stress than in plane strain and the plane-stress fracture toughness is much greater than the plane-strain fracture toughness. Figure 13.8 gives a striking illustration of this effect for aluminium alloy 7075-T6. When $t > 15$ mm $K_c \approx K_{1c}$ and plane-strain conditions apply. As t is decreased below 15 mm, conditions at the crack tip increasingly move towards plane stress and K_c increases. The toughness is a maximum at $t \approx 2$ mm when the plastic zone is all under plane stress. However, thinner sheets have a *lower* toughness. As shown in Fig. 13.7 the plastic work G_c needed to shear the metal apart at the crack tip is proportional to t. As t tends to zero the toughness tends to zero as well. The data in Fig. 13.8 show that, when $t = 0.91$ mm, $K_c \approx 2.6K_{1c}$. In the absence of experimental data for DTD 546 we assume, as a first approximation, that this ratio also holds true for the alloy used in the Comet. Then $K_c \approx 2.6 \times 30$ MPa $\sqrt{m} \approx 80$ MPa $\sqrt{m}$ and

$$a_c = \frac{1}{\pi}\left(\frac{K_c}{Y\sigma}\right)^2 = \frac{1}{\pi}\left(\frac{80 \text{ MPa } \sqrt{m}}{116 \text{ MPa}}\right)^2 \approx 150 \text{ mm}. \tag{13.14}$$

The value of r_y(plane stress) computed from Eqn. (13.13) is 8.3 mm so the length of the real crack is given by

$$a_c(\text{actual}) = a_c(\text{equivalent elastic}) - r_y(\text{plane stress})$$

$$= 150 \text{ mm} - 8 \text{ mm} \approx 140 \text{ mm}. \tag{13.15}$$

The standard requirement for using LEFM in plane stress is that

$$a \geq 50r_y(\text{plane stress}). \tag{13.16}$$

Equation (13.16) gives $a \geq 50 \times 8.3$ mm $= 415$ mm. This is nearly three times the value of a given by Eqn. (13.14), so is it valid to use LEFM? In fact the answer is yes. Figure 13.9 shows how the accuracy of LEFM is affected by the size of the plastic zone. The interesting thing is that the error depends very strongly on the geometry of the specimen. The standard requirement permits an error of $\approx 7\%$ for

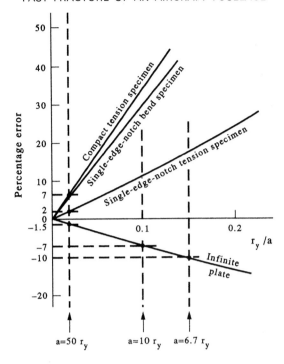

FIG 13.9 How the size of the plastic zone affects the accuracy of LEFM (data from Knott).

two of the standard geometries used for determining values of K_c (the compact tension specimen and the single-edge-notch bend specimen). The error for the single-edge-notch tension specimen is only 2%; an infinite plate containing a crack of finite length is in error by only 1.5%. The standard requirement is drafted to allow for the least favourable geometry and is unduly restrictive for the geometry that we have assumed here. As we can see from Fig. 13.9 an error of 7% in an infinite plate requires that $a \geq 10 \times r_y \approx 83$ mm. In fact $a_c \approx 150$ mm so the error is less than the maximum of 7% allowed.

Obviously, as we approach the windows the stress gets bigger and the critical crack size decreases. For example, when the stress is 195 MPa (de Havilland's figure) our equations give $a_c = 54$ mm. In this case $a = 6.7 \times r_y$ and the error in using LEFM is about 10%. The length of the real crack is ≈ 46 mm, not much more than the length of the fatigue crack shown in Fig. 13.3. If we move into a region of still higher stress the critical crack length will decrease even more. However, as σ approaches σ_y LEFM becomes increasingly shaky and to do an accurate analysis we would need to use the more advanced techniques of elastic–plastic fracture mechanics.

Design Implications

How can we minimise the risk that an aircraft cabin will fail by fast fracture? An obvious approach (used in military transport aircraft) is not to have windows in the first place. This is not likely to be popular with the flying public, but an engineer might well be prepared to trade the view for a less worrying flight. Another change is to do away with holes as much as possible. In fact modern aircraft use epoxy-resin glue instead of rivets for fixing the skin to the framing. Modern finite-element stress analysis can be used to find out where the stress "hot spots" are. The skin in such areas is left thick, while the skin in less highly stressed regions is thinned down by electrochemical machining to save weight. Finally, it is important to design for maintenance: if the critical crack length is large then cracks can be spotted during routine inspections and removed before they become dangerous.

References

T. Bishop, "Fatigue and the Comet Disasters", in *Source Book in Failure Analysis*, American Society for Metals, 1974, p. 209.

R. W. Hertzberg, *Deformation and Fracture Mechanics of Engineering Materials*, 3rd edition, Wiley, 1989.

J. F. Knott, *Fundamentals of Fracture Mechanics*, Butterworth, 1973.

Ministry of Transport and Civil Aviation, *Civil Aircraft Accident: Report of the Court of Inquiry into the Accidents to Comet G-ALYP on 10th January 1954 and Comet G-ALYY on 8th April 1954*, HMSO, 1955.

I. J. Polmear, *Light Alloys*, Arnold, 1981.

R. A. Smith, "Thirty Years of Fatigue Crack Growth – an Historical Review", in *Fatigue Crack Growth – 30 Years of Progress*, edited by R. A. Smith, Pergamon, 1986.

CHAPTER 14

FAST FRACTURE OF AN AMMONIA PRESSURE VESSEL

Background

Figure 14.1 shows part of a steel tank which came from a road tanker vehicle. The tank consisted of a cylindrical shell about 6 m long. A hemispherical cap was welded to each end of the shell with a circumferential weld. The end caps themselves were fabricated from separate "petals" of steel but this detail is not shown on the diagram. The tank was used to transport liquid ammonia. In order to contain the liquid ammonia the pressure had to be equal to the saturation pressure (the pressure at which a mixture of liquid and vapour is in equilibrium). The saturation pressure increases rapidly with temperature. Data are given in Appendix 1H: at 20°C the absolute pressure is 8.57 bar; at 50°C it is 20.33 bar. The *gauge* pressure at 50°C is 19.33 bar, or 1.9 MPa. Because of this the tank had to function as a pressure vessel. The maximum operating pressure was 2.07 MPa gauge. This allowed the tank to be used safely to 50°C, above the maximum temperature expected in even a hot climate.

While liquid was being unloaded from the tank a fast fracture occurred in one of the circumferential welds and the cap was blown off the end of the shell. In order to decant the liquid the space above the liquid had been pressurised with ammonia gas using a compressor. The normal operating pressure of the compressor was 1.83 MPa; the maximum pressure (set by a safety valve) was 2.07 MPa. One can imagine the effect on nearby people and hardware of this explosive discharge of a large volume of highly toxic vapour.

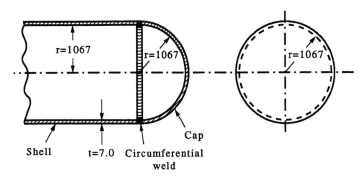

r=1067

r=1067

r=1067

Cap

Shell   t=7.0   Circumferential weld

FIG 14.1   The weld between the shell and the end cap of the pressure vessel. Dimensions in mm.

Details of the Failure

The geometry of the failure is shown in Fig. 14.2. The initial crack, about 2.5 mm deep, had formed in the heat-affected zone (HAZ) between the shell and the circumferential weld. The defect went some way around the circumference of the vessel. The cracking was *intergranular*: the steel had come apart at the grain boundaries. It is well known (Smithells) that ferritic steels are prone to intergranular stress-corrosion cracking (SCC) when they are brought into contact with anhydrous liquid ammonia. The final fast fracture occurred by *transgranular cleavage*: the crack had run along crystallographic planes in the individual grains and there had been essentially no plastic deformation at the crack tip. This indicates that the HAZ must have had a very low fracture toughness. In this case study we predict the critical crack size for fast fracture using linear-elastic fracture mechanics (LEFM).

Material Data

The tank was made from high-strength low-alloy steel to ASTM A517 Grade E. The composition specified by the ASTM standard is given in Table 14.1.

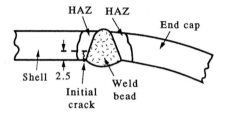

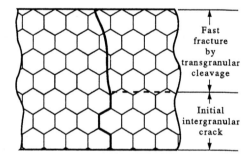

FIG 14.2   The geometry of the failure. Dimension in mm.

TABLE 14.1

Element	Weight %	Element	Weight %	Element	Weight %
Carbon	0.10 to 0.22	Sulphur	0.040 max	Molybdenum	0.36 to 0.64
Silicon	0.18 to 0.37	Phosphorus	0.035 max	Copper	0.17 to 0.43
Manganese	0.36 to 0.74	Chromium	1.34 to 2.06	Titanium	0.03 to 0.11

Samples were cut from the failed tank and were tested with the following results.

Parent plate
Yield strength $\sigma_y = 712$ MPa.
Tensile strength $\sigma_{TS} = 833$ MPa.
Tensile ductility $\varepsilon_f = 22\%$.
Vickers hardness HV $= 280$.
Charpy V-notch impact energy (CVN) $= 22$ lbf ft (30 J) at $-34°$C.

Weld bead
CVN $= 8$ lbf ft (11 J) at $-34°$C.
CVN $= 10$ lbf ft (14 J) at $-3°$C.

Heat-affected zone
HV $= 300$ to 370.
CVN $= 2$ lbf ft (3 J) at $-34°$C.
CVN $= 5$ lbf ft (7 J) at $-3°$C.

We can estimate the plane-strain fracture toughness of the heat-affected zone from the result

$$K_{1c} \approx 15.5 \, (CVN)^{1/2}, \tag{14.1}$$

which is discussed in Appendix 1E. Here K_{1c} and CVN have units of ksi $\sqrt{in}$ and lbf ft. Since CVN ≈ 5 lbf ft, $K_{1c} \approx 35$ ksi$\sqrt{in} \approx 39$ MPa $\sqrt{m}$.

The yield strength of the HAZ can be estimated as follows. We first find the tensile strength using the relation between tensile strength and Vickers hardness given in Appendix 1C. Then

$$\sigma_{TS} \, (MPa) \approx 3.2 \times HV$$

$$\approx 3.2 \times 335 \approx 1070. \tag{14.2}$$

To find the yield strength we assume that the ratio of σ_y/σ_{TS} is the same for both the parent plate and the HAZ. Then $\sigma_y \approx 1070$ MPa $\times (712/833) \approx 915$ MPa.

Stresses Acting on the Crack

Axial Stress

The membrane stresses in the pressure vessel can be found using the results for thin-walled tubes and spheres given in Appendix 1A. The principal directions are shown in Fig. 14.3. For the shell

$$\sigma_1 = \frac{pr}{t}, \qquad \sigma_2 = \frac{pr}{2t}, \qquad \sigma_3 = 0. \tag{14.3}$$

For the cap

$$\sigma_1 = \frac{pr}{2t}, \qquad \sigma_2 = \frac{pr}{2t}, \qquad \sigma_3 = 0. \tag{14.4}$$

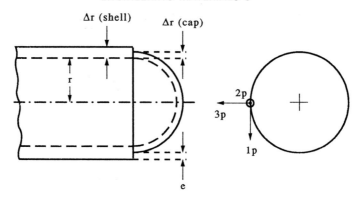

FIG 14.3   When the vessel is pressurised there is a mismatch between the shell and the cap.

The axial stress acting on the crack is therefore given by

$$\sigma = \frac{pr}{2t} = \frac{1.83 \text{ MPa} \times 1067 \text{ mm}}{2 \times 7.0 \text{ mm}} = 140 \text{ MPa}. \tag{14.5}$$

Bending Stress

Equations (14.3) and (14.4) show that the hoop stress in the shell is twice the stress in the end cap. As we saw in Chapter 5 this introduces a discontinuity in the circumferential stress and leads to elastic distortions where the cap meets the shell. The situation is shown in Fig. 14.3. The radial distortions can be found by using the elastic stress–strain relations given in Appendix 1A. The relevant equation is

$$\varepsilon_1 = \frac{\sigma_1}{E} - v\frac{\sigma_2}{E} - v\frac{\sigma_3}{E}. \tag{14.6}$$

Inserting principal stresses from Eqns. (14.3) and (14.4) we obtain

$$\frac{\Delta r(\text{shell})}{r} = \frac{pr}{tE}\left(1 - \frac{v}{2}\right) \tag{14.7}$$

and

$$\frac{\Delta r(\text{cap})}{r} = \frac{pr}{2tE}(1 - v). \tag{14.8}$$

The radial mismatch between the shell and the cap is then given by

$$e = \Delta r(\text{shell}) - \Delta r(\text{cap}) = \frac{pr^2}{2tE}. \tag{14.9}$$

$E = 212 \times 10^3$ MPa (Smithells) so $e = (1067 \text{ mm} \times 140 \text{ MPa})/(212 \times 10^3 \text{ MPa}) = 0.71$ mm.

It is relatively simple to define an upper limit for the stress state produced by this mismatch. The first step is to assume that the cap is infinitely stiff. As shown in

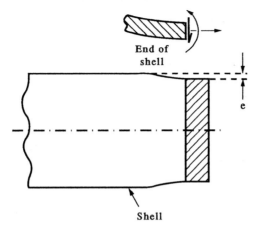

FIG 14.4   The worst case is to assume that all the mismatch is accommodated by the shell. The shell wall at the location of the failure is then subjected to an axial force, a shear force and a bending moment.

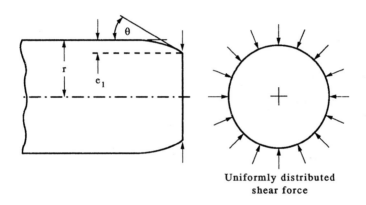

Uniformly distributed
shear force

FIG 14.5   The shear force makes the end of the shell contract.

Fig. 14.4, the end of the shell is then subjected to a bending moment and a shear force in addition to the axial membrane stress. It is the bending moment which is important, because this tries to open up the crack.

The second step is to find out what the shear force does to the end of the shell. The situation is shown in Fig. 14.5. Roark and Young show that the radial displacement and angular rotation at the end of the shell are related by

$$e_1 = \frac{\theta}{\lambda},$$
(14.10)

where

$$\lambda = \left\{ \frac{3(1-v^2)}{r^2 t^2} \right\}^{1/4}.$$
(14.11)

The third step is to find out what the bending moment does to the end of the shell.

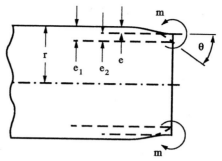

Uniformly distributed edge
moment of m 2πr

FIG 14.6   The bending moment makes the end of the shell expand.

The situation is shown in Fig. 14.6. Roark and Young show that the radial displacement and angular rotation at the end of the shell are related by

$$e_2 = \frac{\theta}{2\lambda}.$$

(14.12)

The bending moment acting on unit length of the circumference is given by

$$m = \theta D \lambda,$$

(14.13)

where

$$D = \frac{Et^3}{12(1 - v^2)}.$$

(14.14)

Then

$$e = e_1 - e_2 = \frac{\theta}{2\lambda},$$

(14.15)

which gives

$$m = 2eD\lambda^2.$$

(14.16)

In order to find the maximum bending stress in the shell wall we apply this moment over a length b of the circumference, as shown in Fig. 14.7. The standard result for elastic beams (Appendix 1B) gives

$$\sigma_{max} = \frac{Mc}{I},$$

(14.17)

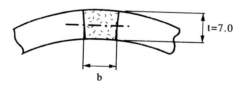

FIG 14.7   Cross-section of part of the shell wall in bending. Dimension in mm.

where $M = mb$, $c = t/2$ and $I = bt^3/12$. Thus

$$\sigma_{max} = \frac{6m}{t^2}. \tag{14.18}$$

Finally, Eqns. (14.18), (14.16), (14.14), (14.11) and (14.9) can be combined to give

$$\sigma_{max} = 1.82\left(\frac{pr}{2t}\right), \tag{14.19}$$

assuming that $v = 0.29$ (Smithells). The term $(pr/2t)$ is simply the axial stress (which is 140 MPa). Thus $\sigma_{max} = 1.82 \times 140$ MPa $= 255$ MPa.

In practice, of course, the cap is not infinitely stiff. It will deflect under the action of the shear force in the same way as the shell, although the equations are different. Because of this the maximum bending stress will not be as great as the value calculated here. In fact, it is probably a reasonable working approximation to adopt a value equal to half the upper limit, giving $\sigma_{max} \approx 130$ MPa.

Stress Intensity of the Crack

As shown in Appendix 1E the stress intensity of the crack in the axial stress field is given by

$$K_1 = Y\sigma\sqrt{\pi a}. \tag{14.20}$$

The geometry factor Y is a function of a/W, as shown in Fig. 14.8. In the present case $a/W = 2.5$ mm/7.0 mm $= 0.36$ and $Y = 1.92$. $\sigma = 140$ MPa and $a = 2.5 \times 10^{-3}$ m, which gives $K_1 = 24$ MPa $\sqrt{m}$.

The stress intensity in the bending stress field is given by

$$K_1 = Y\sigma_{max}\sqrt{\pi a}. \tag{14.21}$$

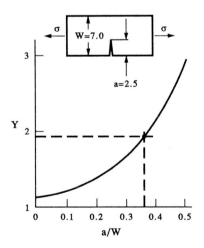

FIG 14.8 Y value for the crack in the tensile part of the stress field. Dimensions in mm.

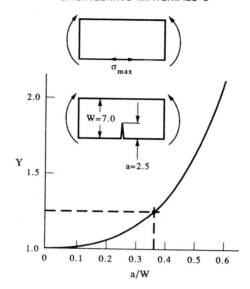

FIG 14.9   *Y* value for the crack in the bending part of the stress field. Dimensions in mm.

As shown in Fig. 14.9, $Y = 1.25$. $\sigma_{\max} \approx 130$ MPa and $a = 2.5 \times 10^{-3}$ m, so $K_1 \approx 14$ MPa $\sqrt{\text{m}}$.

The total stress intensity is 24 MPa $\sqrt{\text{m}}$ + 14 MPa $\sqrt{\text{m}} \approx 38$ MPa $\sqrt{\text{m}}$. This is essentially equal to the value of 39 MPa $\sqrt{\text{m}}$ estimated for K_{1c} and explains why the crack became unstable when it was only 2.5 mm deep.

The analysis does not consider the residual stress left by the welding process. As illustrated in Example 6.1, highly restrained welds contain residual stresses of the order of the yield stress. The circumferential weld was not highly restrained and it had also been subjected to a post-welding heat treatment (PWHT). As a result the residual stresses should have ended up well below yield. However, even a small residual stress (say 0.1 of the yield stress) would have generated an extra stress intensity of ≈ 12 MPa $\sqrt{\text{m}}$.

Validity of Plane-Strain LEFM

The standard requirement for plane strain (see Appendix 1E) is that the length of the crack tip is not less than the term

$$2.5\left(\frac{K_{1c}}{\sigma_y}\right)^2 = 2.5\left(\frac{39 \text{ MPa } \sqrt{\text{m}}}{915 \text{ MPa}}\right)^2 = 4.5 \text{ mm}. \tag{14.22}$$

Since the initial crack went some way around the circumference of the vessel this requirement is satisfied. LEFM is only valid when the plastic zone is small. The standard requirement (see Appendix 1E) is

$$a, (W - a) \geq 50r_y \text{(plane strain)}. \tag{14.23}$$

From Appendix 1E

$$r_y(\text{plane strain}) = \frac{1}{6\pi}\left(\frac{K_{1c}}{\sigma_y}\right)^2 = 0.097 \text{ mm}. \tag{14.24}$$

Thus a and $(W - a)$ must both be greater than 4.8 mm. $(W - a) = 4.5$ mm, which is very close to the minimum allowed value. $a = 2.5$ mm, which is half the allowed minimum. The errors involved in using LEFM when the dimensions are less than the minimum requirement were discussed in Chapter 13. Looking at Fig. 13.9 we can see that the geometry in the axial stress field is similar to that for a single-edge-notch specimen. In our case $r_y/a = 0.097$ mm$/2.5$ mm $= 0.039$. The error for a single-edge-notch specimen is then $\approx 3\%$, well below the error of 7% allowed for a compact tension specimen. The geometry in the bending stress field is more like that in a single-edge-notch *bend* specimen. The error for this situation is $\approx 11\%$. Overall, the error is comparable to that allowed for a compact tension specimen and LEFM is valid.

Design Implications

This case study provides a graphic example of the consequences of having an inadequate fracture toughness. It also shows how the properties of the parent plate can be wrecked by the welding process. If the CVN of the HAZ had been the same as that of the parent plate *at the much lower temperature of* $-34°C$ the fracture toughness would have been about 80 MPa $\sqrt{m}$. Under the axial stress alone the critical crack size would have been around 100 mm. This is much more than the wall thickness of 7 mm. There would have been a leak-before-break situation: the cracking would probably have been detected before disaster occurred. Of course, leak-before-break design assumes that through-thickness cracks will actually be found in practice. This requires constant vigilance and alertness on the part of maintenance and inspection staff. For example, when the vessel is being put through a hydraulic test it is important to make sure that the vessel will hold the pressure for a reasonable time without topping it up from the pump: a slow but persistent drop in the water pressure could well indicate a through-thickness crack.

SCC is an endemic problem where ferritic steels are used to contain liquid ammonia. In a recent survey (see Harrison *et al.*) over half of a sample of seventy-two ammonia storage spheres were found to have stress-corrosion cracks. A standard precaution is to add 0.2% water to the liquid. This inhibits the corrosion process, but has the disadvantage that the steel above the level of the liquid is not protected. High-strength steels are the most prone to SCC and should be avoided if at all possible. The parameters of the welding process should be chosen carefully to reduce the hardness of the weld bead and the HAZ. PWHT should be carried out to minimise the residual stresses. As shown in Appendix 1H ammonia boils at $-34°C$ under atmospheric pressure. One way of storing liquid ammonia is therefore to refrigerate the tank to $-34°C$ so that it does not have to contain any pressure: the driving force for SCC is limited to the residual stresses in the welds and the risk of explosion is removed.

References

E. P. Dahlberg and W. L. Bradley, "Failure Analysis of an Ammonia Pressure Vessel", in *Fracture and Fracture Mechanics: Case Studies*, edited by R. B. Tait and G. G. Garrett, Pergamon, 1985, p. 327.

J. D. Harrison, S. J. Garwood and M. G. Dawes, "Case Studies and Failure Prevention in the Petrochemical and Offshore Industries", in *Fracture and Fracture Mechanics: Case Studies*, edited by R. B. Tait and G. G. Garrett, Pergamon, 1985, p. 281.

R. W. K. Honeycombe, *Steels: Microstructure and Properties*, Arnold, 1981.

R. M. N. Pelloux, "The Analysis of Fracture Surfaces by Electron Microscopy", in *Source Book in Failure Analysis*, American Society for Metals, 1974, p. 381.

R. J. Roark and W. C. Young, *Formulas for Stress and Strain*, 5th edition, McGraw-Hill, 1975.

C. J. Smithells, *Metals Reference Book*, 6th edition, Butterworth, 1984.

R. D. Zip and D. H. Breen, "Correlating Fracture Toughness Curves with Fractographic Features", in *Source Book in Failure Analysis*, American Society for Metals, 1974, p. 371.

CHAPTER 15

FAST FRACTURE OF A MOTORWAY BRIDGE

Background

In November 1954 plans were outlined for a new freeway between the centre of the city of Melbourne, Australia and a position south of the intersection of Hanna Street and City Road in South Melbourne. South of the city centre lies the River Yarra, which is about 300 feet wide. 500 feet to the south of the river is a four-track railway line, which occupies a width of about 100 feet. Other major difficulties in this section included Yarra Bank Road, Whiteman Street, the Queensbridge Road–City Road intersection and a tram line. It was decided to bypass these obstacles by putting the motorway on a bridge. This started at the north bank of the river and ran almost due south for about one-third of a mile to end at Grant Street. The main bridge had two separate parallel decks, referred to as the "eastern" and the "western" carriageways. Each carriageway was carried over a total of twenty-five separate spans. Construction was started in September 1957 and the bridge was opened to traffic in April 1961.

The construction of a typical span is shown in Figs. 15.1 and 15.2. The load-bearing structure consisted of four steel girders which ran parallel to one another along the length of the span. The ends of the girders were supported on short cantilever girders which extended from vertical piers. The girders were secured to a reinforced concrete deck which formed the base of the highway. Each girder came in the form of a large I-beam. The section of the beam was too large for it to be rolled as a single piece. The two flanges and the web were therefore obtained as separate rolled sections and were fabricated into an I-beam by machine welding using continuous single-pass fillet welds. Vertical plates were added at intervals of 1.4 m to stiffen the webs.

Now a uniform I-beam is not an optimum shape for a simply supported span. The applied bending moment is zero at each end of the span but rises to a maximum at mid-span. Since the I-beam has a uniform resistance to bending it will have spare load capacity everywhere except at mid-span. The solution adopted in the bridge girders was to use a basic section that was too weak to take the maximum bending moment, but to add extra metal towards mid-span. Specifically, a flat steel *cover plate* was welded to the lower (tension) flange of the I-beam over a length of ≈ 21 m. The cover plate was narrower than the tension flange and this allowed it to be

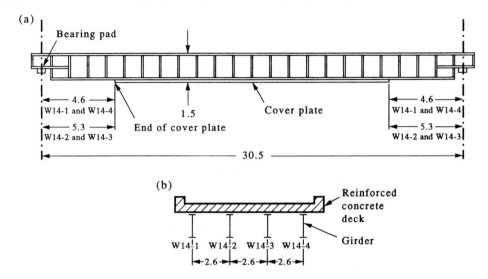

FIG 15.1   Outline drawing of the span. (a) Side elevation of a girder, (b) transverse cross-section through the span. Dimensions in m.

machine welded to the flange using a pair of continuous single-pass fillet welds. In order to reduce the stress concentration factor between the plain and reinforced lengths of the flange the width of the cover plate was reduced progressively over the last 460 mm of its length from its original width of 305/350 mm to ≈ 76 mm. The tapered edges and the end of the cover plate were welded by hand using three-pass fillet welds.

The Collapse

On 10 July 1962 at 11 am a low loader and trailer with a gross weight of 45 tons drove onto the western carriageway of the bridge from the South Melbourne end. The vehicle had only travelled as far as the second span (no. W14) when the deck collapsed. All four girders had cracked and the span was only prevented from falling to the ground by the restraining effect of the concrete deck and by the vertical concrete walls which enclosed the space beneath the bridge. Fortunately there was no serious injury or loss of life, but if the collapse had taken place during the rush hour the situation might have been much more serious. Although the low loader was much heavier than an average vehicle it should not have over-stressed the bridge. Although the air temperature at the time of the collapse was about 7°C the lowest air temperature during the previous night was only 2°C. Because the bridge had a large thermal mass the girders would have responded very slowly to changes in the air temperature and might well have been at ≈4°C when they failed. Only on one previous occasion (July 1961) was the bridge exposed to colder conditions, when an air temperature of 0.5°C was recorded.

As shown in Fig. 15.3, the fractures had all started in the tension flanges at the ends of the cover plates. All four girders had fractured from the southern ends of

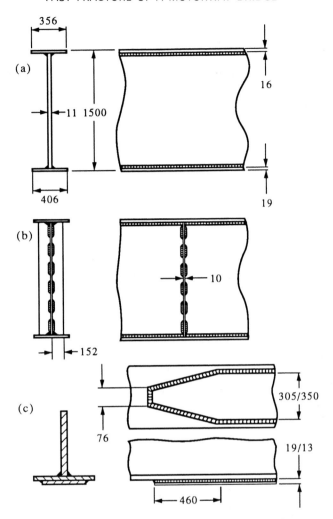

FIG 15.2   Details of the girders. (a) Fabricated I-beam, (b) web stiffeners, (c) end of cover plate. Dimensions in mm.

the cover plates; three girders had also fractured from the northern ends. The fractures at W14-1 south, W14-2 south and W14-3 south had propagated across the complete cross-section of the I-beam. The remaining four fractures had arrested in the webs just below the compression flanges at the top of the girders. The development of the fractures is indicated in Figs. 15.4 and 15.5. The fractures all initiated in the heat-affected zone at the toe of the transverse weld. The initial cracks were ≈ 3 mm deep and extended for the full 76-mm length of the weld. Within hours of the welding operation five of the initial cracks had spread: in three positions (W14-1 south, W14-2 north and W14-3 south) the crack had travelled right through the flange and had even moved into the web. The deepest crack was ≈ 100 mm wide by ≈ 45 mm deep. It was easy to distinguish the enlarged defects on the fracture

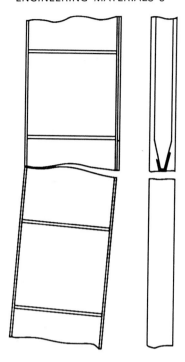

FIG 15.3   Location of the fractures.

surfaces because they were rusty. The welds were cleaned with either phosphoric acid or hydrochloric acid before the girders were painted. Evidently the acid had penetrated the cracks and corroded their surfaces. The surfaces of six of the cracks even showed traces of the red priming paint.

The main fracture surfaces were generally bright and uncorroded: they had obviously propagated as fast fractures from the enlarged defects. However, the fracture at W14-2 south was more complicated. The fracture in the lower flange and in most of the web was rusty, which indicated an earlier fast-fracture event. It was thought that this had probably occurred during the previous cold snap in July 1961. After the fast fracture had occurred the crack continued to spread up the web by fatigue until the final collapse took place. During the year that followed this early failure girder W14-2 would have carried virtually none of the load and this would have increased the stress on the remaining girders.

After the collapse the bridge was closed to traffic and a major investigation was mounted. A total of 168 transverse welds were milled out of the tension flanges and were examined for cracks. Toe cracks were found at 86 of the welds. Most of the cracks were between 2 and 16 mm deep but there were several cases where cracks had propagated right through the flange. These included spans over City Road and Whiteman Street, where a collapse might have caused serious loss of life. The incident was investigated by a Royal Commission, who ended their report with the warning that "we cannot conclude our report without recording our very grave concern about the future of the steel work of the bridge. We know that it contained

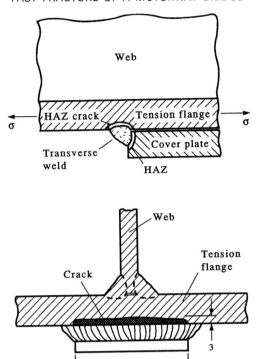

FIG 15.4   Location of the initial cracks. Dimensions in mm.

a great many cracks that have been found and removed but from all we have learnt, we are certain that there must be many more. Some of these may be in critical and highly-stressed regions. Such cracks must be a continuing source of danger, either of brittle fracture or fatigue, unless some means is found to reduce the tensile stress concentrations to negligible amounts. We wish to point out that we do not know of any determinations of the fatigue properties of this steel even at this late stage."

Materials Properties

In the tender documents the contractor was invited to offer a superstructure in reinforced concrete, pre-stressed concrete, mild steel, high-tensile steel or light alloy. In the event, the girders were made from a high-tensile low-alloy steel to BS 968 (1941). This was stronger than carbon–manganese structural steel: girders made from it would be lighter and would put less load on the foundations. However, as we shall see later, the low-alloy steel was much less weldable than ordinary structural steel. After the failure numerous samples of the steel were removed from the girders for mechanical tests and chemical analyses. Some representative results are summarised below.

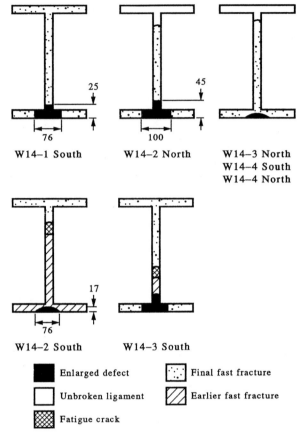

FIG 15.5   Development of the fractures. Dimensions in mm. Not to scale.

TABLE 15.1   *Composition of Flange Plates*

Element	Weight %	Element	Weight %
Carbon	0.22 to 0.26	Phosphorus	≈0.010 to 0.042
Silicon	≈0.03 to 0.34	Chromium	0.24 to 0.25
Manganese	1.70 to 1.80	Nitrogen	0.008 to 0.012
Sulphur	≈0.021 to 0.038		

Tensile strength of plate ≈ 540 to 630 MPa.
Hardness of plate = 180 to 210 HV.
Hardness of weld metal = 205 to 240 HV.
Hardness of HAZ = 255 to 485 HV.
Charpy V-notch impact energy of flange at 21°C = 6 to 22 lbf ft (8 to 30 J).
Charpy V-notch impact energy of flange at 0°C = 2 to 15 lbf ft (3 to 20 J).

Data were not given for the yield stress of the plate. However, similar steels with a tensile strength in the range 540 to 630 MPa typically yield over a range of 325 to 380 MPa (Smithells).

Failure Analysis

Characteristics of the HAZ

The first thing to notice from the data is the large variability in the hardness of the heat-affected zone. The lowest hardness measured in the HAZ (255 HV) is 30% more than the average hardness of the parent plate. However, the highest hardness in the HAZ (485 HV) is 2.5 times the average in the parent plate. Appendix 1G shows that if a steel containing the amount of carbon in the girders (0.22 to 0.26 weight %) is quenched from red heat to give martensite it will have a hardness of ≈ 530 to 575 HV. This is only 10 to 20% more than the maximum hardness in the HAZ. The harder parts of the heat-affected zones presumably had a structure consisting of martensite and bainite. The formation of bainite and martensite in the HAZ is favoured by (a) rapid cooling after the welding operation and (b) a low critical cooling rate (CCR). The presence of untempered martensite makes the HAZ brittle: we will discuss the consequences of this shortly. But why should martensite form in the first place?

Alloying elements such as carbon, manganese, chromium and nickel all decrease the critical cooling rate because they slow down the diffusive decomposition of austenite. Most welding codes assess the effect of the alloying elements from the empirical formula

$$CE = C + \frac{Mn}{6} + \frac{Cr + Mo + V}{5} + \frac{Ni + Cu}{15}. \tag{15.1}$$

CE is the *carbon equivalent* of the steel and the chemical symbols represent the concentration in weight % of each element. The "worst" composition recorded in the flanges was 0.26 C, 1.80 Mn and 0.25 Cr, giving a CE of 0.61. This is high—the cooling rate must be controlled very carefully to produce sound welds when the CE is more than ≈ 0.45.

The lowest cooling rates are obtained when thin sections of steel are welded using a high electrode current. The highest cooling rates are obtained when thick sections are welded using low currents. The cooling rate can also be reduced by *preheating* the steel before welding using the heat from a gas torch. The preheat temperature depends on the steel and the variables of the welding process, but can be as high as ≈ 350°C. In fabricating the bridge the minimum preheat was supposed to be 55°C, but this was not controlled accurately in practice. Sometimes the welds at the ends of the cover plates were made before the flanges of the girders were welded to the webs; sometimes they were made afterwards. In the first sequence of assembly the heat input used to make the short transverse weld would have been small compared to the thermal mass of the flange. In the second sequence the HAZ would have lost heat to the web as well. The combination of low preheat and large thermal mass would have led to a high cooling rate. Given the high CE it is not surprising that the welds had a hard, brittle microstructure.

Cracking in the HAZ

The most likely cause of the HAZ cracks is *cold cracking*. The essential ingredient in cold cracking is hydrogen, and the process is sometimes called hydrogen-induced cracking as a result. The source of the hydrogen is usually H_2O: this can come from water on the surface of the steel, moisture in the atmosphere or dampness in the flux coating of the welding electrode. The water decomposes at the high temperature of the electric arc to give hydrogen, some of which is absorbed by the molten metal in the weld pool. When the weld pool solidifies hydrogen dissolves in the austenite as an interstitital solid solution and diffuses laterally into the austenite in the HAZ. As the weld cools the steel becomes increasingly supersaturated with hydrogen and there is a strong tendency for the gas to come out of solution at defects such as dislocations, particle–matrix interfaces and grain boundaries. Because the hydrogen atom is so small it can diffuse rapidly through the steel even at room temperature. In fact cracking only starts once the weld has cooled to room temperature, hence the term "cold cracking". Cold cracking is encouraged by high residual stresses and the presence of hard, brittle microstructures. This is why it occurs in the HAZ and why it usually starts from a position of stress concentration, such as the toe or root of the weld. The cracks initiate after a specific time interval, or incubation period, and propagate in a slow, erratic fashion. They grow by a cleavage mechanism, which may be either intergranular or transgranular.

Because of the dangers of hydrogen cracking when welding a low-alloy steel the specifications called for welding electrodes of the low-hydrogen type. The flux covering of these electrodes would have had a chemical composition which contained as little associated hydrogen as possible. However, the flux could still have absorbed moisture from the atmosphere and it was important to dry the electrodes before using them by baking them in an oven. The packets in which the electrodes were delivered had instructions that the electrodes were to be dried at 150°C for 30 minutes. However, it was not clear that the welding supervisors understood just how vital it was (a) to dry the electrodes thoroughly, (b) to use electrodes straight from the oven. Indeed, there were times when they had to be told to gather up odd electrodes which had been left lying around the site. The Royal Commission concluded that "laxity in shop organisation in the treatment of electrodes may have contributed to the failure of the bridge". Interestingly, experiments done after the failure showed that it was necessary to bake the electrodes at 350°C to drive off all the water.

Notch Ductility of the Plate

The Charpy V-notch impact test (summarised in Appendix 1E) is the conventional way of measuring the resistance of a steel to brittle fracture starting from stress raisers such as cracks or notches. In contrast to fracture toughness the Charpy impact energy (CVN) cannot be used to predict the stress at which a given flaw should go unstable. However, it is a convenient test for ranking the resistance of different batches of steel to fast fracture. Experience of brittle failures in welded ships showed that structural steels needed to have a CVN of at least 15 lbf ft at the lowest operating temperature to avoid trouble.

The CVN is a structure-sensitive property and the values are affected strongly by the composition, processing and fabrication of the steel. BS 968 (1941) did not lay down requirements for impact testing. However, the County Roads Board added the requirement that impact tests should be carried out as a safeguard against fast fracture. To be approved, each plate had to have an impact energy equivalent to at least 10 lbf ft CVN at 0°C and 16 lbf ft CVN at 21°C. One of the consultants advising the Board even wrote that he "would have nothing to do with the use of BS 968 if adequate assurance of its resistance to brittle fracture was not available". As we have seen, samples taken from the failed flanges gave CVN values as low as 2 and 6 lbf ft at the two test temperatures, indicating a very poor notch ductility. In the event the requirement was not implemented properly and not a single plate was rejected on the grounds of inadequate impact energy. There were, in fact, instances of plates being re-tested several times in an attempt to find an impact value which would pass the specification. This low notch ductility was the main reason why the HAZ cracks grew and eventually propagated as fast fractures.

An interesting illustration of how the CVN is sensitive to the steelmaking process is given in Fig. 15.6. Graph (a) shows the Charpy transition curve for BS 968 steel made at the time of the collapse by an independent steelmaker. The first point to notice is that the CVN goes down as the plate thickness increases. Even so, provided

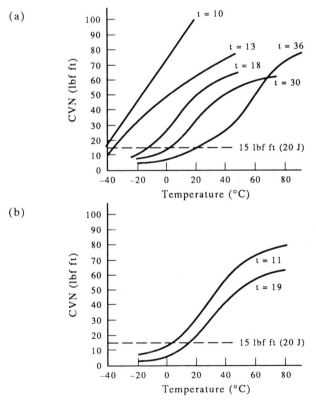

FIG 15.6 Charpy V-notch impact energies (CVN) for (a) BS 968 plate from an independent source, (b) BS 968 plate from the bridge. Plate thicknesses in mm.

the plate is less than 30 mm thick, the 15 lbf ft requirement is satisfied down to 0°C. 18 mm plate (as in the flanges) has a CVN of ≈ 25 lbf ft at 0°C; 10 mm plate (as in the webs) has a CVN of ≈ 75 lbf ft at 0°C. Graph (b) shows the curve for samples removed from the bridge. The impact properties are much worse: the average CVN of the flanges at 0°C is only ≈ 6 lbf ft.

Fast Fracture

Fracture mechanics was not in widespread use when the collapse took place and the Report did not try to quantify the fast-fracture problem. It is interesting to see what a simple LEFM analysis can predict. The fast fracture equation is

$$K_{1c} = Y\sigma\sqrt{\pi a}. \tag{15.2}$$

Equation (15.2) can be used to estimate the stress at which the enlarged defects shown in Fig. 15.5 should go unstable and propagate through the flange by fast fracture. A typical crack geometry is shown in Fig. 15.7. Since $a \ll W$, $Y = 1$ (Appendix 1E, Case 2).

We can estimate the plane-strain fracture toughness of the flange using the result

$$K_{1c} \approx 15.5\,(CVN)^{1/2}. \tag{15.3}$$

Here, K_{1c} and CVN have units of ksi $\sqrt{in}$ and lbf ft. As discussed in Appendix 1E this is an approximate result, valid for CVN values between 5 and 50 lbf ft. Because of the fundamental differences between fracture toughness and impact tests the correlation must not be used for design purposes, but it is useful for analysing failures when experimental data for K_{1c} are not available directly. The relationship

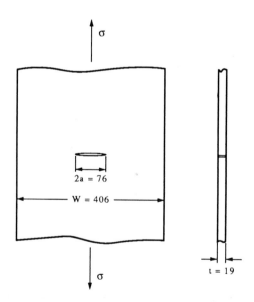

FIG 15.7 Geometry of the through-thickness cracks in the tension flange. Dimensions in mm.

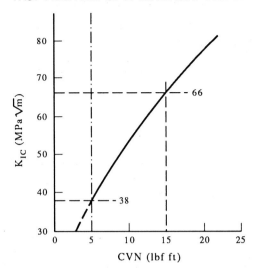

FIG 15.8   Approximate relationship between plane-strain fracture toughness K_{1c} and Charpy V-notch impact energy CVN.

is plotted in Fig. 15.8, with the units of K_{1c} converted to MPa $\sqrt{m}$. When the impact energy is 5 lbf ft, $K_{1c} \approx 38$ MPa $\sqrt{m}$. As we have seen, this impact energy is typical of the more brittle samples of flange plate at the temperature of the failure. Then

$$\sigma_c \approx \frac{38 \text{ MPa } \sqrt{m}}{\sqrt{\pi \times 0.038 \text{ m}}} = 110 \text{ MPa}. \tag{15.4}$$

If the defect is ≈ 100 mm long, as in W14-2 north, then $\sigma_c \approx 96$ MPa. These stresses are small compared to the average yield stress of ≈ 350 MPa and could well have been generated during normal service. They were even more likely after the premature failure of girder no. W14-2.

Is it valid to use plane-strain LEFM in this situation? As shown in Appendix 1E the standard requirement is that

$$2.5 \left(\frac{K_{1c}}{\sigma_y} \right)^2 \leq a, t, (W - a). \tag{15.5}$$

If we assume that $\sigma_y \approx 400$ MPa in the most brittle regions we get a limiting dimension of ≈ 22 mm. This is only slightly more than the thickness of the flange so we make only a small error in using a plane-strain value for the fracture toughness. Both a and $(W - a)$ are greater than the limiting dimension, so LEFM is certainly valid.

When the impact energy reaches the "safe" value of 15 lbf ft the correlation of Fig. 15.8 gives $K_{1c} \approx 66$ MPa $\sqrt{m}$. Using this value in Eqn. (15.4) we get $\sigma_c \approx 191$ MPa, which is still only $\approx 55\%$ of the average yield stress. However, the limiting dimension given by Eqn. (15.5) is now increased to ≈ 68 mm. This is nearly four times the flange thickness so it is not correct to use a plane-strain value for the fracture toughness. As shown in Appendix 1E the actual value of the fracture toughness, K_c, will be significantly higher than K_{1c}. Because the web is thinner than the flange it should be tougher still. If we make a rough guess that $K_c \approx 2 \times K_{1c} = 132$ MPa $\sqrt{m}$ then we get $\sigma_c \approx 382$ MPa. This stress is of the order of the yield stress so LEFM is hardly valid! But the calculation certainly shows why the "safe" Charpy value of 15 lbf ft greatly reduces the chance that the cracks will go unstable.

Although an impact energy of 15 lbf ft may stop a crack going unstable, it may not be enough to stop a *running* crack. This is because the *dynamic* fracture toughness of steel can be much lower than the *static* fracture toughness. Even if the bulk of the material is tough it may not be possible to arrest a crack which has run from a local region of low toughness. It is therefore important to design to the *minimum*, not the *average*, value of the impact energy or fracture toughness.

Design Implications

The collapse provided a convincing illustration of the problems inherent in welded construction. The designers would have designed the structure using a continuum approach. The design deflections would have been calculated using Young's modulus for steel and the design loads would have been limited to some fraction of the load at which the structure would have gone plastic. Since $E \approx 210$ GPa for any ferritic steel we can see that Young's modulus is hardly a structure-sensitive property. Neither is it sensitive to temperature: as a rule-of-thumb E halves between absolute zero and the melting point. σ_y is rather more sensitive to structure, to the extent that it is $\approx 50\%$ greater for BS 968 than it is for a straightforward carbon–manganese structural steel. σ_y does increase as the temperature goes down; it roughly trebles between $0°C$ and $-200°C$. However, the thermal cycle of the welding operation reprocessed the steel in the HAZ and made it hard and brittle; and hydrogen liberated in the electric arc diffused into the embrittled HAZ and formed cold cracks. The integrity of the structure then depended not on E or σ_y but on the fracture toughness, expressed in terms of the Charpy V-notch impact value. This was extremely sensitive to structure (e.g. 2 to 15 lbf ft from one sample to another) and extremely sensitive to temperature (e.g. 6 lbf ft at $0°C$ and 20 lbf ft at $20°C$). To make matters worse, the welds in which the fractures initiated were placed on the surface of the major tension member and the cracks which formed in the HAZ were aligned at right angles to the direction of maximum tensile stress. The welds were subjected to a stress concentration which was caused by the abrupt change of section between the flange and the cover plate. The situation was made even more dangerous by the residual stresses which are inevitable in a large welded structure which has not been stress-relieved.

References

M. F. Ashby and D. R. H. Jones, *Engineering Materials 2*, Pergamon, 1986.
K. E. Easterling, *Introduction to the Physical Metallurgy of Welding*, Butterworth, 1983.
Government of Victoria, *Report of the Royal Commission into the Failure of Kings Bridge*, 1963.
R. W. K. Honeycombe, *Steels: Microstructure and Properties*, Arnold, 1981.
C. J. Smithells, *Metals Reference Book*, 6th edition, Butterworth, 1984.

E. Brittle fracture

CHAPTER 16

FRACTURE OF A CAST-IRON GAS MAIN

Background

Manor Fields is a large private residential estate in the Putney district of London. The thirteen blocks of luxury apartments which make up the estate were built in the early 1930s and are situated in extensive grounds. Each block is three floors high. The walls are built from brick and the flat roofs are concealed behind parapets. The overall appearance is quite attractive and the front elevations are further enhanced by the use of carved mock-Tudor stonework around the entrance doors. The developers must have had a soft spot for higher education as the individual blocks have names such as Harvard House, Magdalene House and Girton House.

Early in the morning of 10 January 1985 a resident of one of the top-floor apartments in Newnham House noticed a slight smell of gas. He opened his front door and found that the smell of gas on the landing outside was much stronger. A few minutes later a huge explosion occurred which demolished the six apartments in the centre of the block and broke windows in most of the surrounding buildings over a quarter mile radius. Of the nine people who were in the apartments at the time, only one survived.

Origin of the Gas Leak

Figures 16.1 and 16.2 show the location of the gas leak. A narrow access road runs along the back of Newnham House. Buried at a depth of about 0.8 m beneath the surface of the road was a gas supply main. The main was made up from lengths of grey cast-iron pipe which were joined together by spigot-and-socket joints. The pipe had a nominal bore of 150 mm and a nominal wall thickness of 11 mm.

The road had originally been laid as reinforced concrete panels of various lengths. The concrete was 100 to 150 mm thick and had a central layer of reinforcing mesh. The road had been dug up on previous occasions and when the surface had been reinstated the concrete had not been reinforced. Because of this there were a number of transverse breaks in the road surface. As shown in Fig. 16.2, the surface of the road had settled by 200 mm in one place. The settlement was first noticed in 1983, and by 1984 the residents were complaining about the pool of water which gathered in the depression when it rained.

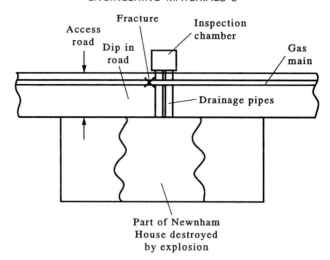

FIG 16.1   Plan view showing the location of the leaking gas main.

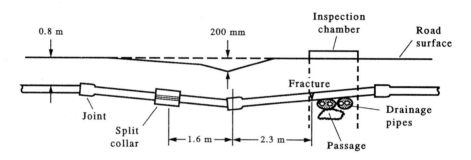

FIG 16.2   Elevation of the gas main in the region of the leak.

The ground beneath the gas main was found to be poorly consolidated. Presumably, when the footings were dug for Newnham House the ground under the access road was filled with loose subsoil and rubble which was not adequately compacted. As shown in Fig. 16.2, the gas main had settled where it ran below the dip in the road. The ground had obviously given way below the pipes and this meant that they had to support the weight of the soil and concrete that lay immediately above.

On the opposite side of the road from Newnham House was a brick-built inspection chamber which took the outflow from two pairs of drainage pipes. Each drainage pipe was made from earthenware sections and the two pipes in each pair were cased in concrete. The underside of the gas main was in firm load-bearing contact with the top of the concrete casings. The pipe was not able to settle in this place and the differential settlement had set up a bending stress in the pipe. The pipe had fractured close to the point of support and this had allowed gas to escape from the main.

Settlement beneath the drainage pipes had opened up a horizontal passage about 75 mm deep and 300 mm wide which allowed the leaking gas to pass from the break in the pipe to the outside wall of Newnham House. The gas then seeped through gaps in the brickwork below road level into the open void which lay beneath the ground floor of the building. Once in the void, the gas was able to spread around the building through the service ducts and the gaps between the wooden floorboards. Eventually an explosive mixture formed and it only needed someone to light a gas ring for the mixture to explode.

Failure Analysis—Details of the Fracture

Figure 16.3 shows how the pipe fractured. The plane of the fracture surface was essentially at right angles to the axis of the pipe. Most of the surface was slightly corroded but near the bottom of the pipe it was bright and clean. The clean fracture had occurred when the pipe was removed from the ground for inspection. The rest of the fracture came from the original failure. The light corrosion indicated that the original failure occurred only a short time before the pipe was removed from the ground.

In grey cast iron most of the carbon occurs as flakes of graphite. The flakes are weak in tension and this makes the iron brittle. The elongation to failure and fracture toughness are typically only 0.4% and 10 MPa $\sqrt{m}$. Failure initiates where the local stress reaches the tensile strength of the iron. The fracture propagates as a flat cleavage crack which runs at right angles to the direction of the maximum tensile stress. The flat facets left by the cleavage fracture reflect the light and the freshly fractured surface looks bright. The plastic zone at the crack tip is small and there are no shear lips at the edges of the fracture surface. There is no gross plastic deformation in the region of the failure and the fractured pieces can be fitted back together again like the pieces of a jigsaw puzzle.

The fracture in the gas pipe showed all of these features. We can see from the way that the crack ran that it must have initiated near the top of the main. This indicates that the tensile stress was a maximum at this position. The fact that the crack ran at right angles to the axis of the pipe indicates that the direction of the tensile stress was parallel to the axis of the pipe. This pattern of stress can be generated by applying a bending moment M to the pipe as shown in Fig. 16.3.

Figure 16.2 shows an unusual feature of the pipework in the region of the leak.

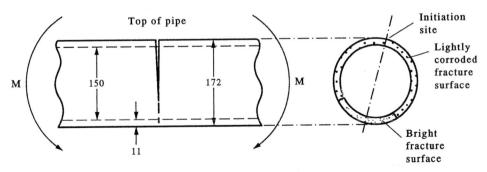

FIG 16.3   Details of the failure. Dimension in mm.

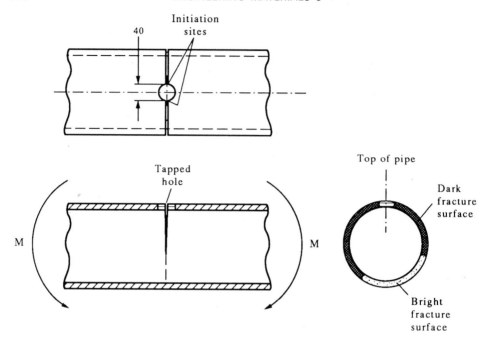

FIG 16.4   Details of the fracture inside the split collar. Dimension in mm.

This is the split collar which had been bolted around the pipe at a location 3.9 m away from the leak. When the collar was removed it was found that the pipe inside was also broken. Figure 16.4 shows the details of this second fracture. The cracks had started from the edge of a tapped hole, 40 mm in diameter, where a branch pipe had been attached to the main. The crack had caused a gas leak in 1982 and the gas board had dug up the road to take remedial action. This involved removing the branch pipe and making the pipe gas tight by installing the split collar. Figure 16.5 shows how the collar functioned. A two-piece rubber gasket was inserted between the outside of the gas pipe and the bore of the collar. The bolts were then tightened and this compressed the gasket and sealed the length of pipe which ran underneath.

This second fracture was very similar to the one which had occurred next to the drainage pipes. However, the original fracture surface was corroded and looked almost black. This showed that the crack had occurred much earlier than 1985. The geometry of the fracture was again consistent with the application of a bending moment to the pipe as shown in Fig. 16.4

Failure Analysis—Properties of the Pipe

Samples were cut from the pipe and were polished for metallography. The micro-structure showed that the metal of the pipe was pearlitic grey cast iron, i.e. the graphite flakes were embedded in a matrix of pearlite. Samples were also analysed chemically. The results in weight % were: 2.80 total C, 2.16 graphitic C, 2.57 Si,

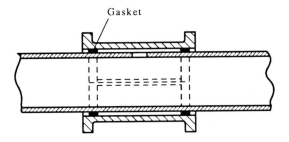

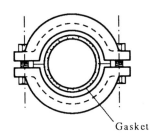

FIG 16.5   Schematic of the split collar.

0.50 Mn, 0.074 S, 1.41 P. This composition is typical of grey cast iron. The silicon was added to encourage the carbon to separate out as graphite rather than as iron carbide when the metal solidified and cooled during the casting process. However, some of the carbon was present as iron carbide in the pearlite and this is why the total carbon content was 0.64% greater than the graphitic carbon. Tensile tests were carried out using specimens cut from the pipe and these gave tensile strengths between 131 and 138 MPa.

The data for tensile strength can be used to estimate the bending moment at which the pipe will break. The cross-section of the pipe in bending is shown in Fig. 16.6. Using the appropriate results from Appendix 1B we can write

$$\sigma_{max} = \frac{Mc}{I}, \tag{16.1}$$

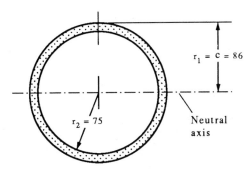

FIG 16.6   Cross-section of the pipe in bending. Dimensions in mm.

where

$$I = \frac{\pi}{4}\left(r_1^4 - r_2^4\right). \tag{16.2}$$

Combining Eqns. (16.1) and (16.2) we get

$$\sigma_{max} = \frac{4Mc}{\pi(r_1^4 - r_2^4)}. \tag{16.3}$$

Finally,

$$M_{max} = \frac{\sigma_{TS}\pi(r_1^4 - r_2^4)}{4c}$$

$$= \frac{138 \text{ N mm}^{-2}\,\pi(86^4 - 75^4)\text{ mm}^4}{4 \times 86 \text{ mm}}$$

$$= 29.1 \times 10^6 \text{ N mm} \approx 29 \text{ kN m}. \tag{16.4}$$

The moment needed to initiate fracture at the 40-mm hole will be less than 29 kN m because the hole acts as a stress raiser. Appendix 1A considers the case of an infinite plate containing a circular hole and subjected to a uniform uniaxial stress σ. The maximum stress occurs at the edge of the hole: its magnitude is 3σ and its direction is parallel to the uniform applied stress. This means that if the top of the pipe were to contain a very small hole then fracture would initiate when the applied stress at the top of the pipe was $\sigma_{TS}/3$. Of course, the 40-mm hole is *not* very small in relation to the size of the pipe: the ratio between the diameter of the hole and the outside diameter of the pipe is in fact 0.23. For this situation the stress concentration factor is reduced to 2.4 (Attewell *et al.*) and fracture should initiate when the applied stress at the top of the pipe is approximately $\sigma_{TS}/2.4$. This requires a bending moment of about 12 kN m.

Failure Analysis—Loading on the Pipe

Although the concrete slabs had settled above the broken section of the gas main it is not clear how much of their weight was actually carried by the pipes. There was probably friction and jamming between adjacent slabs of concrete which would have taken some of the load. We can set a lower limit on the load carried by the pipe by neglecting the weight of the slabs and considering only the weight of the soil held up by the pipe. Figure 16.7 shows the approximate size of the wedge of soil supported by the pipe. The volume of soil above a 1-m length of pipe is roughly $0.5 \times \pi(0.8 \text{ m})^2 \times 1 \text{ m} = 1 \text{ m}^3$. If the density of the soil is $\approx 1.5 \text{ Mg m}^{-3}$, then the weight of the wedge is $\approx 1.5 \text{ Mg m}^{-1}$ and the load on the pipe is $\approx 15 \text{ kN m}^{-1}$.

It is interesting to note that the self-weight of the pipe itself is negligible. The volume of a 1-m length of pipe is $\pi(86^2 - 75^2) \times 1000 \text{ mm}^3 = 5.6 \times 10^{-3} \text{ m}^3$. The density of cast iron is about 7 Mg m^{-3} (Smithells) so the self weight generates a loading of only 0.39 kN m^{-1}.

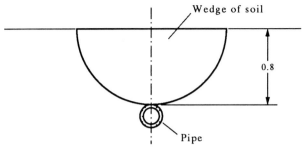

FIG 16.7 The wedge of soil supported by the pipe. Dimension in m.

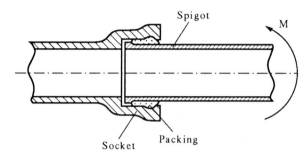

FIG 16.8 Longitudinal section through the spigot-and-socket joint.

Failure Analysis—Estimating the Moments

The moment that the loading generates in the pipe can be estimated using the standard results for elastic beam theory given in Appendix 1B. But there is a complicating factor that we need to look at before we can start on the analysis. As Fig. 16.2 shows, there is a spigot-and-socket joint right in the middle of the damaged section of the gas main. Does this joint function as a pin-jointed or a solid connection? The question is important, because a pin-jointed connection will support the pipe less well than a solid one. A longitudinal section through the joint is shown in Fig. 16.8. The gap between the spigot and the socket is packed with hemp fibres and molten lead is then poured into the joint. The lead solidifies and seals the ends of the pipes together. If a bending moment is applied to the joint the lead on one side of the joint will be squashed; the lead on the other side will either be stretched or will be pulled away from the cast iron. Because lead creeps readily at ambient temperature the joint behaves as a solid connection in response to a short-term load but acts as a pin-jointed connection in response to a long-term load. There is no simple way of deciding how the joint behaved during the failure so we analyse the situation for two cases: (a) where the joint is locked solid, and (b) where the joint is flexible.

Analysis for the Solid Joint

Figure 16.9 shows the situation where the joint is locked solid. The main then behaves as a continuous beam carrying a uniformly distributed load *F per unit*

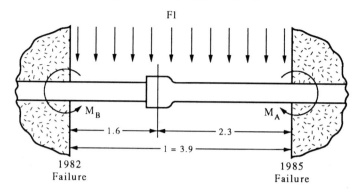

FIG 16.9   Finding the moments when the joint is locked solid. Dimensions in m.

length. As a first approximation we assume that the ends of the beam are built-in. The clamping moments are then given by

$$M_A = M_B = \frac{Fl^2}{12}.$$

(16.5)

$F \approx 15 \text{ kN m}^{-1}$ and $l = 3.9$ m, so $M_A = M_B \approx 19$ kN m.

Analysis for the Flexible Joint

Figure 16.10 shows the situation where the joint is flexible. The main then behaves as two separate cantilever beams connected by a pin-joint which can only transmit a vertical force f. Equilibrium and compatibility give the following equations.

$$M_A = \frac{Fl_1^2}{2} - fl_1,$$

(16.6)

$$M_B = \frac{Fl_2^2}{2} + fl_2,$$

(16.7)

$$\delta = \frac{Fl_1^4}{8EI} - \frac{fl_1^3}{3EI},$$

(16.8)

$$\delta = \frac{Fl_2^4}{8EI} + \frac{fl_2^3}{3EI}.$$

(16.9)

Combining Eqns. (16.8) and (16.9) we get

$$f = \frac{3(l_1^4 - l_2^4)}{8(l_1^3 + l_2^3)} F.$$

(16.10)

Thus

$$f = \frac{3(2.3^4 - 1.6^4)}{8(2.3^3 + 1.6^3)} \text{ m} \times 15 \text{ kN m}^{-1} = 7.4 \text{ kN}.$$

(16.11)

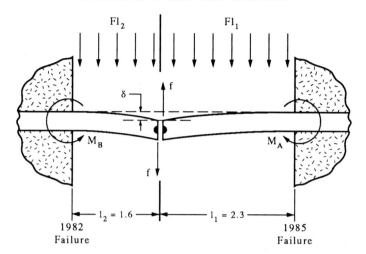

FIG 16.10   Finding the moments when the joint is flexible. Dimensions in m.

This value for f can be put into Eqns. (16.6) and (16.7) to give $M_A \approx 23$ kN m and $M_B \approx 31$ kN m.

Failure Analysis—Discussion

Table 16.1 compares the moments needed to break the pipe with the moments generated by the weight of soil above the pipe. The moments all have units of kN m. Even when the joint is solid the moment at the site of the 1982 crack is 50% more than the moment needed to initiate fracture at the edge of the 40-mm hole. When the joint is flexible the moment is 2.5 × the moment needed to initiate fracture. Although the moments given in Table 16.1 are only estimates, we can see that it was almost inevitable that the main would have fractured at the 40-mm hole.

The situation for the 1985 failure is less clear-cut. Even when the joint is flexible the applied moment is still 20% less than the moment needed to break the pipe. However, additional loads are applied to the pipe when vehicles pass along the access road. The most common heavy vehicle to use the road is the refuse disposal truck. We assume that this weighs 10 tonnes and has four wheels. The load on each wheel is then ≈ 2.5 tonnes. If the whole of this 2.5-tonne load is applied to the 3.9-m length of pipe, the resulting increase in the unit load is ≈ 6 kN m^{-1}. As a result the total unit load goes up from ≈ 15 kN m^{-1} to ≈ 21 kN m^{-1}, an increase of 40%.

TABLE 16.1

Location	Moment to break pipe	Moment generated with solid joint	Moment generated with flexible joint
1982 failure	12	19	31
1985 failure	29	19	23

The bending moment in the pipe at the 1985 failure is increased to ≈ 30 kN m, which is just enough to break the pipe.

Design Implications

Grey cast iron was first used for the distribution of water and gas over 150 years ago. Even now 80% of the water mains in the UK are the old grey cast-iron pipes. However, new water mains are usually made from spheroidal-graphite cast iron ("SG" iron) which was first produced in 1948. Basically, if small amounts of the elements cerium or magnesium are added to molten cast iron just before it is cast the graphite precipitates as spherical particles rather than as flakes. When the graphite is in the form of spheres it has a much smaller weakening effect than when it is present as flakes and SG iron has a much higher strength, ductility and Charpy impact energy than grey iron. For this reason it is often called "ductile iron". A low-strength SG iron typically yields at 230 MPa, breaks at 370 MPa with an elongation of 17% and has a Charpy impact energy of 13 J (Smithells). It is obviously much better at standing up to adverse ground conditions than grey iron.

Old water-distribution systems often develop leaks because the grey iron has fractured or corroded. This wastes water and can lead to flooding but it is rarely dangerous. On the other hand, leaks from broken or corroded gas mains are potentially lethal. In fact the Putney explosion was not an isolated event. Between 1972 and 1984 in the UK leaks from gas mains led to an average of eighteen explosions per year. The total number of fatalities over the period was thirty-nine. These incidents initiated a major programme of replacing old gas mains. Some of the best replacement materials are in fact polymers such as MDPE (medium-density polyethylene), HDPE (high-density polyethylene) and UPVC (unplasticised poly-vinylchloride). These typically yield at ≈ 20 MPa. The Young's modulus is ≈ 50 to 150 times less than it is for cast iron. Because of this the polymer pipes are "whippy" and can absorb substantial ground movements without damage. And polymeric pipes have two further advantages: they do not corrode and they can be joined on site by heat sealing. The ends of the pipes are first machined true using a special rotary planing machine. The two ends to be joined are heated with an electric hotplate and are then forced into contact using a hydraulic ram. The softened thermoplastics fuse together and a high-strength, leak-proof joint is the result.

References

H. T. Angus, *Cast Iron: Physical and Engineering Properties*, 2nd edition, Butterworth, 1976.
M. F. Ashby and D. R. H. Jones, *Engineering Materials 2*, Pergamon, 1986.
P. B. Attewell, J. Yeates and A. R. Selby, *Soil Movements Induced by Tunnelling and Their Effects on Pipelines and Structures*, Blackie, 1986.
Health and Safety Executive, *The Putney Explosion*, HMSO, 1985.
K. J. Pascoe, *An Introduction to the Properties of Engineering Materials*, 3rd edition, Van Nostrand Reinhold, 1978.
C. J. Smithells, *Metals Reference Book*, 6th edition, Butterworth, 1984.

CHAPTER 17

CRACKING OF FOAM INSULATION IN A LIQUID METHANE TANK

Background

Figure 17.1 shows a half-section through a tank used for storing liquid methane at atmospheric pressure. Methane boils at $-162°C$ and this is the temperature at which the tank has to operate. As Appendix 1E shows, ferritic steels go through a transition from ductile to brittle fracture as the temperature is decreased. The temperature at which this transition takes place is generally located between $-40°C$ and $0°C$ depending on the composition and structure of the steel. This is well above the temperature of the tank. Because liquid methane is highly flammable it is vital to ensure that the tank stays sound. Ferritic steels are ruled out because the consequences of a brittle fracture would be catastrophic. In fact the tank is made from an aluminium alloy: because aluminium has a face-centred-cubic structure it does not go brittle at low temperature and limited distortions can be accommodated by plastic flow. Even so, it is considered necessary to have a second line of defence should the tank spring a leak. One way of doing this is to put the tank into a leak-proof aluminium jacket and to put a layer of thermal insulation on the outside of the jacket. However, aluminium is expensive and a second aluminium tank would add a lot to the cost of the installation.

In fact the tank has a jacket made from *mild steel*. The insulation is put into the space between the tank and the jacket and this minimises the cooling effect of the methane. Under normal operating conditions the jacket should not go brittle. But what happens if the tank springs a leak? If the insulation is porous (like fibreglass matting) then the liquid methane will flow through the insulation to the wall of the jacket and will boil off. As a result the jacket will cool down to $-162°C$ and may fail by brittle fracture. If the jacket is gas-tight it will explode under the internal pressure; and if the jacket is open to the atmosphere the escaping gas may ignite. In fact the insulation is *not* porous. The jacket is lined with a layer of *closed-cell foam* made from *rigid polyurethane* (PUR). The foam is applied in layers about 25 mm thick to the inner wall of the jacket. The starting chemicals are mixed with a blowing agent such as carbon dioxide and the resulting froth is sprayed on using a hand-held nozzle. The cell walls then polymerise, producing a light but rigid foam. The operators build up successive layers of foam until the required thickness is achieved. The tank is then lowered into the jacket and the small assembly gap is filled with fibreglass matting. The theory is that if the tank leaks

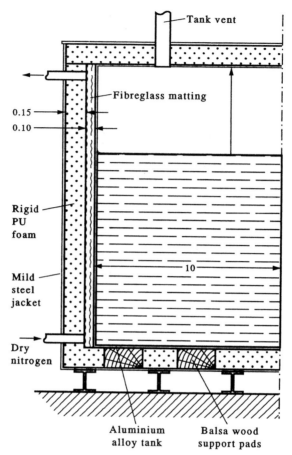

FIG 17.1   Schematic half-section through the liquid methane tank. Approximate dimensions in m.

the methane will be arrested by the closed-cell structure of the PUR and the jacket will be protected.

In fact a number of incidents have occurred where the PUR has suffered from brittle fracture and cracks have propagated right through the thickness of the insulation. This is obviously unacceptable because liquid methane could travel down the cracks and reach the inner wall of the jacket. Repairs are costly, because they involve removing the tank, stripping off the PUR layer and starting again from scratch. In this case study we look at the causes of these failures and outline a design solution to the problem.

Thermal Stresses in the Foam

As we can see from Fig. 17.2, the temperature of the foam decreases linearly with distance through the layer. The foam wants to contract as it gets cold, but it is prevented from doing so by the very rigid steel plate of the jacket to which it is

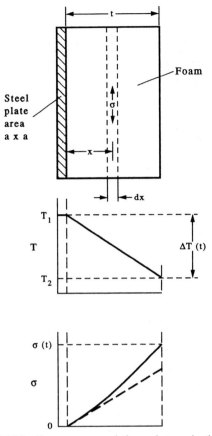

FIG 17.2   Temperature and thermal stress in the foam.

stuck. As shown in Appendix 1A, this generates a biaxial tensile stress in the plane of the layer which is given by

$$\sigma = \frac{\alpha \Delta TE}{(1 - v)}. \tag{17.1}$$

The variation of stress with distance is shown in Fig. 17.2. It is non-linear because Young's modulus for polyurethane (PU) increases as the temperature decreases. The thermal stress is a maximum at the inner surface of the PUR layer.

The obvious explanation for the failures is that the maximum thermal stress was large enough to crack the foam in tension. As shown in Fig. 17.3, polymers behave as elastic–brittle solids provided they are colder than about $0.75T_g$. For PUR $T_g \approx 100°$C, or 373 K. We should therefore expect elastic–brittle behaviour below ≈ 280 K, or $\approx 7°$C. Most of the foam is well below this temperature and there is no plasticity to relieve the thermal stress. The only alternative is brittle fracture. But a good deal of work is needed to confirm this mechanism. We begin by listing data for a typical PUR foam as used in cryogenic applications (Gibson and Ashby).

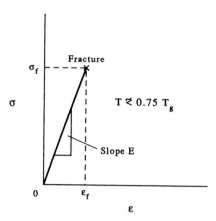

FIG 17.3   Stress–strain relation for rigid polyurethane foam (PUR) in tension.

Density of solid PU $\rho_s \approx 1.20 \text{ Mg m}^{-3}$.
Density of foam $\rho \approx 0.06 \text{ Mg m}^{-3}$.
Relative density of foam $(\rho/\rho_s) \approx 0.05$.
Cell size of foam $l \approx 0.5 \text{ mm}$.
Thermal expansion coefficient of solid PU (and foam) $\alpha \approx 10^{-4} \, {}^\circ\text{C}^{-1}$.
Poisson's ratio of foam $v \approx 0.3$.
Young's modulus of foam $E \approx 34 \text{ MPa}$ at -100°C.
Fracture stress of foam $\sigma_f \approx 1.4 \text{ MPa}$ at -100°C.
Elongation to fracture of foam $\varepsilon_f \approx 4\%$ at -100°C.
Plane-strain fracture toughness of foam $K_{1c} \approx 0.05 \text{ MPa} \sqrt{\text{m}}$ at -100°C.

$T_1 \approx 0^\circ\text{C}$ and $T_2 \approx -100^\circ\text{C}$ so $\Delta T(t) \approx 100^\circ\text{C}$. Equation (17.1) then gives $\sigma(t) \approx$ 0.5 MPa. It is interesting to note that this maximum stress is substantially less than the fracture stress of ≈ 1.4 MPa at -100°C. This means that on a simple basis the foam should not have fractured. In order to understand why the foam *did* in fact break we need to look at the mechanical behaviour of foams in rather greater detail.

Deformation and Fracture of Foams

Figure 17.4 is a schematic section through a foam. When the foam is loaded in tension those cell walls which lie at right angles to the tensile axis deflect in bending like a set of miniature beams. The bending of the walls is restrained to some extent by the end faces of the closed cells, which in turn undergo an extensional strain. Young's modulus E is governed by the overall elastic response of the foam and includes the contributions of both the cell walls and the end faces. Experimental data for PUR foams (Gibson and Ashby) are a good fit to the equation

$$E = E_s \left(\frac{\rho}{\rho_s}\right)^{3/2}, \tag{17.2}$$

where E_s is the modulus of solid PU. The best fit is obtained by choosing values

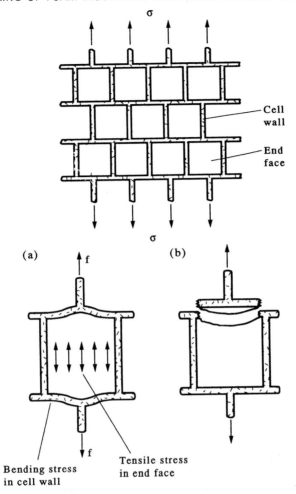

FIG 17.4   Schematic section through a foam showing (a) the elastic behaviour, (b) the fracture behaviour.

for E_s of 3.0 GPa at $-100°C$ and 1.9 GPa at $20°C$; between these two extremes both E_s and E vary linearly with temperature.

When the stress reaches the fracture stress the end faces rupture in tension, transferring load to the transverse cell walls. Fracture is completed when the cell walls break in bending. Experimental data for PUR foams (Gibson and Ashby) are a good fit to the equation

$$\sigma_f = 0.65\sigma_{fs}\left(\frac{\rho}{\rho_s}\right)^{3/2}, \tag{17.3}$$

where σ_{fs} is the fracture stress of solid PU. We get the best fit by choosing values for σ_{fs} of 186 MPa at $-100°C$ and 112 MPa at $20°C$; between these two extremes both σ_{fs} and σ_f vary linearly with temperature.

Resistance to Thermal Shock

The foam should fracture when $\sigma = \sigma_f$. This means that the temperature differential must be at least

$$\Delta T = \frac{(1 - v)\sigma_f}{\alpha E}, \tag{17.4}$$

where ΔT is the *thermal shock resistance* of the foam. If we eliminate σ_f and E using Eqns. (17.3) and (17.2) we get

$$\Delta T = \frac{0.65(1 - v)}{\alpha}\left(\frac{\sigma_{fs}}{E_s}\right). \tag{17.5}$$

According to Eqn. (17.5), the thermal shock resistance is independent of the relative density of the foam and can be changed only by altering the properties of the solid polymer itself. But in real foams this is not the case: variations in the relative density lead to variations in the thermal shock resistance.

We can illustrate this effect with a simple example. Consider a uniform bar of foam which is held between two fixed grips, as shown in Fig. 17.5. The bar is stress-free to begin with but it is subsequently cooled down by ΔT. This generates a tensile stress $\alpha \Delta T E_1$, where E_1 is the modulus of the foam. The thermal shock resistance is

$$\Delta T_1 = \frac{\sigma_{f1}}{\alpha E_1}. \tag{17.6}$$

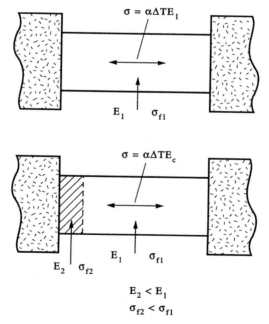

FIG 17.5　A uniform bar of foam is held between two fixed grips and is then cooled by ΔT.

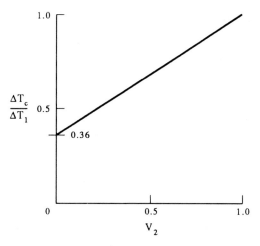

FIG 17.6   How the thermal shock resistance of the composite bar varies with the volume fraction V_2 of the low-modulus region.

Now consider a bar of foam which contains a region of low relative density. The modulus E_2 and fracture stress σ_{f2} are both lower than in the bulk of the foam. The tensile stress is $\alpha \Delta T E_c$, where E_c is the modulus of the composite bar. The thermal shock resistance is

$$\Delta T_c = \frac{\sigma_{f2}}{\alpha E_c}. \tag{17.7}$$

Thus

$$\frac{\Delta T_c}{\Delta T_1} = \left(\frac{\sigma_{f2}}{\sigma_{f1}}\right)\left(\frac{E_1}{E_c}\right). \tag{17.8}$$

The modulus of the composite bar is given by the standard result

$$E_c = \left\{\frac{1 - V_2}{E_1} + \frac{V_2}{E_2}\right\}^{-1}, \tag{17.9}$$

where V_2 is the volume fraction occupied by the weak region. Now the density of the foam is likely to vary quite considerably in practice because it is difficult to exert tight control over the spraying operation. It might be reasonable to expect regions where the relative density was say half that in the bulk of the foam. Equations (17.2) and (17.3) then indicate that both the modulus and the fracture stress should be reduced by a factor of $2^{3/2}$, to give $E_2 = 0.36E_1$ and $\sigma_{f2} = 0.36\sigma_{f1}$. If we substitute these relations into Eqns. (17.9) and (17.8), we can calculate $\Delta T_c / \Delta T_1$ for a given value of V_2. The results of the calculations are plotted in Fig. 17.6. In fact, $\Delta T_c / \Delta T_1$ is linear in V_2 and has a minimum value of 0.36 when V_2 is small. The thermal shock resistance is greatly reduced by the presence of a small low-density region and fracture is likely.

Fast Fracture

Although cracking may initiate in a localised zone of weakness it will not necessarily propagate through the rest of the foam layer. It is easy to estimate the size of the critical crack by using the fast fracture equation

$$K_{1c} \approx \sigma\sqrt{\pi a}. \tag{17.10}$$

Since $K_{1c} \approx 0.05$ MPa $\sqrt{m}$ and $\sigma(t) \approx 0.5$ MPa, then $a \approx 3$ mm. This is only six times the size of a single cell and it is likely that cracks will propagate from all but the smallest regions of weakness.

As soon as a crack forms at the cold side of the layer it runs straight through to the warm side. The situation is shown in Fig. 17.7. The stress field in the foam can be approximated by adding together a bending stress field having a maximum stress $\sigma_{max} = 0.5\sigma(t)$ and an axial stress field $\sigma = 0.5\sigma(t)$. The stress intensity of the crack is obtained by adding together the stress intensities given by Cases 4 and 5 in Appendix 1E to give

$$K_1 = (Y_4 + Y_5) \times 0.5\sigma(t)\sqrt{\pi a}. \tag{17.11}$$

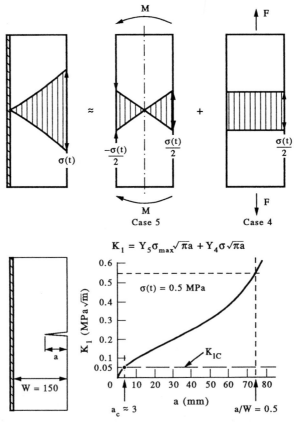

FIG 17.7 Stress intensity of a crack in the foam layer. Dimension in mm.

K_1 is plotted in Fig. 17.7 as a function of the length of the crack. As the crack moves into the foam K_1 increases rapidly: at the stage when the crack has gone halfway through the layer the stress intensity has gone up to 0.54 MPa $\sqrt{m}$, which is about ten times the value of K_{1c}.

Weibull Analysis

Another way of explaining the failures is to use Weibull statistics. The fracture stress of rigid foams is commonly measured by loading specimens in tension until they break. Because of structural variations the breaking stress will vary from one specimen to another. The survival probability of one specimen is given by the Weibull equation

$$P_s(V_0) = \exp\left\{-\left(\frac{\sigma}{\sigma_0}\right)^m\right\}. \tag{17.12}$$

V_0 is the volume of the test specimen (typically $\approx 5 \times 10^{-5}$ m³). The Weibull equation is plotted in Fig. 17.8 for two values of the Weibull modulus m (5 and 10). A reasonable estimate of m for PUR foam is ≈ 8; to get a more accurate value we would need to carry out a very large number of tests on samples taken from the insulation itself. The fracture stress σ_f is the *median strength* — the stress needed to break 50% of the specimens in the test sample. σ_0 is the stress needed to break (1/e) or 37% of the specimens. The precision of the published data for σ_f is rather poor so it is reasonable to assume that $\sigma_0 \approx \sigma_f$.

The survival probability of the insulating layer is given by

$$P_s(V) = \exp\left\{-\frac{1}{V_0}\int_V\left(\frac{\sigma}{\sigma_0}\right)^m dV\right\}, \tag{17.13}$$

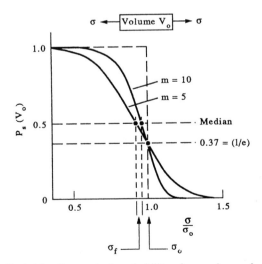

FIG 17.8   Weibull plot for the survival probability of a specimen of volume V_0 loaded to a stress σ in uniaxial tension.

where V is the volume of the layer. As shown in Fig. 17.2, the volume element $dV = a^2\,dx$. From Eqn. (17.1) we have

$$\frac{\sigma}{\sigma_0} = \frac{\alpha\Delta T}{(1-v)}\left(\frac{E}{\sigma_0}\right). \tag{17.14}$$

From Eqns. (17.2) and (17.3) we have

$$\frac{E}{\sigma_0} = \frac{E_s}{0.65\sigma_{fs}}. \tag{17.15}$$

At $-100°C$ we have

$$\frac{E}{\sigma_0} = \frac{3.0\ \text{GPa}}{0.65 \times 186\ \text{MPa}} = 24.8. \tag{17.16}$$

At $20°C$ we have

$$\frac{E}{\sigma_0} = \frac{1.9\ \text{GPa}}{0.65 \times 112\ \text{MPa}} = 26.1. \tag{17.17}$$

The average value of (E/σ_0) is ≈ 25, so we can write

$$\frac{\sigma}{\sigma_0} \approx \frac{25\alpha\Delta T}{(1-v)}. \tag{17.18}$$

The temperature drop at a distance x from the jacket is given by

$$\Delta T = \frac{100x}{t}, \tag{17.19}$$

so

$$\frac{\sigma}{\sigma_0} \approx \frac{25 \times 10^{-4}}{(1-0.3)}\left(\frac{100x}{t}\right) = \frac{0.36x}{t}. \tag{17.20}$$

Finally, we use Eqns. (17.13) and (17.20) to give

$$P_s(V) = \exp\left\{-\frac{1}{V_0}\int_0^t \left(\frac{0.36x}{t}\right)^8 a^2\,dx\right\}$$

$$= \exp\left\{-3.1 \times 10^{-5}\left(\frac{a^2 t}{V_0}\right)\right\}. \tag{17.21}$$

If the survival probability of the layer is $(1/e)$ then

$$3.1 \times 10^{-5}\left(\frac{a^2 t}{V_0}\right) = 1. \tag{17.22}$$

The volume of foam insulation which has a 37% probability of survival is therefore given by

$$a^2 t = 3.2 \times 10^4 V_0 \approx 1.6\ \text{m}^3. \tag{17.23}$$

Since $t = 0.15$ m, $a \approx 3.3$ m. This means that there is a 63% chance that cracks will form in the foam layer at intervals of ≈ 3 m.

Design Solutions

An obvious solution to the problem is to reinforce the foam with fibres of glass or drawn polymer. The fibres will bridge the cracks, helping to keep the crack faces together and increasing the fracture toughness considerably. This is, of course, how solid polymers are toughened in composite materials. Possibilities include (a) mixing short chopped fibres into the foam while it is being sprayed, and (b) incorporating layers of fibre mesh. However, this adds to the cost and complexity of the operation and may not be cost-effective.

References

M. F. Ashby and D. R. H. Jones, *Engineering Materials 1*, Pergamon, 1980.
M. F. Ashby and D. R. H. Jones, *Engineering Materials 2*, Pergamon, 1986.
L. J. Gibson and M. F. Ashby, *Cellular Solids—Structure and Properties*, Pergamon, 1988.

CHAPTER 18

TROUBLE WITH THICK-WALLED TUBES

CASE STUDY 1: A HYDRAULIC RAM EXPLODES AT THE LAUNCH OF THE GREAT EASTERN

Introduction

In December 1853 contracts were signed for the construction of the *Great Eastern*. It was to be the biggest ship that had ever been built, with dimensions of $692 \times 83 \times 58$ feet and a displacement of 22,500 tons. The length of the vessel was not to be exceeded for 50 years, and the displacement was not surpassed until the *Lusitania* was launched in 1906. The ship was powered by three sets of steam engines driving a pair of paddles and a single 20-foot propeller. In addition it could carry up to 6500 square yards of sail on five masts to save fuel on long ocean passages. The *Great Eastern* was designed to carry 4000 passengers and would have been able to make a return voyage from the UK to Australia without refilling the coal bunkers en route. The designer was one of the most gifted and charismatic engineers of all time—Isambard Kingdom Brunel.

The *Great Eastern* was built at Millwall on the River Thames. At this point the water is only 1000 feet wide. If the vessel had been launched in the conventional way it would probably have buried itself in the opposite bank. Brunel originally envisaged that the ship would be constructed in a dry dock but to save money it was decided to lay the keel on the shore 330 feet in from the high-water mark and to go for a sideways launch into the river. Brunel then recommended that the ship should be built on a mechanical slipway which would in turn rest on a large number of rollers. The complete unit could then have been run down the gentle slope of the foreshore in an easily controlled manner. This method was turned down as being too expensive and the ship was built straight on the launching ways; it was to prove a bad mistake.

The *Great Eastern* was to be launched on 3 November 1857, two hours before high water. The operation called for perfect co-ordination between the slipway gangs and had to be done in complete silence. It was a fiasco. The directors of the construction company had sold thousands of tickets to people who had flocked from all over London to see the great event and the noisy throng invaded every part of the shipyard. A gang operating one of the cable drums was distracted and one of them was killed outright by the spinning handles. The launch was aborted after the ship had moved only a few feet. Because the initial movement had been arrested it was to take an immense amount of effort to get the ship off again.

The country was scoured for hydraulic rams which could be set up to apply a horizontal thrust to the side of the hull. Among them was the jack used for lifting the tubular girders of Robert Stephenson's Britannia Bridge at Menai Straits. With a total force of 5500 tons the huge vessel was gradually inched down the slope until, on 31 January 1858, she was floated off by the tide and towed to her fitting-out berth at Deptford. There is always a greater risk of accident when people are working under pressure and the protracted launching operation had its fair share of incidents. *The Times* reported as follows:

> "It was something unheard of in the history of mechanics. In fact the accident to a windlass, when a side of its massive iron drum was crushed like a nut, was not only never known before, but until yesterday such a breakage was considered almost impossible. Through the sides of a hydraulic ram, ten inches in diameter, the water was forced through the pores of the solid iron like a thick dew, until the whole cylinder ripped open from top to bottom with a noise like a dull underground explosion. The iron of this cylinder averaged six inches in thickness and stood a pressure upward of twelve thousand pounds per square inch before it gave way. The massive cast-iron slab against which the base of another ram rested was split like a board, but this of course was a mere bagatelle among the other mishaps, which are not only expensive in themselves but by the delays they give rise to, occasion an expenditure in which the cost of the repair is a mere item."

Failure Analysis

Figure 18.1 is a schematic diagram of the hydraulic ram. The cylinder is an example of a thick-walled pressure-vessel. The relevant parameters (converted to SI units) are as follows.

External radius $a = 279$ mm.
Internal radius $b = 127$ mm.
Internal pressure at fracture $p \approx 83$ MPa.
Force on ram at fracture $F \approx 429$ tonnef.
Ratio $a/b = 2.20$.

The cylinder was almost certainly a casting and the material was probably grey cast iron. The account in *The Times* that "the water was forced through the pores of the solid iron like a thick dew" is consistent with this: if the molten metal was not degassed properly before it was poured into the mould then the final casting could well have contained interconnected porosity through which the water would have been forced under the immense pressure gradient. Casting would have been the obvious method for forming the cylinder: the bore would have been cast to shape by inserting a core into the mould and would only have required a final machining

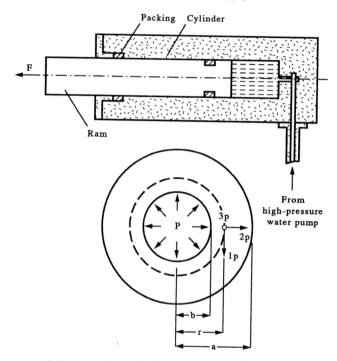

FIG 18.1   Schematic diagram of the hydraulic ram.

operation to bring it to the correct dimensions. The alternative, a hollow wrought-iron forging, would have been very difficult and expensive to produce. The inference is that the hoop stress at the inner surface of the cylinder reached the tensile strength of the cast iron and that the cylinder failed by brittle fracture.

The elastic equations for a thick-walled tube under internal pressure are given in Appendix 1A. The equation for the hoop stress

$$\sigma_1 = \frac{pb^2(a^2 + r^2)}{r^2(a^2 - b^2)} \tag{18.1}$$

can be re-written in the form

$$\sigma_1 = p\left\{\left(\frac{a}{b}\right)^2 + \left(\frac{r}{b}\right)^2\right\}\bigg/\left(\frac{r}{b}\right)^2\left\{\left(\frac{a}{b}\right)^2 - 1\right\} \tag{18.2}$$

to show that the hoop stress is a function of the dimensionless groups a/b and r/b. The ratio of σ_1/p is plotted as a function of r/b in Fig. 18.2. The hoop stress falls off rapidly with distance from the inner wall of the cylinder. Because of this we might think that fracture should initiate at the surface of the bore. Then $\sigma_1 = 1.52p$ and the failure stress is $\approx 1.52 \times 83$ MPa $= 126$ MPa. The reality is not as simple as this.

The strength of grey cast iron is determined largely by the shape and size of the graphite flakes which in turn are controlled by the rate at which the iron cools from the solidification temperature. The cooling rate is limited by the size of the casting:

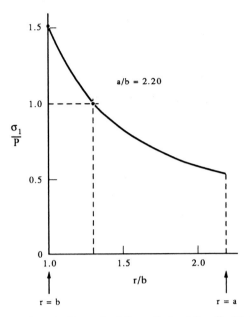

a/b = 2.20

$\frac{\sigma_1}{P}$

FIG 18.2   The hoop stress in the wall of the cylinder falls off with increasing distance from the bore.

thin sections cool faster than thick ones and have a higher tensile strength as a result. As an example a 15-mm bar of Grade 14 iron (nominal strength 14 tons/square inch) has a strength of ≈ 275 MPa; a 150-mm bar of the same material has a strength of only ≈ 110 MPa. The weakest cast iron normally specified is Grade 10 (nominal strength 10 tons/square inch). This has a strength of ≈ 200 MPa in a 15-mm bar and ≈ 50 MPa in a 150-mm bar (Angus, Smithells).

The wall of the cylinder is ≈ 150 mm thick. However, because it has a smaller surface-to-volume ratio than a round bar, it will cool more slowly. We should therefore expect it to have an average strength less than 50 MPa if it had been made of Grade 10 iron. Near to the inner wall of the cylinder, on the other hand, the cooling rate during casting will increase and we would expect strength levels of 200 MPa and more if Grade 10 iron had been used. This is a lot more than the value of ≈ 126 MPa which we estimated at the bore of the cylinder: this means in fact that fracture should *not* have started at the inner wall.

If we move out from the inner wall by 38 mm (a quarter of the wall thickness) the value of r/b increases to 1.30 and, as Fig. 18.2 shows, σ_1/p drops to 1.00. Then $\sigma_1 \approx 83$ MPa. The quarter-wall position is approximately where one would expect the *actual* strength to be equal to the *average* strength of the whole cross-section. This means that if the cylinder wall had been made from Grade 10 iron with an average strength of less than 50 MPa, then failure should have initiated roughly a quarter of the way through the wall with a margin of ≈ 30 MPa to spare. On this basis failure is also likely with a Grade 12 iron (average strength less than 75 MPa). Once fracture initiates, the crack will travel through the full thickness of the cylinder wall. This is because (a) cast iron has a low fracture toughness (≈ 10 MPa $\sqrt{m}$) and (b) the stress intensity of the crack increases rapidly as it grows.

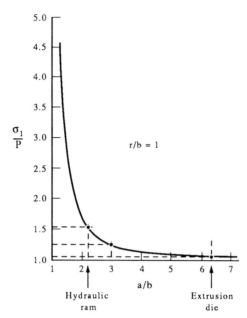

FIG 18.3   The maximum value of the hoop stress falls off as the wall thickness of the cylinder increases.

Could failure have been avoided by having a thicker wall? The answer is, only up to a point. Figure 18.3 shows how the hoop stress at the inner wall depends on the ratio a/b. As we have already seen, when $a/b = 2.20$, $\sigma_1(b) = 1.52p$. If we increase the wall thickness to give $a/b = 3.00$, then $\sigma_1(b) = 1.25p$. This 18% decrease in hoop stress is achieved at the expense of a 108% increase in the volume, weight and cost of the casting. Further increases in a/b are even less attractive and a better solution is to use a material with a higher strength. Of course, this was not a feasible option at the time. The failure could have been avoided by carrying out a hydraulic test on the cylinder and then fitting a pressure relief valve set to blow at $\approx 50\%$ of the test pressure. However, if pressure relief valves *had* been fitted the temptation to blank them off would have been too strong to resist!

CASE STUDY 2: AN EXTRUSION DIE DISINTEGRATES IN A RESEARCH LABORATORY

Introduction

Figure 18.4 is a diagram of a steel die used for extruding solid wires of eutectic lead–bismuth alloy for experiments on flux pinning in superconducting materials. Since the eutectic melts at only 124°C, the die could be filled by pouring molten alloy into the bore and allowing it to solidify. The assembly was then warmed up to a temperature of $\approx 50°C$ and extrusion was carried out using a 50-tonne press. Normally only a small fraction of this load was needed to extrude the product,

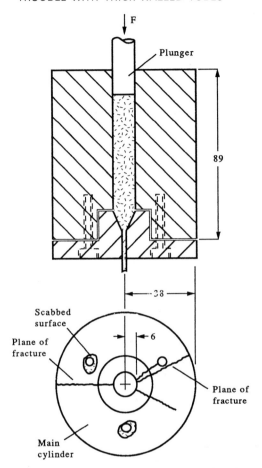

FIG 18.4   Diagram of the extrusion die. Dimensions in mm.

but on one occasion the press was allowed to run away and the die exploded under the internal pressure. As shown in Fig. 18.4, the die had broken in half across the central hole. The fracture surface was flat and there were no shear lips at the edges. Two other radial cracks had initiated at the surface of the bore but these had arrested in the wall. Scabs had also broken away from the surface around two of the tapped holes. The analysis of the failure was complicated by the fact that there was no information about the composition or the heat treatment of the steel. The surface of the die was very hard so it was assumed that the steel had been quenched and tempered. The most likely explanation of the failure was that the steel had not been tempered properly after quenching and had failed by brittle fracture.

Failure Analysis—Materials Properties

In order to check out the theory it was decided to do some simple experiments. To begin with, a small piece of steel was cut from the bottom of the die using a water-cooled abrasive disc. One of the cut surfaces was ground flat and Vickers hardness tests were done on it. Near to the outer surface of the die the hardness was ≈ 660 HV. 10 mm in from the surface it was ≈ 630 HV. A flat was ground in the middle of one of the fracture surfaces: it had a hardness of ≈ 600 HV. The cut-out section was then divided into two pieces using the abrasive saw. One piece was heated to bright red heat using a gas torch and was quenched into cold water. The hardness of the quenched piece varied between 890 HV and 974 HV. The rapid quench would have produced a structure consisting of untempered martensite and probably a little retained austenite. Appendix 1G shows how the hardness of untempered martensite varies with the carbon content of the steel. The maximum hardness of 974 HV lies off the graph but it indicates a carbon content of at least 1.0 weight %. The second piece was heated to bright red with the gas torch and was then cooled very slowly by turning the flame down progressively. When the sample had cooled to room temperature it was polished, etched in 2% nital, and examined using a metallurgical microscope. The former austenite grain boundaries were decorated with a thin white-etching film of primary iron carbide. The rest of the structure was dark-etching and indistinct, and presumably consisted of small nodules of fine pearlite. The hardness was ≈ 310 HV, which is just what one would expect for fine pearlite (Bain and Paxton). The structure was typical of a normalised steel with a carbon content above the eutectoid composition of 0.80% carbon (Ashby and Jones, Pascoe).

The hardness measurements on the die show that the steel has a high hardenability: even though the die is 76 mm in diameter, there is only a small fall-off in hardness away from the surface. This cannot be achieved by carbon alone. As the table in Appendix 1G shows, a plain carbon steel with 0.86% carbon will only transform to 100% martensite in an oil quench if the diameter of the component is less than ≈ 15 mm. However, the table also shows that adding ≈ 1% chromium should give 100% martensite in the middle of the die and it seems likely that a low-alloy steel of some sort was used in its manufacture.

Figure 18.5 is a plot which shows how the hardness of plain-carbon martensite decreases with increasing tempering temperature. Alloying elements tend to reduce the rate at which the hardness falls off, but the data do suggest that the die could have been tempered at a temperature as low as 300°C. This is just about the worst temperature that one could choose. Figure 18.6 shows how the Charpy V-notch impact energy of quenched low-alloy steels depends on the tempering temperature. There is a pronounced dip in the graph at ≈ 300°C and steels tempered at this temperature can have an impact energy that is less than if they had not been tempered at all.

To make a quick estimate of the brittleness of the steel a slice about 1 mm thick was cut from the bottom of the die. The slice was cut up to give a strip ≈ 10 mm wide and ≈ 37 mm long. The strip was tested by attempting to bend it around the surface of a large round bar, as shown in Fig. 18.7. However, it snapped like a carrot before it could be bent by plastic deformation. This "bend test" is commonly used to measure the ductility of normalised steel intended for structural applications: then, one would normally be able to bend a 1 mm strip through an angle of 180° around a bar having a diameter of only 3 mm! Clearly, the steel from the die is extremely brittle and the fracture stress σ_f is less than the yield stress σ_y.

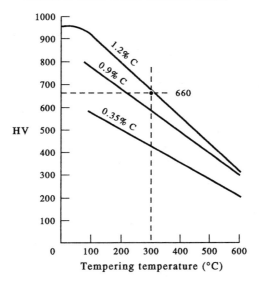

FIG 18.5   How the hardness of plain carbon martensite falls off with increasing tempering temperature (data from Bain and Paxton).

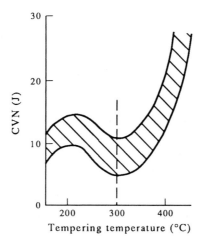

FIG 18.6   How the Charpy V-notch impact energy depends on the tempering temperature (data for a range of low-alloy steels from Bain and Paxton).

The fracture stress was estimated from three-point loading tests, as shown in Fig. 18.8. Four separate tests were done: σ_f varied between 656 and 1321 MPa, and had an average value of 864 MPa. As shown in Appendix 1C

$$\sigma_y \, (\text{MPa}) \approx \frac{H}{3} = \frac{600 \times 9.81}{3} = 1960. \qquad (18.3)$$

σ_y is the yield stress after the extra 8% plastic strain of the hardness test. Heat-treated high-carbon steels typically work-harden by ≈ 160 MPa after a strain of 8%, so the

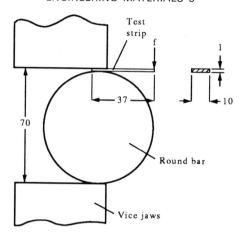

FIG 18.7   Schematic illustration of the bend test. Dimensions in mm.

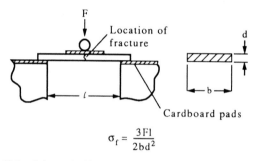

$$\sigma_f = \frac{3Fl}{2bd^2}$$

FIG 18.8   Schematic illustration of the three-point loading test.

yield stress of the die before testing would have been ≈ 1800 MPa. This is obviously well above the fracture stress.

Failure Analysis—Mechanics

To get fracture the pressure in the bore of the die must generate a hoop stress σ_1 which is at least equal to σ_f. The maximum value of the hoop stress is in fact limited by yield. The principal stresses in the wall under elastic conditions are given in Appendix 1A. They are

$$\sigma_1 = \frac{pb^2(a^2 + r^2)}{r^2(a^2 - b^2)},$$

$$\sigma_2 = -\frac{pb^2(a^2 - r^2)}{r^2(a^2 - b^2)},$$

$$\sigma_3 = 0. \tag{18.4}$$

The steel will start to yield when the principal stresses satisfy the Von Mises yield criterion (Appendix 1C). However, it is tedious to use this result for the thick-walled tube because it makes the algebra cumbersome. The Tresca yield criterion is perfectly adequate for our analysis and is much easier to handle. According to Tresca yield will occur when

$$\sigma_1 - \sigma_2 \geq \sigma_y, \tag{18.5}$$

in other words when

$$p \geq \frac{\sigma_y}{2}\left\{1 - \left(\frac{b}{a}\right)^2\right\}\left(\frac{r}{b}\right)^2. \tag{18.6}$$

r/b has a minimum value of 1 at the inner wall of the cylinder and this is where yield starts. Using the dimensions given in Fig. 18.4, we can see that $a/b = 38/6 = 6.33$. The condition for first yield is then found to be $p \geq 0.49\sigma_y$. Interestingly, the external diameter of the die could have been reduced substantially without decreasing the yield pressure much. For example, with an external diameter of 40 mm, $a/b = 3.33$ and yield occurs when $p \geq 0.46\sigma_y$. In other words, decreasing the volume by 3.86 times has reduced the yield pressure by only 6%.

Since $\sigma_y \approx 1800$ MPa, $p \approx 0.49 \times 1800$ MPa ≈ 880 MPa. What value of $\sigma_1(b)$ does this pressure generate? Figure 18.3 shows that, when $a/b = 6.33$, $\sigma_1(b) = 1.06p$. Thus, $\sigma_1(b) \approx 1.06 \times 880$ MPa ≈ 930 MPa. This is slightly more than σ_f, which suggests that brittle fracture might have occurred before the inner wall had a chance to yield. What force must the press deliver to make the inner wall of the die yield? The cross-sectional area of the plunger is 113 mm², so the axial force is 880 N mm^{-2} × 113 mm² = 9.9 tonnef. This is well within the capability of the press.

If the load is increased above 9.9 tonnef, a zone of plastic deformation will spread out from the bore, as shown in Fig. 18.9. It can easily be shown (Calladine) that the pressure needed to make the zone grow to a radius c is

$$p = \frac{\sigma_y}{2}\left\{1 - \left(\frac{c}{a}\right)^2\right\} + \sigma_y \ln\left(\frac{c}{b}\right). \tag{18.7}$$

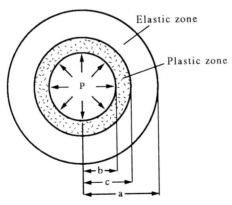

FIG 18.9   Plastic deformation begins at the inner wall of the cylinder. As the pressure is increased a concentric zone of yielded material spreads out from the bore.

Now the pressure which we can apply to the bore is limited by the yield strength of the plunger itself. If we assume that the yield strength of the plunger is the same as the yield strength of the die then $p = \sigma_y$. Equation (18.7) then becomes

$$1 = 0.5\left\{1 - \left(\frac{c}{a}\right)^2\right\} + \ln\left(\frac{6.33c}{a}\right). \tag{18.8}$$

This can be solved numerically to give $(c/a) = 0.27$. Then $c = 0.27 \times 38 \text{ mm} \approx$ 10 mm. The conclusion is that, even at the highest pressure which the plunger can deliver, the zone of plasticity will only spread out from the inner wall of the die by some 4 mm.

As the pressure increases and the zone of yielded material grows the hoop stress at the bore actually *decreases*. Equation (18.4) shows that, at the wall of the bore, $\sigma_2 = -p$. The Tresca yield criterion can then be written as

$$\sigma_1(b) = \sigma_y - p, \tag{18.9}$$

for the metal at the wall of the bore. Obviously, as p goes up, $\sigma_1(b)$ goes down. The maximum hoop stress now acts at the boundary between the elastic and plastic zones and increases progressively as the plastic zone grows. In fact, in the present situation the hoop stress is limited to $\approx 960 \text{ MPa}$ by the small size of the plastic zone.

Conclusions

This case study has shown that quenched steels must be tempered at the correct temperature if brittle fractures are to be avoided. Although the steel of the die had a high hardness as measured in an indentation test it broke in a brittle fashion at the fracture stress which was only half the yield stress. Had the steel been tempered properly the fracture stress would have been greater than the yield stress. Between yield and fracture the steel would have exhibited a limited amount of ductility and this would have prevented the failure.

References

H. T. Angus, *Cast Iron: Physical and Engineering Properties*, 2nd edition, Butterworth, 1976.
M. F. Ashby and D. R. H. Jones, *Engineering Materials 2*, Pergamon, 1986.
E. C. Bain and H. W. Paxton, *Alloying Elements in Steel*, 2nd edition, American Society for Metals, 1966.
C. R. Calladine, *Plasticity for Engineers*, Ellis Horwood, 1985.
K. J. Pascoe, *An Introduction to the Properties of Engineering Materials*, 3rd edition, Van Nostrand Reinhold, 1978.
J. Pudney, *Brunel and his World*, Thames and Hudson, 1974.
C. J. Smithells, *Metals Reference Book*, 6th edition, Butterworth, 1984.

F. Fatigue

CHAPTER 19

FATIGUE OF PIPE-ORGAN COMPONENTS

Background

In this case study we look at a rather unusual example of fatigue which led to the failure of a number of the mechanical connections in a recently rebuilt pipe-organ. The instrument in question is in fact situated in the Chapel at Christ's College, Cambridge (which is the college at which the author holds a Fellowship). In order to describe the failures and their significance we need first to understand the basic workings of the organ.

Figure 19.1 is a general view of the organ. Figure 19.2 is a close-up of the area where the player sits (the "console"). Figure 19.3 is a schematic cross-section of the instrument showing the main parts. The organ has three musical departments: the "Great", the "Choir" and the "Pedal". The Great provides the basic musical tone quality of the organ. The Choir is used for added colour or musical contrast. The Pedal supplies the bass notes for the whole instrument. The organist plays notes on the Great by pressing down keys on a keyboard, or manual, with his or her fingers. The Choir has its own separate manual which is mounted just above the Great manual. The organist plays the Pedal using his or her feet: mounted below the bench on which the organist sits is a "pedalboard" which is essentially a large keyboard designed so it can be played with the feet.

Each of the departments is based on a wind chest which is fabricated from boards of wood. The wind chest is supplied with low-pressure air from an electric fan-blower. On top of each wind chest are rows (or "ranks") of organ pipes made from either wood or metal. Each rank has one pipe for every semitone in the musical scale. All the pipes in a particular rank have a unified acoustic quality which is characteristic of that particular rank. Although Fig. 19.3 only shows a few ranks in each department, there is in fact a total of twenty-three ranks on the whole instrument, with nine on the Great, seven on the Choir and seven on the Pedal. It is possible to play a tune on the Great, for example, using only one rank (which would be rather quiet), or every rank (which would be rather loud) or any combination of ranks in between. The same can be done on the Choir, and on the Pedal. Since all three departments can be played independently the organist can produce a wide variety of sounds and therefore has to use considerable judgement when deciding which combination is best suited to the piece of music being performed.

FIG 19.1   The organ in the Chapel of Christ's College, Cambridge, England.

Each wind chest is divided into a top half and a bottom half. The bottom half is always under air pressure. The top half is divided along the length of the wind chest into a large number of separate boxes, one for each note in the musical scale. Each box is separated from the air in the lower half by a wooden flap valve called a pallet. The pallet is normally kept closed by a spring. The air pressure also helps to press the pallet shut against its seating. Each pallet is connected to either a key on one of the manuals or to a pedal on the pedalboard by a mechanical linkage. The arrangement is shown schematically in Fig. 19.3. We can see from the figure that if a note or a pedal is pressed down the linkage will pull the pallet away from its seating and let air into the box above. This means that air is now available to make the pipes above the box "speak".

In order to play notes on a particular rank the organist pulls out a small wooden knob called a "stop". Because there are twenty-three ranks on the organ there are also twenty-three stops. These are positioned on either side of the manuals as shown in Fig. 19.2. Each stop is connected by a mechanical linkage to a long,

FIG 19.2   Close-up of the organ showing the two keyboards, or "manuals". The wooden knobs on either side of the manuals are the stops, used to connect individual ranks of pipes to the wind supply. The pedalboard is out of sight below the bench on which the player sits.

narrow strip of wood or "slider" which slides underneath the rank of pipes. The slider has a row of holes bored through it, one for each pipe in the rank. When the rank is shut off the holes in the slider are exactly out of phase with the holes at the feet of the pipes: air cannot get to the pipes even if the pallets are opened. To activate the rank the stop must be pulled out. This moves the slider a small distance along the length of the chest so that every hole in it lines up with the hole at the foot of a pipe.

The failures had all occurred in the linkages between the pedalboard and the pallets in the Pedal wind chest. A total of about five linkages had failed, affecting five separate musical notes. An unexpected failure during an organ recital or a Chapel service can be very off-putting for both the organist and the listeners and it was important to establish the cause of the breakages.

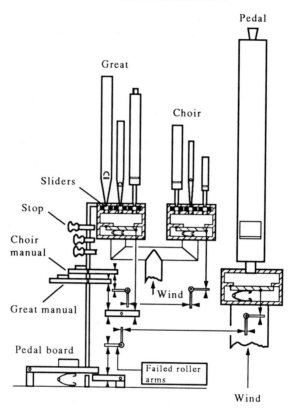

FIG 19.3   Schematic section through the organ. Note the mechanical action which links
each pallet to either a key or a pedal.

Operational History

There has been an organ in Christ's Chapel since at least 1510, for the College
accounts in that year record a payment of six pence for "the mendying of thorgans".
In the early 1530s an organ builder called William Beton installed "ye newe
orgayns" at a cost of £10. This instrument lasted until 1705, although it had to be
dismantled during the Civil War.

In 1705 a new and updated organ was installed by the famous organ builder
Bernard Smith at a cost of £140. This instrument had a Great similar in size to the
present one, but had no Choir or Pedal departments. The original carved organ
case survives, as do some of the pipes. Soon after 1785 the organ became derelict
and some of the pipes were stored away in boxes for safe-keeping. A writer in the
College magazine in 1893 recalled that "we had no organ; only what some facetious
student called 'organic remains', in the shape of some dusty pipes stowed away
behind the empty organ-case".

The organ was rebuilt in 1865 by Hill and Son at a cost of £397. Choir and Pedal
departments were added since these had become essential to playing the musical
repertoire of the period. By 1909, however, musical fashions had changed yet again.

In order to keep the organ up to date it was rebuilt by Norman and Beard who added a third manual and increased the number of pipes to the vast total of 1900. It was only with the greatest difficulty that all the pipes could be crammed into the space in the organ loft and the author recalls that one had to behave like a pot-holer to gain access to the machinery of the instrument. The only way of connecting the manuals and pedals to the pallets was by using the rather bizarre "pneumatic action" in which the signal was transmitted by sending puffs of high-pressure air along rubber tubes!

By 1983 the organ had become totally unreliable and the opportunity was taken of rebuilding it along the lines of the much more economical Hill version of 1865. The complex pneumatic action was discarded and a brand-new version of the original mechanical action was used instead. Naturally, there had been many developments in materials since 1865 and it was cheaper to make some parts from aluminium alloys and plastics. Of course, innovation in design nearly always involves an element of risk and the failures were traced to an inappropriate use of an aluminium–copper alloy in a small part of the mechanism. It was in fact a simple operation to replace the offending parts with ones made from steel and no trouble has been experienced since the modifications were made.

Details of the Failures

The general location of the failures is marked in Fig. 19.3. Figure 19.4 shows the details of the mechanical action in this area. The pedal is connected to the end of the roller arm by a vertical rod of wood called a tracker. When the pedal is pushed down the tracker pulls the end of the roller arm down and this makes the roller rotate. The arm at the other end of the roller rotates as well, actuating the horizontal tracker. Both the arms and the roller were made from aluminium alloy.

Figure 19.5 is a cross-section of the roller-arm assembly which shows the position of the fractures. The arm was made from round rod 4.75 mm in diameter. The end of the rod had been turned down to a diameter of 4 mm to fit the hole in the roller and had been riveted over to hold it in place. The turning operation had obviously been done with a sharp-ended lathe tool because there was no radius where the 4-mm section met the main body of the rod. The fatigue crack had initiated at this sharp change of section.

Figure 19.6 is a general view of the fracture surface taken in the scanning electron microscope (SEM). The top of the arm is at the top of the photograph. This is the region that is put into tension when the tracker is pulled down. The photograph shows that the fatigue crack initiated in this region of maximum tensile stress and then propagated downwards across the 4-mm section. The progress of the crack is marked by "clam-shell" marks. These show the tip of the crack at various stages in its growth and are caused by changes in either the loading pattern or the environment. Eventually the crack reached its critical length and the remaining cross-section failed by fast fracture.

Figure 19.7 is a close-up taken from the surface of the fatigue crack just before it went unstable. The picture shows three further features of fatigue: surface smearing, fatigue striations and multiple cracking. The smearing is caused by the

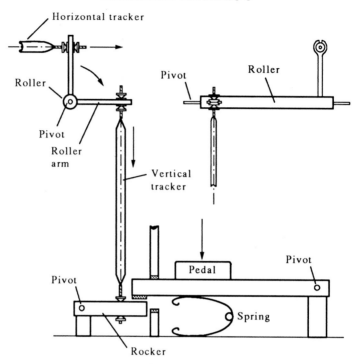

FIG 19.4   Schematic of the mechanical action in the region of the failed components.
Note that the vertical tracker and the rocker are both made from light wood and their
weight can be neglected.

opposite faces of the fatigue crack rubbing over one another as the load on the
component cycles. Flat, burnished areas are characteristic features of a fatigue
surface. Fatigue striations are groups of very fine, closely spaced lines which indicate
the progress of the crack tip with each cycle of loading. They are usually obvious
in aluminium alloys but are often almost invisible in steels. Multiple cracking is a
product of the fatigue processes that occur in the plastic zone at the crack tip. Both
the smearing and the striations are shown more clearly in Fig. 19.8, which is taken
at higher magnification.

   Figure 19.9 is a close-up taken from the region of fast fracture. The surface shows
the classic features of ductile failure by microvoid coalescence which are summarised
in Appendix 1C.

Failure Analysis—Materials Data

   An approximate chemical analysis of the aluminium-alloy roller arm was done in
the SEM. This showed that the arm contained several percent copper. The metal
was probably Duralumin, a common age-hardening alloy of the 2000 series. Typical
properties are given by Smithells. The alloy contains about 4.4% Cu, 1.0% Mg,
0.75% Mn and 0.4% Si. The Vickers hardness of the arm was found to be 69.

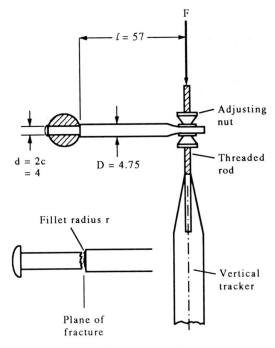

FIG 19.5   Location of the fractures. Dimensions in mm.

Age-hardened Duralumin has a Vickers hardness of about 110. This is 60% more than the hardness of the arm, which is therefore in quite a soft condition. A fatigue cycle of ±140 MPa will make a smooth, un-notched specimen of age-hardened Duralumin break after 5×10^8 cycles. Annealed Duralumin has a fatigue strength of only ±95 MPa for the same life. The material of the roller arm should therefore have a fatigue strength that is somewhere between these two values. The yield and tensile strengths of annealed Duralumin are approximately 130 MPa and 230 MPa; for age-hardened Duralumin they are approximately 255 MPa and 430 MPa.

Failure Analysis—Loading History

Tests were carried out with the organ running to find the force F that was needed to depress the end of the roller arm. At the top end of the pedalboard (the end that plays the highest notes in the Pedal department) the force was about 0.5 kgf. At the bottom end of the pedalboard (which plays the lowest notes in the Pedal) the force was about 1 kgf. The difference is caused mainly by a systematic decrease in the size of the pallets along the length of the wind chest. The pipes at the bottom end of the wind chest are much bigger than those at the top and they need more wind to make them speak properly. To get enough air through to the big pipes the pallets at the bottom end of the wind chest must be larger than those at the top end. A larger force is needed to open the big pallets against the pressure of

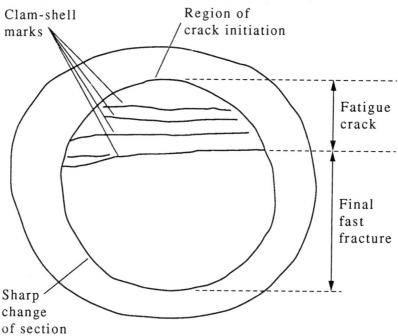

Clam-shell marks

Region of crack initiation

Fatigue crack

Final fast fracture

Sharp change of section

FIG 19.6   General view of one of the fracture surfaces taken in the scanning electron microscope (SEM). Magnification: ×15. The identification diagram points up the main features of the fracture surface.

the air underneath them, and this increases the force required to make the trackers move.

When we described the workings of the organ there was one complication which we did not mention. This is that the instrument, like nearly all pipe organs of any size, is fitted with "couplers". For example, when the organist pulls out the wooden

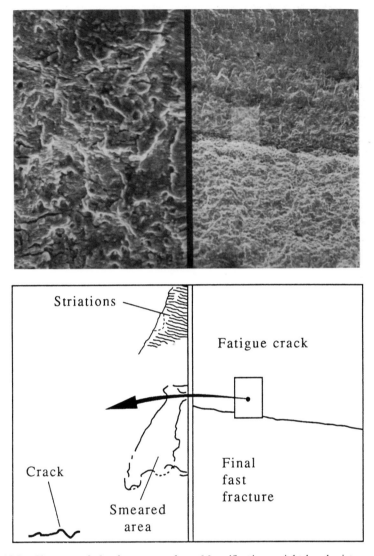

FIG 19.7   Close-up of the fracture surface. Magnifications: right-hand picture ×35, left-hand picture ×270. The identification diagram points up typical examples of cracks, striations and smeared areas.

knob labelled "Choir to Great" he or she can press a key down on the Great manual and see the corresponding key on the Choir manual move down of its own accord. This is not meant to send shivers down the spine of an uninitiated observer, however! It is done so that the organist can select pipes from the Choir and play them alongside pipes from the Great when using only the Great manual. Naturally, because the keys on the Great manual are then driving two sets of mechanisms and pallets instead of just one the force needed to depress the keys is approximately doubled. The couplers which involve the pedalboard are Great to Pedal, and Choir to Pedal. When these are both selected and a "key" on the pedalboard is pressed

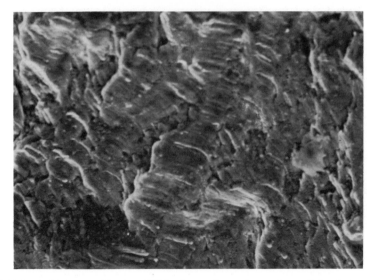

FIG 19.8   A high-magnification SEM picture from the surface of the fatigue crack. Magnification: ×390. Note striations and smeared areas.

down the corresponding keys on both the Great and Choir manuals move down automatically. The forces measured above are for the worst case where both pedal couplers are "on". In fact, uncoupling the pedals makes little difference to the forces required to move the trackers because the Pedal pallets are much bigger than those on either the Choir or the Great.

It is not easy to obtain an accurate figure for the number of cycles of loading experienced by the failed roller arms. The following estimate is based on reasonable guesses.

Age of action = 2 years ≈ 100 weeks.
Organ played on average for 10 hours in each week.
Popular pedal note sounded once every 10 seconds.
Number of cycles ≈ $(100 \times 10 \times 3600)/10 \approx 3.6 \times 10^5$.

Failure Analysis—Stress Calculations

Using the results given in Appendix 1B, we find that the maximum stress in the reduced section of the roller arm is given by

$$\sigma = \frac{Mc}{I},$$
(19.1)

where

$$I = \frac{\pi c^4}{4}$$
(19.2)

and

$$M = Fl.$$
(19.3)

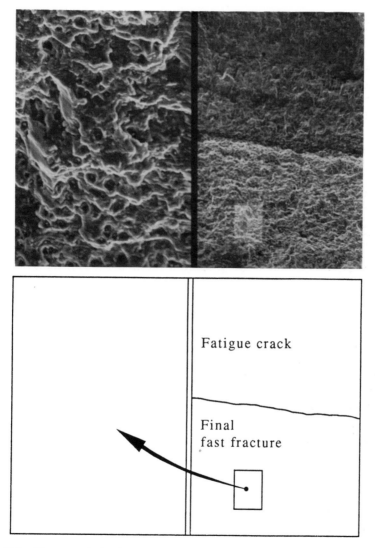

Fatigue crack

Final
fast fracture

FIG 19.9   Close-up of the fracture surface. Magnifications: right-hand picture ×35, left-hand picture ×270. The left-hand picture shows the typical features of ductile failure by microvoid coalescence.

Combining these equations we get

$$\sigma = \frac{4Fl}{\pi c^3}. \tag{19.4}$$

The maximum value of F is 1 kgf, or 9.81 N. Figure 19.5 shows that $l = 57$ mm and $c = 2$ mm. With these values, Eqn. (19.4) gives $\sigma = 90$ MPa. When $F = 0$, the stress is obviously zero across the whole cross-section. The fatigue cycle is therefore ± 45 MPa about a mean tensile stress of 45 MPa.

The effect of the non-zero mean stress can be allowed for by using Goodman's rule. This is

$$\Delta\sigma = \Delta\sigma_0\left(1 - \frac{\sigma_m}{\sigma_{TS}}\right), \tag{19.5}$$

where $\Delta\sigma_0$ is the stress range with a zero mean stress which is equivalent to a stress range of $\Delta\sigma$ with a mean stress of σ_m. Then

$$\Delta\sigma_0 = \frac{\Delta\sigma}{\left(1 - \frac{\sigma_m}{\sigma_{TS}}\right)} \approx \frac{90\,\text{MPa}}{\left(1 - \frac{45\,\text{MPa}}{230\,\text{MPa}}\right)} = 112\,\text{MPa}. \tag{19.6}$$

This gives an equivalent fatigue cycle of ± 56 MPa about a mean stress of zero. This is well below the fatigue strength of ± 95 MPa for annealed Duralumin so the reduced section of the roller arm would appear to be safe. However, this analysis ignores the effect of the stress concentration at the sharp change of section and, as we now show, this has a dominant effect.

Appendix 1A gives a table of stress–concentration factors for the case of a round, shouldered bar in bending. The SCF depends on the ratios D/d and r/d. As shown in Fig. 19.5, D and d are the diameters of the two parts of the bar and r is the radius of the fillet at the change in section. The ratio $D/d = 4.75\,\text{mm}/4\,\text{mm} = 1.19$. The fillet radius is not specified, but a rough guess might be 0.025 mm. Then $r/d = 0.025\,\text{mm}/4\,\text{mm} = 0.006$. The nearest values of D/d and r/d in the SCF table are 1.10 and 0.025, which give an SCF of 2.25. By inspecting the trends in the table we can see that our SCF should be considerably more than 2.25 and could in fact be as large as 4.

On the face of it, this SCF indicates that the maximum local stress at the sharp corner is $4 \times 90\,\text{MPa} = 360\,\text{MPa}$. Of course, this is well above the yield stress of 130 MPa: plastic flow should therefore occur at the corner, which will redistribute the stresses and invalidate the elastic theory used to determine the SCF. However, this is not as simple as it seems. The material in the sharp corner is constrained by the surrounding material and the local yield stress is greater than the uniaxial yield stress as a result. A rather severe example of this *plastic constraint* can be found at the tip of a sharp crack in plane strain. This situation is treated in Appendix 1E. The Von Mises equivalent stress at the crack tip is only 40% of the normal tensile stress. Since yield can only occur when the equivalent stress reaches the uniaxial stress, the material at the crack tip can only yield when the normal stress is equal to $1/0.4 = 2.5$ times the yield stress. Of course, the material at the root of the sharp corner is less constrained than this. But the analysis for the crack does allow us to put an upper limit of about $2.5 \times 130\,\text{MPa} \approx 325\,\text{MPa}$ on the stress at the corner. In view of this it is not surprising that the arms failed by fatigue in a relatively small number of cycles.

Design Implications

This case study has shown that stress concentrations need to be avoided wherever possible in components which are subjected to fatigue loading. The arm could have been made from rod with a uniform diameter of 4 mm fixed into the holes with epoxy-resin glue: in this case it is probable that failure would not have occurred. A bigger safety factor can be achieved by using alloy in the fully age-hardened condition, giving a better fatigue strength. However, even normalised mild steel has a fatigue strength of ± 193 MPa for 10^7 cycles; quenched-and-tempered medium-carbon steel has a fatigue strength of ± 293 MPa for the same life. If the design life of an organ is taken to be 100 years, which involves about 2×10^7 cycles, then a uniform rod of ordinary mild steel should be perfectly satisfactory.

References

C. Clutton, "The Organ", in *Musical Instruments through the Ages*, edited by A. C. Baines, Penguin, 1961, p. 55.

R. W. Hertzberg, *Deformation and Fracture Mechanics of Engineering Materials*, 3rd edition, Wiley, 1989.

R. M. N. Pelloux, "The Analysis of Fracture Surfaces by Electron Microscopy", in *Source Book in Failure Analysis*, American Society for Metals, 1974, p. 381.

C. J. Smithells, *Metals Reference Book*, 6th edition, Butterworth, 1984.

B. G. Whalley, "The Analysis of Service Failures", *Metallurgist and Materials Technologist*, **15**, 21 (1983).

CHAPTER 20

MORE TROUBLE WITH SHARP CORNERS

CASE STUDY 1: COLLAPSE OF A SWING DOOR

Background

Figure 20.1 shows the general arrangement of a swing door which was positioned at one of the entrances to a public theatre. The door was a heavy one—it weighed about 100 kg—and instead of being hung on conventional hinges it was supported by pivots. Unfortunately, after the theatre had been in use for about a year, the top pivot fractured and the door fell out of the frame, narrowly missing a person who was passing through the doorway.

The details of the top pivot assembly are shown in Fig. 20.2. The pivot itself was machined from a round bar of steel. Separate die-cast housings were recessed into the top of the door frame and the top of the door. These contained vertical holes into which the pivot fitted. The pivot was held captive in the door frame by an adjusting lever which fitted into a deep groove machined in the pivot. The extent to which the pivot overlapped the hole in the door could be adjusted by rotating the head of the adjusting screw using a screwdriver. There was nothing to stop the pivot turning in the holes, but because the hole in the door housing was greased the pivot tended not to turn in the frame housing.

The location of the fracture is shown in Fig. 20.3. The pivot had broken right across at one end of the groove. As a result the lower part of the pivot had dropped into the bearing hole and this had allowed the top of the door to come free. It was estimated that the door had swung back and forth about 250,000 times since it had been installed and this pointed to fatigue as the probable cause of the failure.

Material Properties

A cross section was cut from the failed pin and this was mounted and polished for metallographic examination. The microstructure consisted of pearlite in a ferrite matrix. The amount of pearlite indicated that the steel contained about 0.15 weight %

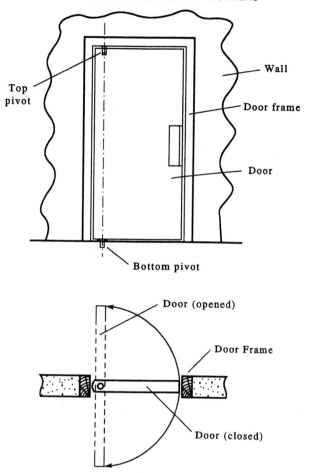

FIG 20.1   The general arrangement of the swing door.

carbon. The Vickers hardness was measured at various positions on the cross section. The hardness was reasonably uniform and had an average value of about 200. The correlation given in Appendix 1C shows that the tensile strength of the steel was about $3.2 \times 200 = 600$ MPa. Normalised 0.20% C steel typically has a tensile strength of 430 MPa and a fatigue strength for 10^7 cycles of ± 200 MPa (Smithells). The much higher tensile strength of the pivot probably indicates that it was made from cold-drawn bar. A steel with a tensile strength of 600 MPa typically has a yield strength of 460 MPa and a fatigue strength for 10^7 cycles of around ± 250 MPa (Smithells).

Stresses in the Pin

Figure 20.4 shows the static forces acting on the door. The weight of the door is supported by the bottom pivot but the top pivot is responsible for keeping the door

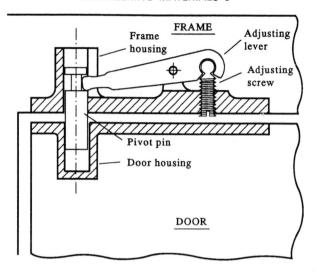

Elevation of pivot assembly

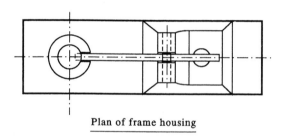

Plan of frame housing

FIG 20.2   Details of the top pivot assembly.

upright. In order to do this it must deliver a horizontal force F to the top of the door. The value of F can be found by balancing moments about the bottom pivot, giving

$$F \times 2500 = 100 \text{ kgf} \times 500. \tag{20.1}$$

Thus $F = 20 \text{ kgf} = 196$ N.

Figure 20.5 shows the static forces acting on the pivot. The reaction to F acts horizontally on the lower part of the pivot. We cannot specify the precise point where the force is delivered, but a fair guess is to assume that it is applied at the centre of the nominal area of contact between the pivot and the door housing. The pivot is a running fit in the frame housing as well: because of this it will rotate slightly and will tend to contact the frame housing at only two places. We assume, as before, that the force is delivered to each contact at the centre of the nominal area. Taking moments we get

$$f \times 15 = F \times 20. \tag{20.2}$$

Thus $f = 260$ N. Because f acts horizontally at the top of the pivot it exerts a

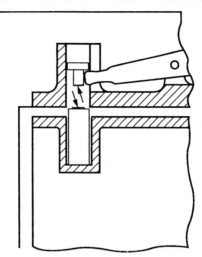

FIG 20.3   Details of the failure. Each fracture surface is marked with an arrow.

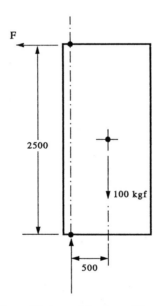

FIG 20.4   Static equilibrium of the door. Dimensions in mm.

bending moment on the reduced section. The bending moment at the location of the fracture is given by

$$M = 260 \text{ N} \times 9 \text{ mm} = 2340 \text{ N mm}. \tag{20.3}$$

The bending moment obviously puts one side of the reduced section into tension and the other side into compression. The maximum tensile or compressive stress

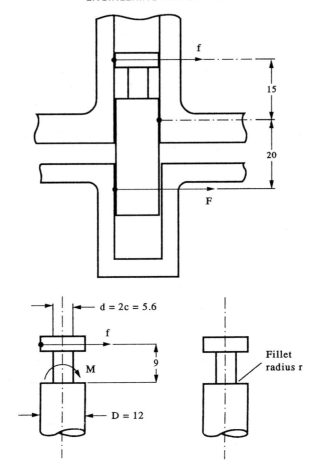

FIG 20.5   Static equilibrium of the pivot. Dimensions in mm.

can be calculated using the standard result given in Appendix 1B. Thus

$$\sigma = \frac{Mc}{I},$$ (20.4)

$$I = \frac{\pi c^4}{4},$$ (20.5)

and

$$\sigma = \frac{4M}{\pi c^3}.$$ (20.6)

$M = 2340$ N mm and $c = 2.8$ mm so $\sigma \approx 140$ MPa. Of course, as the door swings back and forth the reduced section is subjected to an alternating load and the metal is subjected to a nominal stress cycle of ± 140 MPa at the point of failure.

The *actual* stress cycle at the point of failure is very much more than ± 140 MPa because of the stress concentrating effect of the sharp change in section. We have already come across this situation in Chapter 19. The stress-concentration factor (SCF) for a round, shouldered bar in bending depends on the ratios of D/d and r/d. Using the dimensions given in Fig. 20.5 we see that $D/d = 2.14$. If we guess that $r \approx 0.05$ mm then $r/d \approx 0.01$. The SCF table in Appendix 1A shows that the SCF $= 2.59$ when $D/d = 1.50$ and $r/d = 0.025$. The trends in the table show that our SCF is probably in the region of 3.5 because of the more acute change in section. This would imply a maximum stress cycle of around ± 490 MPa at the fillet. This is only $\approx 7\%$ more than our estimate for the uniaxial yield strength of the pivot. Given that there will be some plastic constraint in the sharp corner it is likely that the elastic SCF of ≈ 3.5 will actually apply and that the stress cycle will be close to ± 490 MPa.

Failure Mechanism

The failure is a good example of *high-cycle fatigue*. This is described by Basquin's law

$$\Delta\sigma N_{\mathrm{f}}^{\mathrm{a}} = C, \qquad (20.7)$$

which applies above $\approx 10^4$ cycles. Our estimate for the fatigue strength of the steel gave $\Delta\sigma = 500$ MPa for $N_{\mathrm{f}} = 10^7$. For the door $N_{\mathrm{f}} = 2.5 \times 10^5$ and $\Delta\sigma > 500$ MPa. The increase in $\Delta\sigma$ as N_{f} decreases from 10^7 to 2.5×10^5 is typically of the order of 60 MPa for mild steel (Frost *et al.*). We would therefore expect $\Delta\sigma \approx 560$ MPa for $N_{\mathrm{f}} = 2.5 \times 10^5$, giving a stress cycle of ± 280 MPa. This is much less than our estimated stress cycle of ± 490 MPa, so why did the pivot not break much sooner than it did?

In order to arrive at an SCF of ≈ 3.5 we assumed a fillet radius of 0.05 mm at the end of the groove. A stress concentration of this severity creates a large *stress gradient* in the metal under the fillet. This means that only a tiny volume of metal actually sees a stress cycle which is anything like as large as ± 490 MPa. It is therefore less likely that a fatigue crack will initiate at the fillet—and even if a crack did form, there would be no guarantee that it would propagate down the stress gradient into the bulk of the pivot. In fatigue terms the *effective* SCF will be less than the elastic SCF. In the present case the effective SCF could be around half the elastic SCF (Hertzberg), giving an effective stress cycle which is approximately equal to the fatigue strength for 2.5×10^5 cycles.

Design Implications

The failure provides another graphic illustration of the folly of having a sharp change of section in any component which has to take a fatigue loading. It would have been quite simple to specify a fillet radius of say 2 mm at the ends of the groove, giving $r/d = 0.36$. The table in Appendix 1A then gives an SCF of ≈ 1.3. This reduces the maximum stress cycle from ± 490 MPa to ± 180 MPa. The stress cycle is now 28% less than the fatigue strength for 10^7 cycles, so the pivot should last at least 10^7 cycles (40 years at the current rate of use). However, given the uncertainties in the calculations it would be wise to get the stress amplitude down even more. This can be done very effectively by increasing the diameter of the groove. As Eqn. (20.6) shows the stress varies inversely as the cube of the groove diameter. If we increase the diameter from 5.6 mm to 7 mm we almost halve the stress. The stress cycle at the corner is then reduced to ± 90 MPa, which is well below our estimate for the fatigue strength.

Design modifications of this kind cannot necessarily be done in isolation. One consequence of increasing the diameter of the groove is to decrease the depth of engagement between the pivot and the adjusting lever. If the groove is made too shallow relative to the tolerances on the parts it might be possible for the pivot to slide past the end of the adjusting lever on a proportion of manufactured assemblies. Even if this did not affect the integrity of the unit it could certainly make it difficult to install or remove the door. If a deep groove has to be retained for location purposes then the bending moment on the reduced section can be designed-out by making sure that the top of the pivot cannot touch the bore of the hole in the frame housing. This can be achieved by a combination of the following simple modifications:

(a) decrease the diametrical clearance between the shank of the pivot and the bore of the hole;
(b) increase the diametrical clearance between the top of the pivot and the bore of the hole;
(c) increase the length of shank which lies inside the upper housing.

CASE STUDY 2: FAILURE OF A LIFTING EYE

Background

Figure 20.6 shows a pulley block which was used for lowering an inspection platform down the side of a marine jetty and hoisting it up again. The mass of the platform was around 1800 kg. The block was supported from the boom of a hydraulic crane which was mounted on the deck of the jetty. The platform was lowered and hoisted by the wire rope which was taken from a power winch mounted on the deck near to the base of the crane. The pull of the crane boom was transmitted to the head of the block through a lifting eye. During the hoisting operation the crane exerted a straight pull of 2760 kgf along the axis of the eye. When the platform had been winched up to the level of the deck it was secured directly to the side of the block as shown in Fig. 20.7. The winch was slackened off, the crane was slewed around, and the platform was lowered onto the deck. During this operation the crane boom exerted an oblique pull of 1800 kgf on the eye.

Figure 20.8 shows the detailed construction of the lifting eye. The eye itself was a steel forging, supplied as a standard item. It was fixed to the pulley block by a steel bolt which had been manufactured for the purpose. The threaded end of the bolt was screwed into the tapped hole in the eye until the shoulder on the bolt was in contact with the flat surface at the bottom of the eye. No specifications were

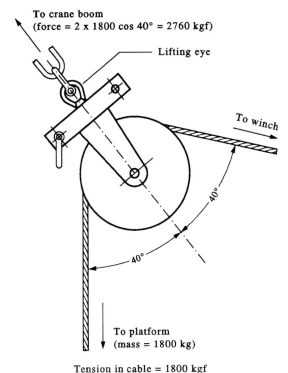

To crane boom
(force = 2 x 1800 cos 40° = 2760 kgf)

Lifting eye

To winch

40°

40°

To platform
(mass = 1800 kg)

Tension in cable = 1800 kgf

FIG 20.6   Schematic of pulley block during hoisting operation.

given for the assembly torque but the bolt was secured against unscrewing by a bissel pin. The shank of the bolt was a running fit in the block.

After about 200 hoisting operations the eye broke away at the top of the block and the platform fell down the side of the jetty. The details of the fracture are shown in Fig. 20.9. The failure had occurred in the bolt at the position where the thread met the shank. In order to allow the screw-cutting tool to run out at the end of the thread a sharp groove had been machined next to the shoulder. The failure had initiated at this sharp change of section as a fatigue crack. The crack had obviously propagated in from the surface of the groove until the remaining cross section became unable to support the load and failed by fast fracture. The fatigue surface was flat and had circumferential "clam-shell" markings. When it was examined in the scanning electron microscope (SEM) it had a burnished appearance. The fast-fracture surface looked bright and examination in the SEM confirmed that it consisted of cleavage facets.

Material Properties

A chemical analysis was carried out on a piece cut from the bolt. The results are given in Table 20.1. The composition is typical of a medium-carbon low-alloy steel capable of reaching a tensile strength of 1500 MPa after heat treatment. A cross

TABLE 20.1

Element	Weight %	Element	Weight %
Carbon	0.40	Phosphorus	0.011
Silicon	0.32	Chromium	1.05
Manganese	0.65	Molybdenum	0.25
Sulphur	0.025	Nickel	1.45

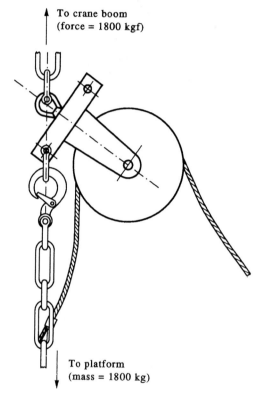

To crane boom
(force = 1800 kgf)

To platform
(mass = 1800 kg)

FIG 20.7   Schematic of pulley block after hoisting operation.

section was cut from the bolt and was ground flat for Vickers hardness testing. The hardness was uniform across the section and had an average value of ≈ 220. The correlation given in Appendix 1C shows that the tensile strength of the bolt was about $3.2 \times 220 \approx 700$ MPa. This figure is a good deal lower than the tensile strength which can be produced by heat treatment. It is therefore probable that the steel was supplied in the soft condition and was not heat treated after machining. A steel of this type with a tensile strength of 700 MPa should have a yield strength of ≈ 540 MPa (Smithells).

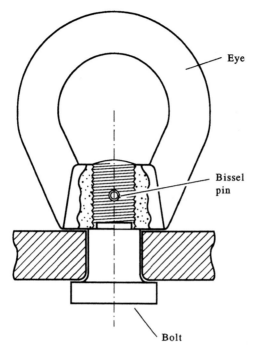

FIG 20.8   Details of lifting eye.

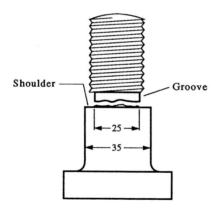

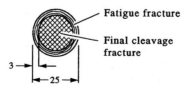

FIG 20.9   Details of the failure. Dimensions in mm.

Stresses in the Bolt

As shown in Fig. 20.10 the straight pull of 2760 kgf simply generates a uniform tensile stress of 55 MPa in the bolt at the position of the groove. In contrast the oblique pull of 1800 kgf generates three stress states, as Fig. 20.11 shows. The vertical component of 1160 kgf generates a uniform tensile stress at the groove of $(1160 \times 9.81)/491$ MPa = 23 MPa. The horizontal component of 1380 kgf generates a uniform shear stress of $(1380 \times 9.81)/491$ MPa = 28 MPa. The overall force of 1800 kgf also applies a *bending moment* to the groove. Because the line of action of the force lies 60 mm away from the centre of the groove the moment is simply $1800 \times 9.81 \times 60$ N mm = 1.06×10^6 N mm.

How important is this bending moment? It is easy to calculate the moment M_{el} at which the cross section starts to yield. Following Eqn. (20.6) we can write

$$M_{el} = \frac{\pi \sigma_y c^3}{4}. \tag{20.8}$$

Setting $\sigma_y = 540$ MPa and $c = 12.5$ mm we get $M_{el} = 0.83 \times 10^6$ N mm. This is 22% less than the actual bending moment, so the bolt should have experienced plastic deformation in service. The uniform tensile and shear stresses are negligible compared to the maximum bending stress and the bending moment is the dominant term.

Could the groove have gone fully plastic? From Appendix 1C the fully plastic moment of the cross section is

$$M_p = \frac{4 \sigma_y c^3}{3}. \tag{20.9}$$

If we compare Eqns. (20.8) and (20.9) we see that $M_p = 1.70 M_{el}$. The bending

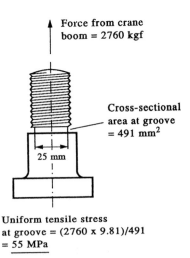

Force from crane
boom = 2760 kgf

Cross-sectional
area at groove
= 491 mm²

25 mm

Uniform tensile stress
at groove = (2760 x 9.81)/491
= 55 MPa

FIG 20.10   Force acting on the bolt during hoisting.

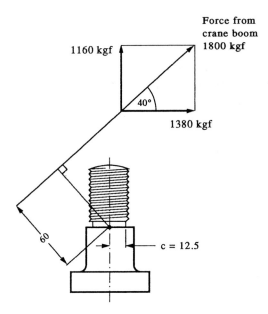

FIG 20.11   Force acting on the bolt after hoisting. Dimensions in mm.

moment needed to create a plastic hinge in the groove is then $1.70 \times 0.83 \times 10^6$ N mm $= 1.41 \times 10^6$ N mm. Because this is 33% more than the actual moment the groove should not have gone fully plastic.

In estimating the bending moment which is applied to the bolt we have of course ignored any load-bearing contact between the bottom of the eye and the turned shoulder. Any contact should serve to reduce the bending stress in the groove, but since the situation then becomes statically indeterminate it also becomes impossible to analyse.

Failure Mechanism

The failure is a classic example of *low-cycle fatigue*. This is described by the Coffin–Manson law

$$\varDelta \varepsilon^{pl} N_f^b = C, \tag{20.10}$$

which applies below $\approx 10^4$ cycles. The plastic strain range at the corner of the groove was obviously large enough for a crack to form and grow to the critical depth for fast fracture after only 200 cycles. Certainly the burnished look of the fatigue surface suggests that the fatigue crack was subjected to a large compressive stress during part of each stress cycle. And the small depth of the critical crack gives extra confirmation that the bolt failed at a high stress.

Design Implications

While the platform was being lowered and hoisted the lifting eye was subjected to a straight pull from the crane. The tensile stress in the bolt was only 10% of the yield stress. However, at the end of the hoisting operation the lifting eye was loaded at an angle. This applied a bending moment to the bolt and generated a maximum tensile stress equal to the yield stress. The failure certainly provides us with a good illustration of the dangers of loading components in bending. The bolt might have lasted rather longer if the steel had been heat treated after machining, or if the groove had been radiused, but the only safe solution was to design the bending load out entirely and this is what was done. It is left to the ingenuity of the reader to suggest a suitable modification!

References

N. E. Frost, K. J. Marsh and L. P. Pook, *Metal Fatigue*, Oxford University Press, 1974.
R. W. Hertzberg, *Deformation and Fracture Mechanics of Engineering Materials*, 3rd edition, Wiley, 1989.
C. J. Smithells, *Metals Reference Book*, 6th edition, Butterworth, 1984.

CHAPTER 21

FATIGUE OF A ROTATING CHEMICAL VESSEL

Background

Figure 21.1 is a schematic diagram of a process vessel in a chemical plant. The vessel is fabricated from mild steel and is operated as follows. First, the chemicals are poured into the vessel through a valve at the top. The valve is then closed and the vessel is rotated about a horizontal axis to agitate the chemicals. The process needs heat and this is supplied by circulating hot water through a hollow jacket which is welded to the outside of the vessel. The vessel rotates at a speed of about 2 revolutions per minute. When the reaction has finished the vessel is stopped with the valve at the bottom. The valve is then opened and the chemicals are poured out. When the vessel is empty it is turned through another half-turn so it can be filled with a fresh batch of chemicals. The cycle of events is then repeated.

The average gross weight of the vessel when it is operating is about 26 tonnes. This means that each trunnion shaft is subjected to an upward reaction force of 13 tonnes applied at the centre-line of the bearing. The bearings are self-aligning so the reaction force applies a bending moment to the shaft. In order to resist this moment the attachment between the shaft and the vessel is stiffened with gusset plates, as shown in Fig 21.2. The gussets are fixed to the vessel and to the shaft by

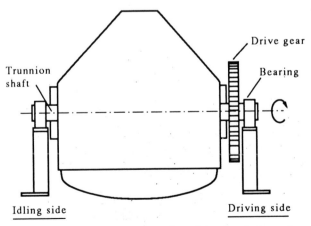

FIG 21.1   Schematic side elevation of the chemical vessel.

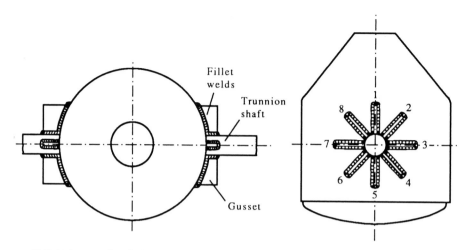

FIG 21.2   Details of the attachments between the trunnion shafts and the side of the
vessel.

fillet welds. Each shaft is attached using a total of eight gussets spaced equally around the circumference of the shaft. After $4\frac{1}{2}$ years in service cracks were found in most of the welds between the shaft and the gussets on the driving side of the vessel.

We can see that if the vessel is rotated through 180° the stresses in the welds will be reversed. If the vessel is rotated by another 180° the stresses will return to their initial values. This means that, with each revolution, the stresses in the welds will undergo a complete cycle of fatigue loading. The total number of fatigue cycles experienced by the welds during the period of operation is given approximately by

$$N = 2 \times 60 \times 24 \times 365 \times 4.5 \times 0.9 = 4.3 \times 10^6. \tag{21.1}$$

The vessel was in continuous use for about 90% of the time and this is allowed for by the "utilization factor" of 0.9. This is a large number of cycles and the welds presumably failed by high-cycle fatigue. In this Chapter we analyse the failure by obtaining an estimate of the stress cycle and relating this to data for the fatigue properties of welds in structural steelwork.

Details of the Failure

Figure 21.3 shows the geometry specified for the welds between the shaft and the gussets. The gussets were made from plate 25 mm thick and the fillet welds had a leg of 12 mm. The weld at gusset number 1 on the driving side was cracked, as shown in Fig. 21.4. The crack almost certainly started at the toe of the weld. This is because (a) the change in section at the toe generates a stress concentration, (b) the surface of the weld bead is rough, (c) there is a tensile residual stress at the toe, (d) the heat-affected zone next to the weld creates a discontinuity in the micro-structure. Once the crack had initiated it spread through the whole cross-section. As a result the gusset became detached from the shaft and only a compressive

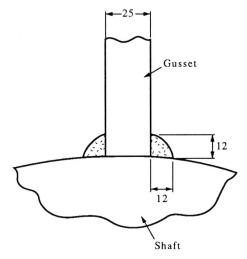

FIG 21.3   Specified geometry for the welds between the shaft and the gussets. Dimensions
in mm.

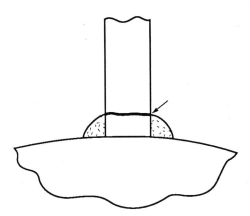

FIG 21.4   Geometry of the fatigue crack in the gusset. The arrow marks the position
where the crack probably started.

load could be transferred from the shaft to the gusset. The welds at gussets 2, 4, 5, 6 and 8 on the driving side were cracked differently. As Fig. 21.5 shows, the cracks had passed through the throat of each weld bead. The cracks almost certainly started at the root of the weld. This is because there is both a high residual stress and a very bad stress concentration at this location. The welds at gussets 3 and 7 on the driving side were not cracked. There were no cracks in the welds on the idling side.

Classifying the Welds

Appendix 1E gives data for the fatigue strengths of welds in structural steelwork. These are obtained from the British Standard for the design of bridges, BS 5400.

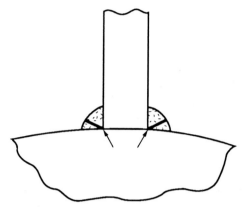

FIG 21.5   Geometry of the fatigue cracks in the gusset-shaft welds. The arrows mark the
positions where the cracks probably started.

The fatigue strength of a weld is very sensitive to its geometry: because of this
BS5400 divides the common types of welds into a number of weld classes, each of
which has a different fatigue curve. The welds between the shaft and the gussets are
responsible for transferring a direct load between the two components and are
therefore classified as "Weld Details on End Connections of Member" by BS 5400.
Because the crack at gusset 1 initiated at the toe of the weld and ran through the
member itself the welded connection is a Class F2 (see Appendix 1E). On the other
hand, the connections at gussets 2, 4, 5 6 and 8 are Class W because the cracks ran
through the throat of the fillet weld.

Fatigue Strengths of the Welds

The fatigue curves for the Class F2 and W welds are shown in Figs. 21.6 and
21.7. The data have been selected from the appropriate diagrams in Appendix 1E,
where the curves for all the weld classes are shown. The data are plotted as log $\Delta\sigma$
against log N_f, where $\Delta\sigma$ is the stress range and N_f is the number of stress cycles.
The curves in Fig. 21.6 give a 97.7% chance that the weld will survive, i.e. a 2.3%
chance that the weld will break. The curves in Fig. 21.7 give a 50% chance of
survival, i.e. a 50% chance that the weld will break. Naturally, for a given N_f, the
value of $\Delta\sigma$ that gives a 2.3% chance of fracture is less than the value which gives
a 50% chance of fracture.

The 50% survival curves are usually chosen for analysing welds which have
cracked. If we look at Fig. 21.7 we can see that a Class F2 weld needs to be subjected
to a stress range of 67 MPa to make it break after 4.3×10^6 cycles: the stress range
at gusset number 1 on the driving side must therefore be at least 67 MPa. The stress
range needed to break a Class W weld after 4.3×10^6 cycles is 40 MPa: the welds
at gussets 2, 4, 5, 6 and 8 on the driving side must have a stress range of at least
40 MPa.

The 97.7% survival curves are usually chosen for designing welds against fatigue.
If we look at Fig. 21.6 we can see that a Class F2 weld will survive 4.3×10^6

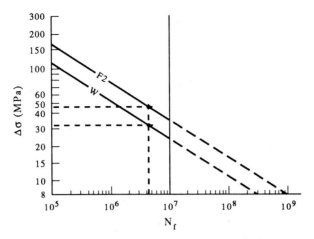

FIG 21.6   Fatigue curves for 97.7% survival.

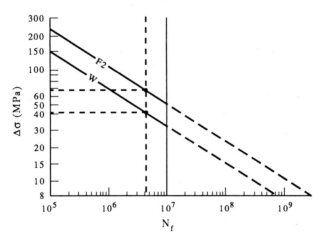

FIG 21.7   Fatigue curves for 50% survival.

cycles as long as the stress range is kept below 46 MPa; a Class W weld will survive 4.3×10^6 cycles if the stress range is kept below 30 MPa. Since there were no cracks at gussets 3 and 7 on the driving side, or at any of the gussets on the idling side, the stress ranges in these welds must have been less than 46 MPa and 30 MPa depending on the geometry of failure.

Where in the welded connection is the stress range defined? In all classes of weld except Class W $\Delta\sigma$ is the stress range in the member. The stress range for the Class F2 weld is therefore the stress range in the gusset itself, which we call $\Delta\sigma_1$. In a Class W weld, however, $\Delta\sigma$ is the stress range in the throat of the fillet weld. We call this stress range $\Delta\sigma_2$. The two stress ranges are related by the equation

$$\Delta\sigma_2 = \Delta\sigma_1\left(\frac{b}{2g}\right). \tag{21.2}$$

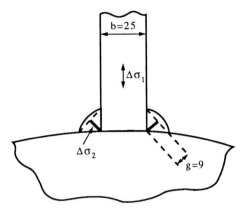

FIG 21.8   Stress ranges in the welded connection. Dimensions in mm.

g is the "effective throat dimension" of the fillet weld and is defined in Fig. 21.8. As shown in Fig. 21.8, $b = 25$ mm and $g = 9$ mm, so $\Delta\sigma_2 = 1.38\Delta\sigma_1$.

Estimating the Stresses

Figures 21.9 and 21.10 show in detail how the trunnion shaft is joined to the vessel. The geometry is complex and it is not easy to determine the value of $\Delta\sigma_1$.

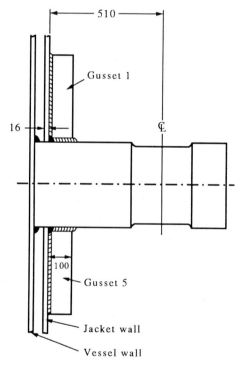

FIG 21.9   Details of the attachment on the driving side. Dimensions in mm.

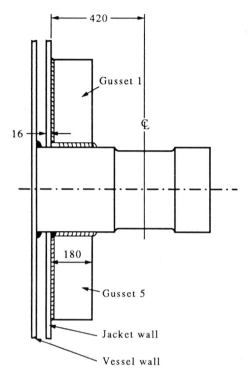

FIG 21.10 ᐳ Details of the attachment on the idling side. Dimensions in mm.

There are two ways of doing this with a reasonable degree of accuracy. The first is experimental. It involves fixing strain gauges to the surface of the gusset near each weld and measuring the change in the electrical resistance as the vessel is rotated through 180°. The second is to analyse the elastic behaviour of the assembly using a finite-element software package. Both methods are time consuming and require specialist resources. However, a rough analysis can be done to show that it is possible to get stresses of the right magnitude.

We begin by making two fairly crude assumptions. The first is that the vessel and the jacket do not take any of the bending moment applied to the shaft. This is obviously wrong: the jacket is attached to the vessel by a complex array of ribs which are welded into the space between the two plates and this double-skinned construction is probably quite effective in restraining the shaft. The second, which is quite reasonable, is that all eight gussets take an equal share of the bending moment. The arrangement for modelling the stress distribution in a single gusset is shown in Fig. 21.11. Here, a force equal to one-eighth of the reaction from the bearing is applied to the end of the shaft and the bending moment that this generates is applied to the end of the gusset. Using elastic equations from Appendix 1B we can write

$$\sigma_{max} = \frac{Mc}{I}, \tag{21.3}$$

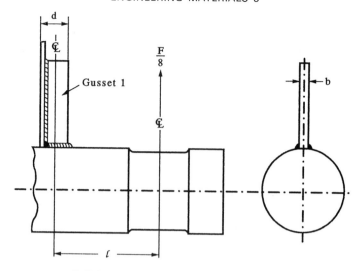

FIG 21.11   Modelling the stress in the gusset.

where

$$M = \frac{Fl}{8},\qquad(21.4)$$

$$c = \frac{d}{2},\qquad(21.5)$$

and

$$I = \frac{bd^3}{12}.\qquad(21.6)$$

Combining these equations gives us

$$\sigma_{max} = \frac{3Fl}{4bd^2}.\qquad(21.7)$$

As we mentioned earlier, $F = 13$ tonnes, or 1.28×10^5 N. For gussets 1 and 5 on the driving side $l = 468$ mm, $d = 116$ mm and $b = 25$ mm. Putting these values into Eqn. (21.7) gives a maximum stress in the gusset of 133 MPa. The stress range in the Class F2 weld is then given by $\Delta\sigma_1 = 2 \times 133$ MPa $= 266$ MPa. The stress range in the Class W weld is given by $\Delta\sigma_2 = 1.38 \times 266$ MPa $= 367$ MPa. These values are well above those of 67 MPa and 40 MPa needed to break Class F2 and Class W welds. Of course, these stress ranges are an upper estimate because we have neglected the stiffening effect of the double-skinned construction. The actual stress ranges were obviously much smaller but were still large enough to break the majority of the welds on the driving side.

For gussets 1 and 5 on the idling side $l = 338$ mm, $d = 196$ mm and $b = 25$ mm. These values give a maximum stress in the gusset of 34 MPa. Thus $\Delta\sigma_1 = 2 \times 34$ MPa $= 68$ MPa and $\Delta\sigma_2 = 1.38 \times 68$ MPa $= 94$ MPa. These upper-bound values are much less than those estimated for the driving side. This can be traced to two factors: (a) the gussets on the idling side have a much larger d than those on the driving side, (b) the bearing on the idling side is closer to the wall of the vessel than is the case on the driving side. Since the actual stress ranges were much smaller than these upper-bound values it is not surprising that the welds on the idling side remained intact.

Design Implications

How can failures of this sort be avoided? One obvious way of strengthening a Class W weld is to increase the penetration, as shown in Fig. 21.12. This increases g and decreases $\Delta\sigma_2$. However, the best solution is to go for a full-penetration weld (see Fig. 21.12). This gets rid of the crack initiation site at the weld root and a Class W failure should not occur. An added bonus of a full-penetration weld is that it decreases the stress concentration at the weld toes on the surface of the gusset. As shown in Appendix 1E this allows us to re-classify the connection as a Class F weld instead of a Class F2, giving slightly better fatigue properties.

The fatigue strength of the Class F connection can be improved even further by grinding the surface of the weld as shown in Fig. 21.12. This removes the rough finish of the weld and gives a smooth transition at the change of section, both of which lower the stress concentration and reduce the chance that a crack will initiate at the toe. The edges of the gussets also act as stress raisers and these, too, should be rounded off by grinding. Finally, the surface of the weld can be shot peened. In this process a jet of small metal particles is fired at the weld using compressed air. Each particle makes a small plastic indentation in the surface which tries to

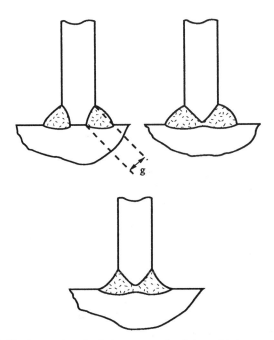

FIG 21.12   Improving the fatigue strength of the welded connection.

make the metal spread out sideways. It is prevented from doing this by the constraint of the bulk of the weld metal beneath the surface. The net effect is that the surface is put into a state of residual compression so small surface cracks cannot grow.

Using the Fatigue Curves

Below 10^7 cycles the fatigue curves given by BS 5400 are based on experimental data for actual welds. But care is needed when using the curves above 10^7 cycles. The situation is summarised in Fig. 21.13. Provided the stress range of the fatigue cycle is constant and the environment is clean, dry air a fatigue limit operates: the welds will survive indefinitely as long as the stress range is less than the stress range at 10^7 cycles. But if the environment is corrosive (e.g. sea water) there is no fatigue limit. In fact, the limited information available at present (see Fry and Greenway) suggests that the curves should be extrapolated following the dashed line in Fig. 21.13. This is the conservative approach to designing welds against fatigue and is the reason why the fatigue curves in BS 5400 are all extrapolated as straight lines above 10^7 cycles.

Even when the environment is not corrosive there will generally not be a fatigue limit. This is because real fatigue cycles vary with time. The loading on a highway bridge, for example, will vary depending on the number of vehicles on the bridge at any time and their weights. There will be occasional peaks in the stress range and these can trigger the growth of defects that would be stable at the average value of the stress range. BS 5400 allows for this by reducing the slope of the fatigue curve above 10^7 cycles as shown in Fig. 21.13. The extent of this reduction depends on the class of weld. The slope of the fatigue curve below 10^7 cycles is defined as $1/m$ (over one decade of stress range the number of cycles goes through m decades). $m = 3$ for Classes D to W, but rises to 3.5 for Class C and 4 for Class B. Above 10^7 cycles the slope of the fatigue curve is set equal to $1/(m + 2)$.

Fatigue data for unwelded components are usually obtained under conditions of zero mean stress in "push–pull" tests. The same is true of the data for welds.

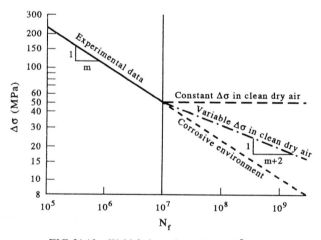

FIG 21.13   Weld fatigue data above 10^7 cycles.

When the mean stress is not zero, Goodman's rule shows that the fatigue strength must be corrected by a factor which depends on the value of the mean stress. No such correction is needed for welds: their strength depends only on the applied stress range and the mean stress is not relevant. This greatly simplifies the design of welded connections, but why it is valid to ignore the mean stress? As Example 6.1 shows, welds contain residual tensile stresses which are generally equal to the yield stress. We first look at what happens when we apply a stress cycle which has a zero mean stress. In the first quarter of the fatigue cycle the weld is loaded in tension and the weld strains plastically: to a first approximation the stress cannot rise above the yield stress. When the load is decreased in the second and third quarters of the cycle the weld will deform elastically in compression: the elastic strain will be proportional to the stress range. Now consider what happens when the peak stress in the cycle is zero, so the weld only sees a compressive loading. As before, the weld starts off at the yield stress. The load is applied and the weld deforms elastically in compression. The elastic strain is again proportional to the stress range. To summarise, the only difference between the two situations is that the weld "shakes down" plastically if it is subjected to a tensile load in the first quarter of the fatigue cycle. Once shakedown has occurred the deformation is entirely elastic: the amplitude of the elastic strain depends only the stress range and is not affected by the value of the mean stress.

References

British Standards Institution, BS 5400: 1980, "Steel, Concrete and Composite Bridges": Part 10: "Code of Practice for Fatigue".

P. R. Fry and M. E. Greenway, "An Approach to Assessing Structural Integrity and Fatigue Failures in Vibrating Equipment", in *Fracture and Fracture Mechanics: Case Studies*, edited by R. B. Tait and G. G. Garrett, Pergamon, 1985, p. 159.

G. Environmental failures

CHAPTER 22

CORROSION OF CENTRAL HEATING SYSTEMS—1: BASIC MECHANISMS

Background

Figure 22.1 is a schematic diagram of a typical closed recirculating heating system. The heat-exchange medium is water and this is circulated between the heat source (the boiler) and the heat sink (the radiators). The maximum water temperature at the exit from the boiler is typically $\approx 80°C$. The construction of the radiators is shown in Fig. 22.2. The radiator is made from two separate sheets of mild steel which are press-formed to produce the shapes of the waterways. The two halves are then joined by electrical-resistance welding. The wall of the radiator is usually between 1 and 2 mm thick. The water pipes are generally made from thin-walled copper tube. The heating modules in the boiler are made from steel, copper or aluminium alloy. Finally, plumbing fittings such as valves are usually made from brass.

Oxygen Reduction—Basic Mechanisms

Mild steel has a strong tendency to rust when it is exposed to water and oxygen. Because the mains water which is used to fill the system is saturated with air one might think that mild steel was an unwise choice of material for radiators and heat exchangers. As shown in Appendix 1F the conditions under which metals corrode in water can be summarised using the electrochemical equilibrium (or Pourbaix) diagram. Figure 22.3 shows the Pourbaix diagram for iron at 25°C. Mains water usually has a pH of 6.5 to 8 so as long as the potential of the iron is kept below −0.6 volt (standard hydrogen-electrode scale) the iron will be immune from corrosion. The Pourbaix diagram also shows the line for the oxygen-reduction reaction. The open-circuit potential for this reaction in mains water is ≈ 0.8 volt. This means that if iron is immersed in mains water saturated with oxygen the voltage difference available to drive the corrosion process will be $0.8 - (-0.6) = 1.4$ volts.

In practice, the oxygen content of the water in the heating system rapidly falls towards zero. The water is only saturated with air when the system is first filled. The solubility of oxygen in water decreases as the temperature increases: it is 8 mg/litre at 25°C but only 3 mg/litre at 80°C. When the water is heated to the operating temperature air is driven out of solution: it escapes through the vent and is bled off through the valve at the top of each radiator. The steel only has to

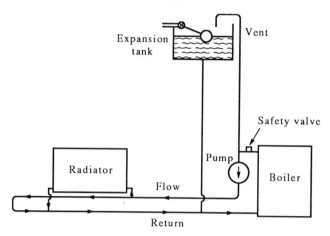

FIG 22.1   Schematic diagram of a closed recirculating heating system.

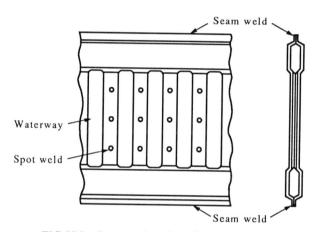

FIG 22.2   Construction of a mild-steel radiator.

corrode to a tiny extent to consume the remaining oxygen and the water becomes de-aerated. It takes typically 75 days for this to happen; at steady state the oxygen concentration is only ≈ 0.3 mg/litre. The oxygen-reduction reaction effectively stops and further corrosion is negligible. Under these conditions mild steel radiators and heat exchangers can last for well over thirty years without rusting through. However, if fresh oxygen gets into the system in any quantity it is possible for radiators to perforate after only two years in service.

Oxygen Pick-up

There are a number of ways in which the water can pick-up oxygen. Obviously, some air will get in when the circuit is drained down (e.g. for repairs, or when there is a risk of freezing) and then re-filled. A limited amount of fresh water is taken

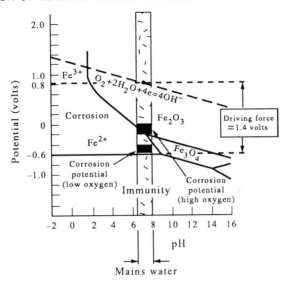

FIG 22.3   The Pourbaix diagram for iron at 25°C showing the oxygen-reduction reaction.

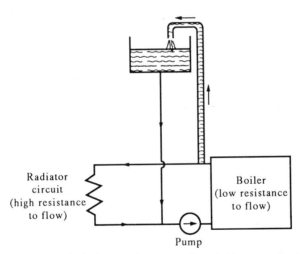

FIG 22.4   Design fault 1—pumping water through the expansion tank.

into the system to compensate for evaporation from the expansion tank or losses due to leaks and this, too, is a source of oxygen. But a major intake of oxygen is usually caused by design faults in the pipework.

A typical fault is shown in Fig. 22.4. If the flow resistance of the radiator circuit is large compared to that of the boiler then, depending on the location of the pump, water can be forced up the vent and discharged into the expansion tank. A particularly bad location for the pump is shown in Fig. 22.4—in the return line between the cold feed and the boiler. It is easy to see that when a large pressure

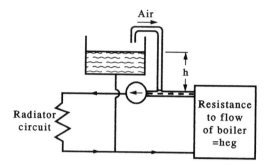

FIG 22.5   Design fault 2—drawing air through the vent.

drop is needed to get the water to flow around the radiator circuit then the water will tend to take the easier route and there will be a circulation loop between the boiler and the expansion tank. Naturally, as the water pours out of the vent it continually picks up air. This "pumping over" would not have happened if the pump had been positioned to the *left* of the cold feed instead of to the right. Figure 22.1 shows the standard position for the pump in a modern installation—in the flow line after the vent. When the pump is put in this position pumping over cannot occur.

Figure 22.5 shows another fault—when air is drawn in down the vent pipe. In order to pull water through the boiler the pump must generate a low pressure in the flow line from the boiler. This low pressure depresses the head of water in the vent pipe. If the expansion tank is not high enough the surface of the water in the vent pipe will be pulled down to the level of the flow line and bubbles of air will be swept into the circuit. Oxygen concentrations of ≈ 5 ppm have been found in systems suffering from this defect. To avoid air intake the static head h must be greater than the flow resistance of the boiler by a suitable safety margin; a typical value for h is 1.5 m.

Corrosion Rates

When the water is kept topped up with oxygen the steel will corrode rapidly. The rate of corrosion is controlled by the oxygen-reduction reaction which is in turn controlled by the rate at which oxygen diffuses through the water to the surface of the steel. Because diffusion is a thermally-activated process the rate of diffusion will increase exponentially with temperature. So also will the corrosion rate, as shown by the rising curve in Fig. 22.6. As we approach 80°C the curve divides into two branches. If the system is pressurised (so none of the dissolved oxygen can escape) the curve carries on upwards. But if the system is vented (as most central heating systems are) the oxygen can come out of solution as the water warms up. As the oxygen concentration falls below 3 mg/litre the oxygen-reduction reaction slows considerably and at 100°C the steel is corroding no faster than it would have done in aerated water at 25°C.

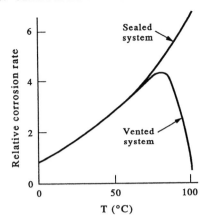

FIG 22.6   How the rate of corrosion of mild steel in aerated mains water depends on
the temperature.

Diagnosis

It is relatively easy to find out whether excessive oxygen has been getting into a
heating system. When oxygen is plentiful the iron corrodes rapidly at a relatively
high potential. The reaction takes place in the Fe_2O_3 field of the Pourbaix diagram
(see Fig. 22.3) and a red sludge of hydrated ferric oxide collects at the bottom of
the radiators. When oxygen is scarce the iron corrodes slowly at a relatively low
potential. The reaction generates ferrous ions and a thin black coating of hydrated
magnetite (magnetic iron oxide) forms instead. Both oxides are five times as dense
as water: although oxide particles can be carried around with the circulating water
most of the sludge remains in the radiators. Contrary to the view commonly held
by many heating "engineers" oxide sludge cannot be removed by flushing water
through the radiators, although it *can* be removed by circulating a warm solution
of inhibited phosphoric acid through the system.

Oxygen Reduction—Some Complications

There are several things which can make steel corrode faster in the presence of
oxygen. These are:

Galvanic Corrosion

The mixture of copper pipes and steel radiators makes up a *galvanic couple*. The
Pourbaix diagram for copper is shown in Appendix 1F. Copper will be immune
from corrosion in mains water as long as its potential is kept below 0 volt. This is
0.6 volt higher than the corrosion potential of iron. Now the copper is in excellent
electrical contact with the steel through soldered joints or compression fittings.
Because of this the iron corrodes in preference to the copper and acts as a *sacrificial
anode*. The copper is protected by *cathodic protection* and provides an inert surface

at which the cathodic oxygen-reduction reaction takes place. In practice the extent of galvanic corrosion is limited by the *ionic conductivity* of the water. If the conductivity is low the electrical resistance of the galvanic cell is high and the steel will only corrode *galvanically* near to a copper surface. On the other hand, if the conductivity is high galvanic attack will occur some distance away from the copper.

There is another way in which the copper pipes can lead to galvanic corrosion. The bore of the tubes corrodes slowly and releases copper ions into the circulating water. At steady state it is quite common to find concentrations of Cu^{2+} of 0.1 to 0.2 mg/litre. Because steel is more reactive than copper the following reaction takes place where the water passes over the steel surfaces:

$$Cu^{2+} + Fe = Fe^{2+} + Cu. \tag{22.1}$$

The copper metal produced by the reaction deposits on the surface of the steel as an extremely thin layer. Galvanic cells are then set up between the islands of deposited copper and the steel in between. However, opinions vary as to how important this mechanism really is.

Dissolved Ions

Mains water typically contains ≈ 50 ppm Cl^- and SO_4^{2-} in association with Na^+, Ca^{2+} and Mg^{2+}. The ions increase the conductivity of the water and help the corrosion processes. The metal ions migrate to the cathodic surfaces where they neutralise the OH^- ions produced by the oxygen-reduction reaction. The Cl^- ions are small and highly mobile: they migrate rapidly to the anodic areas and neutralise the Fe^{2+} ions produced when the iron dissolves. When mains water enters the system to make up for evaporation it brings dissolved ions with it and over a period of time the concentration of the ions in the circulating water will increase. Both Cl^- and SO_4^{2-} are aggressive ions: they help to stop passive oxide films forming on the surface of steel and this encourages corrosion even more.

Pitting

The surface of the steel tends to divide itself into anodic areas (where the iron corrodes) and cathodic areas (where the oxygen-reduction reaction takes place). This separation is encouraged by anything which makes the environment of the steel *non-homogeneous*. In most places the metal is exposed to the flowing water (and the oxygen it contains) and behaves cathodically. However, steel in crevices has almost no exposure to oxygen and behaves anodically. Crevices are present in many places: under deposits of sludge, next to welds and in screwed connections. Obviously, the current of electrons produced by the anodic areas must balance the current of electrons delivered to the cathodic area. Since the anodes are small compared to the cathodes the current *density* at the anodes will be large compared to the current density at the cathodes. The steel will corrode rapidly over a small area and localised *pits* will form.

Reduction of Hydrogen

Figure 22.7 shows the line for the hydrogen-reduction reaction superposed on the Pourbaix diagram for iron. The open-circuit potential for the hydrogen-reduction reaction in mains water is -0.4 volt. The voltage difference available to drive the corrosion process is $-0.4 - (-0.6) \approx 0.2$ volt. Now the hydrogen-reduction reaction *polarises* rapidly: a large voltage difference is needed to make the reaction go at a reasonable rate. The extent of the polarisation depends strongly on the metal or alloy concerned, the condition of the surface and the nature of the environment. Polarisations of ≈ 0.2 to 0.3 volt are common when hydrogen is reduced on the surface of iron. As shown in Fig. 22.7, the practical line for the hydrogen-reduction reaction probably lies *below* the corrosion potential of iron in mains water. In theory, therefore, steel radiators should not corrode by the hydrogen-reduction reaction.

In spite of this hydrogen is often given off in central heating systems. The usual symptom is that one or more of the radiators suddenly goes cold for no apparent reason. This is because hydrogen gas has collected at the top of the radiator and has blanketed the top ends of the vertical waterways. Heating "engineers" often claim that this is caused by dissolved air coming out of solution. In fact when the gas is vented off it can usually be lit with a match! Provided there is no risk of explosion having to vent radiators in this way is just a nuisance. However, the corrosion process produces magnetite which can deposit in the waterways and clog the pumps. Why does this hydrogen reduction occur?

If the temperature is high enough a reaction called the Schikorr reaction can take place. This has two steps as follows.

$$Fe + 2H_2O = Fe(OH)_2 + H_2, \tag{22.2}$$

$$3Fe(OH)_2 = Fe_3O_4 + 2H_2O + H_2. \tag{22.3}$$

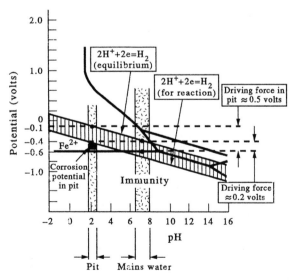

FIG 22.7   The Pourbaix diagram for iron at 25°C showing the hydrogen-reduction reaction.

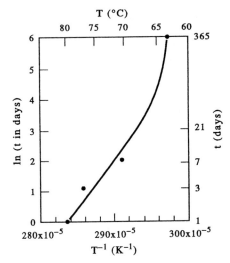

FIG 22.8   How the rate of hydrogen evolution depends on temperature. Radiators have
to be vented after a time interval t. From Fraunhofer—typical data only.

The Schikorr reaction is catalysed by copper ions: when these are present it can
take place above $\approx 60°C$. In systems where the water is aggressive to copper the
concentration of Cu^{2+} which is released into the circulating water can be large
enough to provide the required catalyst. A survey carried out on one central heating
system gave the results shown in Fig. 22.8. Below $\approx 63°C$ there was no detectable
evolution of hydrogen. However, at $70°C$ the radiators had to be vented every week;
and at $80°C$ they had to be vented every day. An obvious way of suppressing the
reaction is therefore to keep the temperature in the system below $60°C$.

Hydrogen and Bacterial Corrosion

Heating systems can become contaminated by bacteria. The most common are
the anaerobic sulphate-reducing bacteria (SRB) such as *Desulfovibrio* which live in
oxygen-starved conditions. As part of the metabolic cycle the organisms convert
sulphate ions into sulphide ions. Sulphide greatly speeds up the hydrogen-reduction
reaction and this allows steel to corrode even in neutral oxygen-free solutions.
Desulfovibrio grows in the pH range 5 to 10 and the temperature range 5 to $50°C$.
Some SRB can survive to even higher temperatures. The usual way of preventing
bacterial corrosion in heating systems is to add a biocide to the water.

Hydrogen and Pitting Corrosion

Once pitting has started it is unlikely to stop. In order to maintain electrical
neutrality in the pit the Fe^{2+} ions attract Cl^- ions from the water outside. The pit
becomes concentrated with $FeCl_2$ and this hydrolyses according to the reaction

$$FeCl_2 + 2H_2O = Fe(OH)_2 + 2HCl. \tag{22.4}$$

The corrosion product forms a crust which covers the mouth of the pit and isolates it from the water outside. The HCl which is trapped inside the pit lowers the pH to ≈ 2. At this low value of pH the hydrogen-reduction reaction gives a voltage differential of ≈ 0.5 volt (see Fig. 22.7) and this is enough to make the iron corrode quite rapidly. Hydrogen gas is given off inside the pit and the crust bursts to let it escape. Of course, the new Fe^{2+} produced by this attack will suck in fresh Cl^- ions and the cycle of events will repeat itself. The pitting process is therefore said to be *autocatalytic*. Because the corrosion is intense and localised it can perforate the radiator wall after only a short time. Hydrogen evolved at low temperatures (where the Schikorr reaction is suppressed) is almost certainly a sign of rapid pitting corrosion.

Using Inhibitors

These problems can be avoided by using *inhibitors*. A common inhibitor for steel is sodium nitrite which is dissolved in the water to give a concentration of ≈ 800 mg/litre of NO_2^- ions. The nitrite functions as an *oxidising agent*. It increases the potential at the surface of the steel and if the concentration is high enough the steel is moved up into the Fe_2O_3 field on the Pourbaix diagram (see Fig. 22.3). A thin stable film of γ-Fe_2O_3 forms and this acts as a very effective barrier to further corrosion. The pH is kept above ≈ 9 in order (a) to avoid the corrosion field on the diagram and (b) to reduce the potential at which the Fe_2O_3 field starts. Because sodium nitrite interferes with the anodic reaction (the oxidation of Fe to Fe^{2+}) it is called an *anodic inhibitor*.

It is very important to have a large enough concentration of nitrite in the water. Figure 22.9 shows how the rate of corrosion is affected by the nitrite concentration. Above a critical concentration there is no corrosion at all. However, as the concentration is decreased the corrosion rate increases enormously; the maximum corrosion rate is much greater than it is when there is no inhibitor present at all. When the

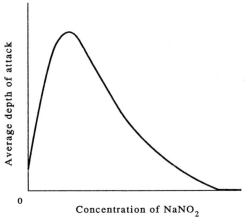

FIG 22.9   How the concentration of sodium nitrite affects the rate of corrosion of mild steel in mains water.

concentration is below the critical level the surface of the steel is not fully passivated. The electron current taken by the cathodic reaction is concentrated at the weak points in the film and these are subjected to rapid localised attack. The nitrite also increases the ionic conductivity of the water and this increases the corrosion rate even more. Sodium nitrite is therefore classified as a *dangerous* inhibitor.

The actual value of the critical concentration depends on several factors. It is difficult for the nitrite to get at the steel when it is covered by deposits or shielded inside crevices: the concentration of inhibitor in the bulk of the water has to be increased considerably to stop the shielded areas corroding. Cl^- and SO_4^{2-} ions attack the passive film so the concentration of inhibitor must be increased to compensate. The minimum amount of extra sodium nitrite is given by

$$\text{weight } NaNO_2/(\text{weight } NaCl + \text{weight } Na_2SO_4) \approx 1. \qquad (22.5)$$

Because it is so important to keep the concentration safely above the critical level the water in the system must be analysed at regular intervals and topped-up with extra inhibitor as required. Inhibitor will obviously be lost if the system leaks or if it is drained down. What is less obvious is that sodium nitrite can also be removed by the action of bacteria such as *Nitrobacter* which are very efficient at oxidising nitrite to nitrate. Because of this it is particularly important to use a biocide in any system which relies on sodium nitrite for its protection.

The best inhibitors for copper are specific organic chemicals such as benzotriazole. These react with the copper at the surface of the metal to produce a uniform adherent film. Benzotriazole is classed as a *cathodic* inhibitor because it interferes with the cathodic oxygen-reduction reaction. Around 5 mg/litre is usually required to stop the copper corroding. Benzotriazole can also deactivate (chelate) copper ions which have already dissolved in the water. The Schikorr reaction cannot occur (because there are no active copper ions to catalyse it) and galvanic attack is inhibited (because copper ions are not reduced at steel surfaces). Finally, benzotriazole is a *safe* inhibitor. Because it is an organic compound it cannot increase the ionic conductivity of the water. And because it inhibits the *cathodic* reaction it does not cause pitting.

CHAPTER 23

CORROSION OF CENTRAL HEATING SYSTEMS—2: CASE STUDIES

CASE STUDY 1: RAPID RUSTING OF STEEL RADIATORS

Background

The subject of this case study is a large central heating system in an office block. When the system was about twenty years old the radiators started to leak and had to be replaced. However, because of the age of the system it was decided not to clean the existing pipework with a chemical descaling agent. The new radiators had a wall thickness of only 1.25 mm and it was decided to protect them against rusting by adding a corrosion inhibitor to the water. The inhibitor was supplied as a concentrated solution of sodium nitrite and sodium borate. The solution was added to the system until the concentration of sodium nitrite in the water was 1300 ppm (parts per million, or mg/litre). The sodium nitrite was the active ingredient in the inhibitor package; the sodium borate was added as a buffer to keep the pH of the water at about 9. However, after only two years the new radiators started leaking. The offending radiators were removed and one was cut open. As shown in Fig. 23.1, a layer of black sludge had formed in the water space at the bottom of the radiator. The corrosion had originated beneath this deposit and had developed into deep pits. One of these had penetrated right through to the outside surface. But there was no evidence of any pitting above the sludge.

The water had been analysed every two months to make sure that the inhibitor concentration was up to the recommended level. In most cases it was found that the concentration had fallen significantly since the previous service and extra inhibitor had to be put in to make up the shortfall. The average concentration reading was ≈ 500 ppm. However, on several occasions the nitrite level was as low as 100 ppm. An analysis of the mains water gave: pH ≈ 8; Cl^- ≈ 50 ppm; SO_4^{2-} $\approx$ 100 ppm; high hardness. However, samples of water taken from the system itself gave: pH ≈ 10; Cl^- ≈ 250 ppm (max); SO_4^{2-} ≈ 300 ppm (max); strong indications of nitrite-oxidizing bacteria.

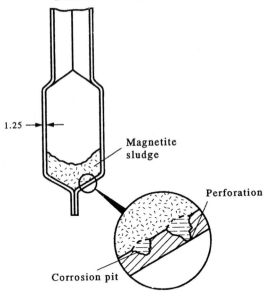

FIG 23.1 A layer of magnetite sludge had formed in the water space at the bottom of the radiator. Pitting corrosion had occurred where the sludge covered the steel. Dimension in mm.

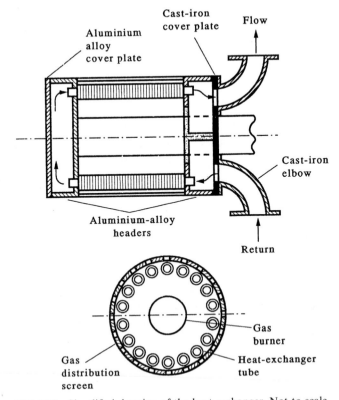

FIG 23.2 Simplified drawing of the heat exchanger. Not to scale.

Failure Analysis

The first thing that one notices about the system is the high concentration of Cl^- and SO_4^{2-}. The minimum concentration of sodium nitrite which must be added to compensate for these aggressive ions can be estimated from Eqn. (22.5). Now the ionic mass of Cl^- is 35 and the molar mass of NaCl is 58. The 250 ppm of Cl^- is therefore equivalent to $(58/35) \times 250 = 414$ ppm of NaCl. The ionic mass of SO_4^{2-} is 96 and the molar mass of Na_2SO_4 is 142. The 300 ppm of SO_4^{2-} is therefore equivalent to $(142/96) \times 300 = 444$ ppm of Na_2SO_4. The combined mass of NaCl and Na_2SO_4 is 858 ppm. Equation (22.5) shows that this must be balanced by at least 858 ppm of $NaNO_2$. As we saw earlier, the average concentration of inhibitor was ≈ 500 ppm and the minimum concentration was ≈ 100 ppm. These levels are far below the minimum requirement so it is hardly surprising that the radiators corroded rapidly. The situation was still worse in practice because the sludge would have screened the metal from the inhibitor. The routine analyses show that the inhibitor level regularly fell from the recommended dose of 1300 ppm to as little as 100 ppm after only two months. Presumably nitrite was consumed by the nitrite-oxidizing bacteria. This would have been avoided if a biocide had been added to the water when the system was first filled.

CASE STUDY 2: HOLES IN ALUMINIUM HEAT-EXCHANGER TUBES

Background

Figure 23.2 is a simplified drawing of a heat exchanger from a large gas-fired central heating boiler. It consists of a number of externally finned aluminium tubes arranged concentrically around a tubular gas burner. The tubes are enclosed by a perforated sleeve of stainless steel; this is meant to spread the hot gases out uniformly as they flow away from the burner. Figure 23.3 shows a cross-section through one of the heat-exchanger tubes. The ends of the tubes are expanded into headers cast from eutectic aluminium–silicon alloy LM6. The cover plate on the return header is also cast in LM6. The cover plate on the inlet/outlet header is cast in grey cast iron and so too are the pair of curved elbows. The boiler contains between two and

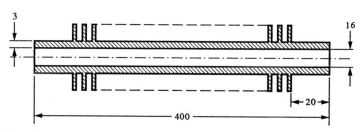

FIG 23.3   Cross-section through the heat-exchanger. Dimensions in mm. Not to scale.

TABLE 23.1   *Composition of Aluminium Alloy 6063*

Element	Weight %	Element	Weight %	Element	Weight %
Silicon	0.20–0.6	Magnesium	0.45–0.9	Tin	0.05 max
Iron	0.35 max	Zinc	0.10 max	Lead	0.05 max
Copper	0.10 max	Chromium	0.10 max	Aluminium	balance
Manganese	0.10 max	Titanium	0.10 max		

six of these units depending on the heat output required. Technical data for a single heat exchanger are as follows:

Maximum heat output = 100 kW.
Heat flux at bore of tube $\approx$ 230 kW m^{-2}.
Flow rate $\approx$ 2 litres/minute.
Flow velocity inside tube $\approx$ 0.5 m s^{-1}.
Maximum temperature at return elbow $\approx$ 80°C.
Maximum temperature at flow elbow $\approx$ 90°C.
Maximum temperature drop across tube wall $\approx$ 10°C.
Reynolds number at 80°C $\approx$ 20,000 (fully turbulent).

The tubes were made from wrought Al–Mg–Si alloy 6063. The specified composition is shown in Table 23.1.

A number of failures were reported where the heat-exchanger tubes had perforated from the inside. The water had been treated with a liquid inhibitor package containing sodium nitrite, sodium borate, sodium silicate, sodium hydroxide and "organic scale and corrosion inhibitors". When the water contained less than 50 mg/litre of chloride the recommended dose of sodium nitrite was 1250 mg/litre. This was achieved by adding 5 parts by volume of the inhibitor liquid to 1000 parts by volume of water. The organic corrosion inhibitor was presumably added to protect any copper in the system; it could well have been benzotriazole. The neat inhibitor had a pH of 11.5 so the pH of the diluted solution was probably around 9.

Failure Analysis

Corrosion Resistance of Aluminium

The Pourbaix diagram for aluminium is given in Appendix 1F. It shows that a protective film of aluminium oxide can form on the surface of the metal as long as the pH of the water is between 4 and 8.5. Of course, the diagram does not give any information about the effectiveness of this barrier layer. When aluminium is exposed to oxygen in the atmosphere it immediately forms a thin invisible film of aluminium oxide. The film is bonded firmly to the surface and is an excellent electrical insulator. Because of this aluminium (a very reactive metal which would otherwise oxidise very rapidly) is widely used as a corrosion-resistant material. Even when the surface is damaged by mechanical abrasion the oxide film reforms immediately. Although this repaired film is very thin to begin with it is still a good barrier to corrosion. On further exposure to the air the film thickens, typically by 100 times, and the barrier becomes more effective. Because of this film aluminium does not corrode in water where the pH is in the range 4 to 8.5.

Outside this pH range the oxide film is unstable and the aluminium corrodes: the rate of uniform corrosion increases by roughly ten times for every unit increase in pH above 8.5. The film can, however, be stabilised by adding chemicals to the water. At low pH aluminium is hardly affected by dilute or concentrated nitric acid or dilute sulphuric acid because the oxidising action of these chemicals provides a strong film-forming tendency. At high pH the film is stabilised by *silicates*— compounds having the variable composition $n\text{Na}_2\text{O}.m\text{SiO}_2$. These are most effective on aluminium when the *module* of the silicate (the ratio m/n) is high (typically ≈ 3 to 3.5). But even sodium disilicate ($\text{Na}_2\text{Si}_2\text{O}_5$; module = 2) can stop uniform corrosion at a pH of ≈ 11.5.

Aluminium Alloy 6063

6063 alloy is one of a series of wrought aluminium alloys containing silicon and magnesium. They have good corrosion resistance, especially in alkaline solutions. Alloying elements, mainly iron and copper, can lead to local weaknesses in the oxide film and can cause pitting. 6063 resists corrosion well because the concentration of the alloying elements (Si and Mg) is small and the harmful elements (Fe and Cu) are present only as low-level impurities. In contrast the corrosion resistance of the 2000 series alloys (which contain 2 to 7% Cu as an alloying element) is relatively poor. Minute particles or films of copper can deposit on these alloys as a product of corrosion and the galvanic cells which are created attack the film. 6063 alloy was therefore a good choice for the heat-exchanger tubes. High-purity aluminium would have been better still although it might not have had the necessary mechanical properties.

Function of Inhibitor Package

It is not easy to get the optimum protection of both steel and aluminium in the same system. The situation is summarised in Fig. 23.4. To protect the steel best the pH needs to be buffered to at least 9 so that it misses the corrosion field on the Pourbaix diagram. However, this is above the pH at which the oxide film on aluminium breaks down. The approach which was adopted here was presumably to protect the steel in the usual way with sodium nitrite and an alkaline solution but to add sodium silicate to stop the film of aluminium oxide breaking down. An alkaline pH is also compatible with any copper in the system: the Pourbaix diagram for copper (see Appendix 1F) shows that a protective oxide film can form in the pH range 7 to 12.5. The protection of the copper can be completed by adding benzotriazole.

The Pourbaix diagrams show that iron is a more reactive metal than copper. Because of this copper ions can be reduced to copper metal by a steel surface (see Eqn. (22.1)). Aluminium is more reactive than both copper and iron. When water containing dissolved copper flows over an aluminium surface there is a tendency for copper metal to precipitate out. If the water contains dissolved iron there is a

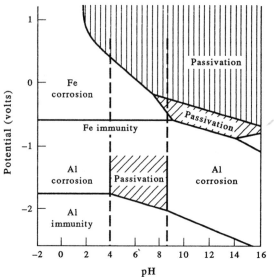

FIG 23.4   The Pourbaix diagrams for iron and aluminium at 25°C.

tendency for metallic iron to precipitate out as well. We have already seen that metallic copper and iron are likely to damage the aluminium oxide film. It is therefore necessary to keep the concentration of copper and iron in the water as low as possible when there is aluminium about. This is another (and very important) function of the inhibitors.

Conclusions

The most straightforward explanation of the failures is an inadequate concentration of inhibitor. The situation is made critical by the special environment of the heat-exchanger tubes. As we saw earlier the bore of the tube is subjected to conditions of high heat flux and temperature (230 kW m^{-2} and 90°C). For comparison conditions at the surface of the heating element in an electric kettle are ≈ 130 kW m^{-2} and 100°C. The chemical environment at the aluminium surface (pH, concentration of ionic species) may be radically different from that in the bulk of the water. The oxide film can also be worn away by suspended particles of corrosion product circulating with the water. When the rate of mechanical damage becomes greater than the rate of chemical healing then rapid corrosion will result.

CASE STUDY 3: EXTERNAL CORROSION OF STEEL WATER PIPES

Background

In our final case study we look at a central heating system which failed from corrosion on the *outside*. The system was installed in 1970 in a large new student

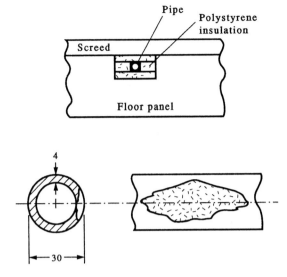

FIG 23.5   The corroded steel pipe. Dimensions in mm.

residence. The building was made from columns and slabs of reinforced concrete. The water pipes were made from mild steel and were laid in the concrete during the construction. By 1985 the pipes had started to leak. A typical failure is shown in Fig. 23.5. The pipe had an outside diameter of ≈ 30 mm and a wall thickness of ≈ 4 mm. It had been cast into the floor panel of one of the student rooms and was insulated with slabs of expanded polystyrene. When the floor was dug up it was found that the pipe had rusted from the outside. The wall of the pipe had perforated and the water inside the heating system had started to leak out. The corrosion deposits were reddish-brown in colour.

Failure Analysis

The roof of the building consisted of a series of horizontal pre-cast concrete slabs. These had started to take in rainwater at the joints. The structure of the building was complex and the rainwater travelled large distances before leaking into the rooms or the public areas. It was concluded that the most likely explanation for the corrosion was that the pipes had come into contact with the leaking rainwater. The horizontal runs of piping on each floor were connected by vertical runs laid in open ducts and it was thought that these could have channelled the rainwater from one floor to another.

Since air was able to get to the surface of the pipe the cathodic oxygen-reduction reaction would have taken place easily. In addition the incoming rainwater would have been saturated with oxygen. The corrosion product was presumably red rust (hydrated Fe_2O_3). The Pourbaix diagram for iron (Fig. 22.3) shows that this oxide is produced when the oxygen content is high. Rainwater is saturated with carbon dioxide: it contains ≈ 1300 mg/litre of the gas. The carbon dioxide converts to carbonic acid and the pH falls to ≈ 4.5 as a result. As we can see from the

Pourbaix diagram rainwater is aggressive to mild steel: iron cannot start to produce an oxide film at a pH of 4.5 unless the potential is taken up to +0.2 volt. A strong oxidising agent would be needed to get the potential this high.

Mild steel typically corrodes at the rate of 0.05 to 0.15 mm/year in slowly moving soft water saturated with air at $\approx 15°C$. However, the pipes were not always at $15°C$: when the heating system was operating they were closer to $70°C$. Figure 22.6 shows that the corrosion rate should go up by ≈ 2.7 times over this temperature interval. The heating system was generally used only for six months out of each year; on average the corrosion rate was probably ≈ 1.8 times greater than it would have been at $20°C$. This would give a corrosion rate of ≈ 0.09 to 0.27 mm/year. If we assume that rainwater had been leaking into the building for ≈ 7 years the depth of attack would be ≈ 0.6 to 1.9 mm. If the steel develops pitting corrosion then the rate of attack in the pit can be up to ten times the rate of general corrosion. In this case the pits would easily be capable of perforating the wall of the pipe after seven years. Localised attack of this sort would have been encouraged by the wet crevices which probably formed between the polystyrene and the outside of the pipe.

Design Implications

Many central heating installations function perfectly well without corrosion inhibitors. But what must be done to avoid corrosion when inhibitors are not used? The first requirement is to make sure that a uniform, stable film of oxide forms on the metal to act as a barrier to corrosion. The pH of the water must be in the range where a passive layer is thermodynamically possible. pH ranges can be estimated from the Pourbaix diagrams (see Appendix 1F): they are 7 to 12.5 and 9 to 14 for copper and iron respectively. Hard water (pH ≈ 8) is less aggressive to copper and steel than soft water (pH ≈ 6.5). The initial condition of the system has an important influence on the state of the protective films. It is bad to test a system with water and then leave it drained down. Corrosion will occur in the aerated pool of water at the bottom of each radiator and the film may be penetrated by corrosion pits. It is best to put the system into working order straight away and to keep it filled with water. Because chloride and sulphate ions attack the oxide films the water should be as free from them as possible. Copper pipes are sometimes attacked by residues of soldering flux or graphite and these, too, should be avoided.

Once a stable film has formed the rate of corrosion depends on the cathodic reaction. The oxygen-reduction reaction can be prevented by keeping the oxygen content of the water as low as possible. Pumps should be placed so they do not pump water through the expansion tank or draw air into the system. The water in the expansion tank should be kept cold to minimise evaporation. Obviously, the system should not be drained if this can be avoided. The hydrogen-reduction reaction can be minimised by keeping the pH high and the temperature low.

Because many systems depart from these ideals it is common to add inhibitors. But these must be used and specified correctly. They are generally recommended for systems which contain aluminium because its oxide film breaks down under mildly alkaline conditions. To get the best results the internal surfaces should first be cleaned by circulating a chemical descaling agent through the system. The system is rinsed and the inhibitor is added immediately afterwards. The inhibitor will react strongly with the bright metal surfaces and will penetrate well into the crevices formed at the welds.

Inhibitors are often supplied as a multi-component "package". Sodium nitrite can attack the lead–tin solder in soldered joints; nitrite-based inhibitors often contain sodium *nitrate* to prevent this. Sodium benzoate is often added because it resists pitting better than sodium nitrite. When the water contains more than one inhibitor the overall effect is usually better than the sum of the effects of the individual inhibitors (there is usually a *synergistic* effect). The pH of the water is controlled by adding a buffer such as sodium borate. Many packages contain specific copper inhibitors such as benzotriazole; this is especially important when the system contains aluminium. In hard-water areas a scale inhibitor is also added to stop hardness deposits building up in the boiler. Finally a biocide is needed to stop bacterial corrosion. The design of inhibitors is a complex business which draws on long experience of laboratory tests and field performance.

References

B. P. Boffardi, "Control of Environmental Variables in Water-Recirculating Systems", in *Metals Handbook*, 9th edition, Vol. 13: *Corrosion*, American Society for Metals, 1987, p. 487.

G. Butler, H. C. K. Ison and A. D. Mercer, "Some Important Aspects of Corrosion in Central Heating Systems", *British Corrosion Journal*, **6**, 32 (1971).

J. B. Cotton and W. R. Jacob, "Prevention of Corrosion in Domestic Hot Water Central Heating Systems", *British Corrosion Journal*, **6**, 42 (1971).

U. R. Evans, *An Introduction to Metallic Corrosion*, 2nd edition, Arnold, 1975.

U. R. Evans, *The Corrosion and Oxidation of Metals*, 2nd supplement, Arnold, 1976.

M. G. Fontana, *Corrosion Engineering*, 3rd edition, McGraw-Hill, 1986.

J. A. von Fraunhofer, "Corrosion in Hot Water Central Heating", *British Corrosion Journal*, **6**, 23 (1971).

E. H. Hollingsworth and H. Y. Hunsicker, "Corrosion of Aluminium and Aluminium Alloys", in *Metals Handbook*, 9th edition, Vol. 13: *Corrosion*, American Society for Metals, 1987, p. 583.

F. L. LaQue and H. R. Copson, *Corrosion Resistance of Metals and Alloys*, 2nd edition, Reinhold, 1963.

I. J. Polmear, *Light Alloys*, Arnold, 1981.

I. L. Rozenfeld, *Corrosion Inhibitors*, McGraw-Hill, 1981.

J. C. Scully, *The Fundamentals of Corrosion*, 2nd edition, Pergamon, 1975.

L. L. Shreier (editor), *Corrosion, Volume 1: Metal/Environment Reactions*, 2nd edition, Newnes-Butterworth, 1976.

J. F. D. Stott, "Assessment and Control of Microbially-Induced Corrosion", *Metals and Materials*, **4**, 224 (1988).

H. Uhlig, *The Corrosion Handbook*, Wiley, 1948.

CHAPTER 24

CORROSION OF STAINLESS STEEL

CASE STUDY 1: RUSTING OF A WATER FILTER AFTER WELDING

Background

Figures 24.1 and 24.2 are photographs of part of a water filter. The filter consisted of a perforated tube with a diameter of 200 mm. The end of the tube was welded to a short screwed section and this was used to couple the filter to a length of ordinary non-perforated pipe. The filter was intended for use in a water-supply network on an irrigation project. The components were made from an austenitic stainless steel of type AISI 304. The perforated tube was made by assembling a tubular cage of steel rods and welding the end of each rod to the screwed coupling. A helix of steel wire was then wound around the outside of the cage to complete the perforated wall. The wire was fixed to the support rods by electrical-resistance spot welding. For some reason the connection between the end of the helix and the coupling was not satisfactory. In order to correct this an extra weld had been made on the outside of the coupling as shown in Fig. 24.1. Finally the weld and the adjoining helix were levelled by grinding.

The filters were transported to their destination by sea. When they were unloaded it was noticed that some of the repair welds had corroded. As shown in Fig. 24.1 a closer inspection revealed that corrosion had occurred not just on the weld bead itself but also on the parts of the helix which had been ground flat. The surface was not pitted, but was covered with a thin uniform deposit of red rust. Light rusting was also observed on the surface of the main attachment weld and on the neighbouring heat-affected zone (HAZ). The rusting in the HAZ can just be seen as a thin dark line on the inside of the coupling in Fig. 24.2, but the rusting on the weld itself only shows up in a colour photograph. Naturally, concern was expressed that the steel might have had the wrong chemical composition. Indeed, it had failed the traditional artisans "test": although an austenitic steel should be non-magnetic some of the filters had shown a marked affinity for a magnet!

FIG 24.1   Exterior view of part of the water filter.

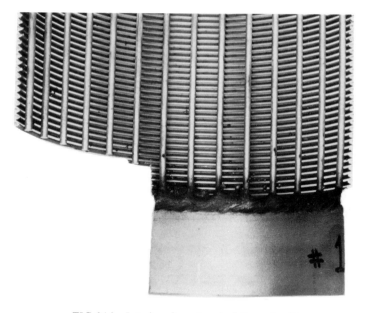

FIG 24.2   Interior view of part of the water filter.

Magnetism—A "Red Herring"

Table 24.1 gives data for the relative magnetic permeability of 304-type stainless steel. When the steel is in the fully annealed condition the permeability is very close to 1 and the metal is effectively non-magnetic. However, the permeability

TABLE 24.1   *Magnetism of 304-Type Stainless Steel as a Function of Cold Work*

Reduction in thickness, $(t_0 - t)/t_0$	Vickers hardness, HV	Relative permeability for $H = 4000$ A m^{-1}	Relative permeability for $H = 16000$ A m^{-1}
0	175	1.0037	1.0040
0.14	218	1.0048	1.0050
0.32	315	1.0371	1.062
0.65	390	1.540	2.12
0.85	437	2.20	4.75

increases with cold work: up to a hardness of about 320 Vickers the effect is small, but larger amounts of cold work result in a significant increase in permeability.

Why can austenitic stainless steels go magnetic in this way? The thermodynamic stability of the austenite phase is controlled by the chemical composition of the steel. The range of composition over which austenite is stable at room temperature can be found from the Schaeffler diagram (Appendix 1G). The vertical axis of the diagram plots the *nickel equivalent*, which is given by $Ni + 30C + 0.5Mn$. The horizontal axis plots the *chromium equivalent*, given by $Cr + Mo + 1.5Si + 0.5Nb$. Nickel, carbon and manganese all stabilise austenite, but we need a nickel equivalent of at least 11 to get into the austenite field on the diagram.

The compositions of AISI 304 and the nearest BS equivalents are given in Table 24.2; the corresponding nickel and chromium equivalents are 10 to 13, and 17 to 20. These values put us right at the bottom of the austenite field on the Schaeffler diagram. Depending on the precise composition the austenite in 304-type stainless is unstable at room temperature; severe cold work can then trigger a displacive transformation where some of the austenite flips over into the magnetic martensite phase instead.

Site engineers tend to be sceptical of complicated explanations so it was decided to stage an experimental demonstration of this theory. A number of samples were removed from the filter and were tested with both a magnet and a hardness tester. The results, which are shown in Table 24.3, were remarkably consistent with the data quoted in Table 24.1. An even better demonstration was to show that all

TABLE 24.2   *Compositions of 304-Type Stainless Steels in Weight %*

Element	AISI 304	BS 304 S15	BS 304 S31
Carbon	0.08 max	0.06 max	0.07 max
Silicon	0.75/1.00 max	1.00 max	1.00 max
Manganese	2.00 max	2.00 max	2.00 max
Sulphur	0.030 max	0.030 max	0.030 max
Phosphorus	0.045 max	0.045 max	0.045 max
Chromium	18.00–20.00	17.5–19.0	17.0–19.0
Molybdenum	–	–	–
Nickel	8.00–10.50	8.0–11.0	8.0–11.0
Titanium	–	–	–
Niobium	–	–	–

TABLE 24.3   *Magnetism of Filter Components*

Component	Vickers hardness, HV	Magnetic attraction
Coupling	180	Very weak
Rod/wire	350	Weak
Rod/wire	430	Strong

traces of magnetism could be removed by heating the samples to red heat and cooling them down to room temperature again.

Weld Decay

When ordinary stainless steels are held in the temperature range 550 to 850°C the carbon reacts with the chromium to produce chromium carbide, $Cr_{23}C_6$. Carbide particles nucleate at the austenite grain boundaries, and as the carbides grow they reduce the concentration of dissolved chromium. The steel becomes *sensitised* and is liable to corrode preferentially along the grain boundaries. The reaction is an example of a *diffusive transformation*: the time required for the carbides to form varies with temperature in the classic "C-curve" fashion. Sensitisation is a potential problem in welded stainless steels because the metal in the HAZ experiences a thermal cycle which takes it into the temperature range for the carbide reaction. A standard way of preventing weld decay is to use *stabilised* steels which contain either titanium or niobium (typically 0.5 or 1.0 weight %). These elements are much stronger carbide-formers than chromium and they remove the carbon from solution before it can form chromium carbide. However, as Table 24.2 shows, 304 stainless is devoid of both niobium and titanium. Is it possible, then, that the filter corroded because of weld decay?

An increasingly common way of avoiding weld decay is to keep the carbon content low (as in the special low-carbon "L" grade which has a maximum carbon content of only 0.03%). In fact, provided the plate is relatively thin, a steel containing as much as 0.06% carbon can be immune from sensitisation. This is because the position of the C-curve depends strongly on the amount of carbon which is initially present in the steel. Table 24.4 gives the data for 304-type stainless steels: with a carbon content of 0.08% the nose of the C-curve lies at 800°C and the reaction only takes 0.5 minutes; with a carbon content of 0.06% the nose lies at 730°C and

TABLE 24.4   *Approximate Kinetics of Sensitisation for 304-Type Stainless Steels*

Weight % carbon in steel	Temperature T^* (°C) for fastest sensitisation	Sensitisation time (minutes) at T^*
0.08	800	0.5
0.06	730	2
0.05	660	10
0.04	620	60
0.03	600	600

the reaction takes 2 minutes; and with a carbon content of 0.03% the nose lies at 600°C and the reaction takes 600 minutes. We know that the filter was not made from 304 L, but was the carbon content high enough to cause weld decay?

In order to answer this question samples were taken from a number of couplings, weld beads, rods and wires and were analysed to find out their chemical composition. In every case this was within the specification for BS 304 S15 given in Table 24.2. The carbon content of the samples varied between 0.03 and 0.04%, which is significantly less than the maximum of 0.08% allowed for AISI 304. As we can see from Table 24.4 a 304 stainless containing 0.04% carbon takes at least 60 minutes to become sensitised. The steel has an average carbon content which is almost as low as that specified for the L grade and weld decay should not have been a problem.

The Passive Film

If the stainless steel was up to specification, and if weld decay was not a factor, then why should the filter have corroded at all? In fact, the standard range of austenitic stainless steels is prone to a number of corrosion problems—the rather vague description "stainless" cannot necessarily be taken to mean "immune to corrosion". The iron in stainless steel wants to react with the environment and the metal depends for its corrosion resistance on a very thin protective film of chromium oxide. If the passive film breaks down for any reason then stainless steel can corrode very rapidly indeed. When the filters were unloaded from their containers they were found to be running with condensation. The film of condensed water would have been saturated with air and would probably have contained a significant concentration of chloride ions picked up from the salty atmosphere. The solution should then have been an ideal medium for corrosion, with an ample supply of oxygen and a reasonable electrical conductivity. We saw in Chapter 22 that chloride ions are also very effective at breaking down the protective films which form on most metals. In view of this it is not surprising that the weak areas were identified in the surface film on the filters.

When a fresh, dry surface of stainless steel is exposed to the oxygen in the atmosphere the passive oxide film rapidly forms of its own accord. For critical applications (e.g. water pipes in the nuclear industry) the film can even be thickened artificially by treating the surface with an oxidising agent such as nitric acid. However, problems can arise when stainless steel is welded. Because the surfaces of the weld bead and the HAZ are exposed to the atmosphere at high temperature they oxidise. A layer of black oxide scale forms on the surface of the weld bead and a layer of "heat tint" oxide forms on the metal immediately alongside. Unfortunately these high-temperature oxides protect the metal much less well than the normal passive film. This is probably why corrosion occurred at the main attachment welds. The problem can be solved by removing the oxide with a pickling solution of nitric and hydrofluoric acids. This produces a fresh clean surface which is rinsed and allowed to passivate naturally in air.

What seems less obvious is why the outside of the filter rusted even though the oxide scale had been ground off. The answer is that the rough, cold-worked surface produced by grinding is more liable to corrode than a smooth stress-free surface.

Indeed, stainless steel components for critical applications are often "cleaned" by *electropolishing*. This dissolves away the cold-worked layer, producing a surface which is smooth, clean and stress-free and which forms an optimum base for the passive film. Electropolished stainless is much in demand in the medical, pharmaceutical and food-handling industries where freedom from contamination is essential.

CASE STUDY 2: STRESS CORROSION CRACKING OF AN EXPANSION BELLOWS

Background

Figure 24.3 is a schematic diagram of an expansion bellows. The bellows formed part of a main which carried saturated steam at a pressure of 10 bar gauge and a temperature of 184°C. The unit had been in use for a couple of years when it was given a routine hydraulic test to 15 bar gauge. A small leak was detected and this was traced to a number of fine cracks in the outer surface of the bellows. The cracks were all aligned in a radial direction and were typically 1 mm long. No material specification was available, but since the metal was non-magnetic it was presumed that it was an austenitic stainless steel. When the bellows was cut up for examination it was found that the wall had been laminated from a total of ten separate thicknesses of metal. The axial section is shown in Fig. 24.4—the individual sheets are 0.35 mm thick and the overall thickness of the wall is 4 mm. This form of construction was obviously intended to combine an adequate strength in the circumferential direction with a low stiffness in the axial direction. The cracks had penetrated through the outermost sheet but had not gone any further. So why had the bellows sprung a leak?

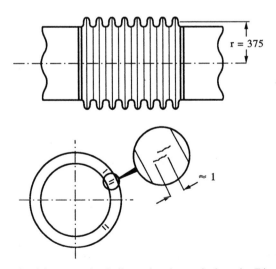

FIG 24.3   Schematic of the expansion bellows showing typical cracks. Dimensions in mm.

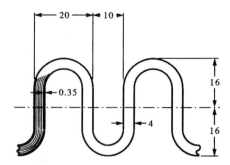

FIG 24.4   Axial section through the bellows wall. Dimensions in mm.

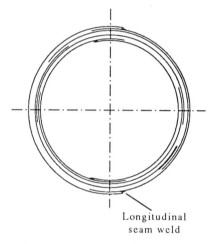

Longitudinal
seam weld

FIG 24.5   Schematic transverse section through the bellows wall.

Figure 24.5 is a schematic transverse section through the bellows wall. Each sheet extended slightly more than half-way around the circumference and the wall was built up by overlapping pairs of sheets. The overlaps were staggered around the circumference in order to distribute the free space which they created. The bore of the bellows was made pressure-tight by welding the ends of the innermost pair of sheets, and the same method was used to seal the outer skin of the bellows. When the bore was inspected the inner skin was found to be torn open. Obviously, when the bellows had been under pressure the water was able to reach the outer skin by flowing through the tear and seeping around the ends of the overlapped sheets. As a result the inner surface of the outer skin would have been exposed directly to the pressure inside the bellows.

Estimating the Hoop Stress

The first step in analysing the failure is to estimate the hoop stress which the internal pressure generates in the outer skin. The situation is shown in Fig. 24.6.

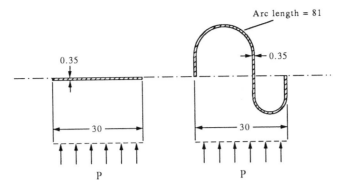

FIG 24.6   Calculating the hoop stress in the outer skin. Dimensions in mm.

We first form the skin into a straight thin-walled tube. The hoop stress in this tube is given by

$$\sigma = \frac{pr}{t},\tag{24.1}$$

with $r = 375$ mm and $t = 0.35$ mm. Because the operating and test pressures are 1.0 and 1.5 MPa respectively, the corresponding hoop stresses are 1071 and 1607 MPa. We next form the skin into a convoluted tube. This increases the amount of metal in the tube wall in the proportion $81/30 = 2.70$. The hoop stress is reduced by the same factor, giving values of 397 and 595 MPa. There was no evidence of any permanent deformation of the bellows after the hydraulic test, which suggests that the yield stress of the skin was greater than 595 MPa. This is not unreasonable as the yield stress of cold-drawn 304 stainless is around 700 MPa. Under operating conditions the hoop stress could have been as high as $397/595 = 0.67$ of the yield stress.

Stress Corrosion Cracking

When some metals are subjected to the combination of a tensile stress and a specific chemical environment they can fail by *stress corrosion cracking*. The cracks tend to initiate and grow at right angles to the maximum tensile stress. Austenitic stainless steels, especially the 304 and 316 types, are susceptible to SCC when they are exposed to solutions containing the chloride ion. The cracks usually follow a transgranular path, but if the steel has become sensitised the cracks often exploit the reduced corrosion resistance of the grain boundaries and follow an intergranular path instead. In general the threshold stress for crack growth decreases as the temperature and the concentration of chloride ion increase. Wilde quotes the example of a heat exchanger made from 321 stainless which cracked under a tensile stress of 90 MPa when exposed to water containing only 1 mg/litre of Cl^- at 205°C.

The failure of the bellows seems to have the hallmarks of SCC. Certainly, both the temperature and the stress were high, and the cracks were aligned at right angles to the hoop stress. When the inner surface of the outer skin was examined in the SEM it was noticed that the cracks coincided with small circular areas of corrosion debris ≈ 1 mm in diameter. This strongly suggests that the cracking started from the steam side. As we saw in Chapter 22 mains water can contain 50 mg/litre Cl^{-1} and, unless this is removed from the water before it is pumped into the boiler, traces of Cl^{-1} can be carried over with the steam. Condensation beneath the outer skin would have provided the aqueous medium needed to support the corrosion processes.

Improving Stainless Steels

AISI 304 is the most common and most basic stainless steel. But its resistance to corrosion (especially pitting and crevice corrosion) can be improved considerably by adding molybdenum, which helps to stabilise the passive film. In fact, Mo is 3.3 times as effective as Cr in resisting pitting corrosion. The most common molybdenum steel is AISI 316. This differs from 304 in containing 2.00–3.00% Mo, 16.00–18.00% Cr and 10.00–14.00% Ni. The Cr content is decreased by 2% to keep the chromium equivalent at about 20, and the nickel content is increased by 2–3.5% to stabilise the austenitic structure.

Recent advances in steelmaking have led to the development of high-molybdenum austenitic steels (the "super austenitics") such as 254 SMO and AL-6XN. These typically contain 20% Cr, 6% Mo and 20% Ni, together with $\approx 0.2\%$ nitrogen. The high nickel content is needed to balance the high chromium equivalent on the Schaeffler diagram. However, the major innovation is the controlled addition of nitrogen (absorbed as an interstitial solid solution). The nitrogen increases the yield strength by $\approx 25\%$, is 16 times as effective as Cr in resisting pitting, and is a cheap and potent alternative to nickel as a stabiliser of austenite (nickel equivalent ≈ 25 N). These remarkable alloys are also highly resistant to stress corrosion cracking.

Just as resistant to pitting and stress corrosion cracking are the new "super duplex" steels such as SAF 2507. This has a composition of 25% Cr, 4% Mo, 7% Ni and 0.3% N. The chromium and nickel equivalents are 29 and 14, giving a structure of 80% austenite plus 20% ferrite on the Schaeffler diagram. Practical alloys have at least 30% ferrite. This two-phase microstructure is much stronger than single-phase austenite and the yield strength is twice that of 304 or 316. Because nickel is expensive the super duplexes are also cheaper than the super austenitics. Super duplexes are finding increasing use in applications as diverse as propellers for ships, pipes in breweries and implants for the human body.

References

R. S. Brown *et al.*, "Arc Welding of Stainless Steels", in *Metals Handbook*, 9th edition, Vol. 6: *Welding, Brazing and Soldering*, American Society for Metals, 1987, p. 320.

R. M. Davison and J. D. Redmond, "A Guide to Using Duplex Stainless Steels", *Materials and Design*, **12**, 187 (1991).

R. M. Davison, T. DeBold and M. J. Johnson, "Corrosion of Stainless Steels", in *Metals Handbook*, 9th edition, Vol. 13: *Corrosion*, American Society for Metals, 1987, p. 547.

T. G. Gooch and D. C. Willingham, *Weld Decay in Austenitic Stainless Steels*, The Welding Institute, 1975.

R. W. K. Honeycombe, *Steels: Microstructure and Properties*, Arnold, 1981.

K. F. Krysiak *et al.*, "Corrosion of Weldments", in *Metals Handbook*, 9th edition, Vol. 13: *Corrosion*, American Society for Metals, 1987, p. 344.

J. O. Nilsson, "Overview: Super Duplex Stainless Steels", *Materials Science and Technology*, **8**, 685 (1992).

J. Olsson, "Stainless Steels for Harsh Environments", *Materials World*, **1**, 220 (1993).

L. L. Shreir (editor), *Corrosion, Volume 1: Metal/Environment Reactions*, 2nd edition, Newnes-Butterworth, 1976.

C. J. Smithells, *Metals Reference Book*, 6th edition, Butterworth, 1984.

B. G. Whalley, "The Analysis of Service Failures", *The Metallurgist and Materials Technologist*, **15**, 193 (1983).

B. E. Wilde, "Stress Corrosion Cracking", in *Metals Handbook*, 9th edition, Vol. 11: *Failure Analysis and Prevention*, American Society for Metals, 1987, p. 203.

CHAPTER 25

DEGRADATION OF A NATURAL MATERIAL

Background

In 1838 Samuel Clegg, Joseph Samuda and Jacob Samuda were granted a patent for an "atmospheric" system of rail traction. The basic concept is outlined in Fig. 25.1. A long cylindrical "vacuum tube" was laid on the sleepers between the running rails. To the "locomotive" (or "piston carriage") was fixed a piston, arranged so that its axis was in line with the axis of the tube. The piston carriage was pushed up to the end of the tube and the piston was inserted into the bore. The carriage was held on its brakes while the length of tube was exhausted of air through a large vacuum pump. The pump was driven by a stationary steam engine which was supplied with steam from a coal-fired boiler. The pumping machinery was housed in an engine house at the side of the track. When the vacuum in the tube had reached ≈ 0.5 bar the carriage was released and was pulled along the line by the piston. At this point the observant reader may wonder precisely how the drive was transmitted from the piston to the carriage, given that the wall of the vacuum tube was in the way. In fact the piston was attached to the piston carriage with a thin vertical plate rather like the centreboard of a sailing dinghy. The tube was slotted along its whole length at the top and the centreboard (or "coulter") ran along this slot just behind the piston. The slot was covered over by a longitudinal flap valve to make the tube air-tight. The valve was opened and closed automatically by the passage of the piston carriage: as we shall see later this entailed a piece of very complex machinery which only a nineteenth-century engineer could have devised!

The arrangement used in normal train working was rather more elaborate. As shown in Fig. 25.2, the vacuum tube was divided up into three-mile lengths. An engine house was erected alongside each break in the tube. The pump could be used to exhaust the section of tube on either side of the break. At each end the length of tube was closed with a pair of entry/exit valves. The length of tube immediately ahead of the piston carriage was pumped-out before the piston was inserted. The carriage was then pulled forward by a windlass and the piston was driven through the entry/exit valve at the start of the section. The valve was pushed open automatically by the force of the piston and closed itself again when the piston had moved past. The pump at the far end of the section was kept running while the train was moving along the section in order to maintain the required vacuum. When the carriage reached the end of the section the piston ran out of the tube

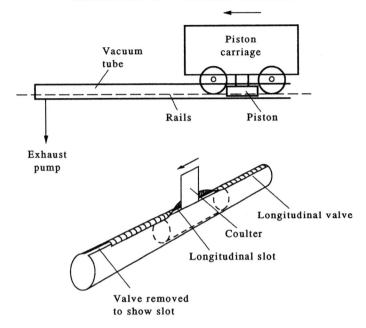

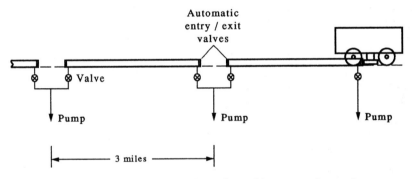

FIG 25.1   Conceptual illustration of the atmospheric system of rail traction.

FIG 25.2   Arrangement used for train working on continuous lines.

through the entry/exit valve. The train could then be diverted into a siding and brought to a halt using the brakes. Alternatively, if the next section of tube had already been evacuated the train could run straight through from one section to the next under its own momentum.

The atmospheric system had a number of unique qualities. The trains ran smoothly and were quiet and clean. Acceleration was brisk and top speeds were high. 40 mph was common when the load behind the piston carriage was 80 tons; when the piston carriage ran light it could reach 75 mph. In the age of electric trains we take these things for granted, but they were revolutionary in the 1840s. The system was especially well suited to commuter lines and underground systems because the intensive working and short distances offset the capital and running costs of the stationary pumps. And yet, more than 40 years later, trains were still being worked through the tube tunnels of the Metropolitan line in London behind condensing steam locomotives. So what went wrong? As we shall see in this case study the atmospheric system suffered from fundamental material problems which badly affected the reliability of the longitudinal valve. But in order to understand what happened we need first to look at the detailed construction and working of the valve.

Technical Details

The vacuum tube, longitudinal valve and piston unit are shown in the following sequence of drawings. Figure 25.3 shows that the running rails are secured to longitudinal baulks of timber to raise the axles of the train above the top of the vacuum tube. Stiffening fins are added to the tube at intervals because the longitudinal slot greatly reduces the ability of the tube to resist the external gauge pressure. Figure 25.4 shows that the tube is assembled from separate pieces joined together by spigot-and-socket joints. Each piece is ≈ 3 m long and is cast to shape in grey cast-iron. The joints allow the sections of tube to expand and contract in response to changes in temperature. Figure 25.5 shows the slot (≈ 65 mm wide) and also the channel (≈ 190 mm wide) in which the valve is mounted. Figure 25.6 shows the integral bracket which is used to secure the tube to the sleeper.

Figure 25.7 is a cross-section through the valve itself. The pressure membrane consists of a continuous leather strip which runs along the length of the slot. The strip is ≈ 180 mm wide and is ≈ 5 mm thick. It is clamped to one side of the valve channel by an iron bar. On its own the leather would get pushed through the slot by the external pressure. To stop this happening the strip is sandwiched between iron plates which are fixed in position by rivets. The bottom plate also closes the

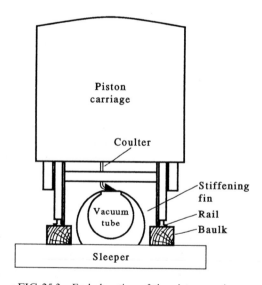

FIG 25.3   End elevation of the piston carriage.

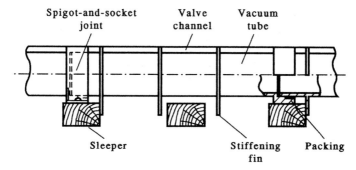

FIG 25.4   Side elevation of the vacuum tube.

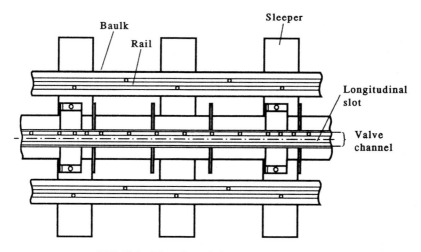

FIG 25.5   Plan view of the railway track.

gap between the leather and the inner circumference of the tube and stops air leaking past the piston. The plan view in Fig. 25.8 shows that the reinforcing plates are ≈ 200 mm long and are separated by short gaps. Because of this it is easy to bend, twist and stretch the valve along its length. Figure 25.9 shows that the valve can be opened by hinging it next to the clamping bar. The coulter is then able to pass between the underside of the valve and the edge of the channel.

The *pièce-de-résistance* of the whole mechanism—the piston unit—is shown in Fig. 25.10. The coulter is fixed to a horizontal frame which runs inside the tube. The piston itself is bolted to the leading end of the frame. Between the piston and the coulter are mounted a pair of wheels (the valve-opening wheels) which run on the bottom plates of the valve and push it open. A second pair of wheels is fixed behind the coulter to support the valve as it falls shut again. To make sure that the valve is properly re-seated behind the piston carriage the top plates are pushed down by a re-seating wheel mounted outside the tube. In order to open and close in this way the valve must be bent, twisted and stretched along its length and this is why the plating is discontinuous. The mind boggles at the idea of this contraption rushing along the tube at high speed, but by all accounts the mechanism worked

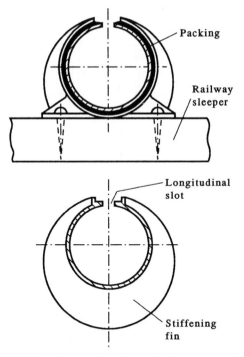

FIG 25.6   Cross-sections through the vacuum tube.

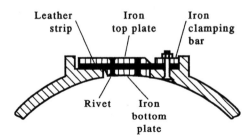

FIG 25.7   Cross-section through the valve.

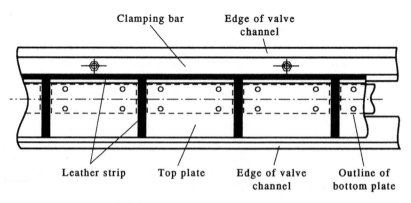

FIG 25.8   Plan view of the valve.

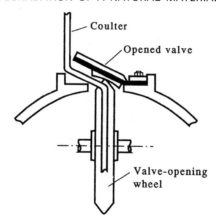

FIG 25.9   Cross-section through the valve in the open position.

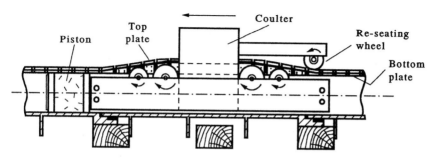

FIG 25.10   Longitudinal section of the vacuum tube showing the piston unit.

well and was virtually silent. In fact the quietness of the system was a safety hazard: there were several cases of maintenance staff being run down by atmospheric trains because they did not hear them approaching.

Operational History

The railways which used the atmospheric system are listed in Table 25.1. Clegg and the Samudas had already shown that their idea worked by setting up a miniature railway with a tube only 92 mm in diameter. This drew two men each weighing 115 kg up a gradient of 1 in 10 at 7 mph. However, the first full-size line was a short demonstration track which was laid on land at Wormwood Scrubs (of prison fame) near London. This had a tube 230 mm in diameter and used old rails from the now historic Liverpool and Manchester Railway. On initial test the system was able to move 11 tons up the gradient of 1 in 120 at 36 mph. The first commercial atmospheric line was opened in 1844, a year after the demonstration line had been lifted. This was a short and steeply graded extension at the far end of the railway from Dublin to Kingstown in Ireland. Trains ran every half-hour between 8 am and 9.30 pm. During the first eleven months of operation 17,506 trains were run at an average speed of 26 mph. The service was apparently reliable although a few trains

TABLE 25.1   *Lines Worked by the Atmospheric System*

Line (atmospheric section)	Length (miles)	Maximum gradient	Diameter of vacuum tube (mm)	Period of atmospheric working
Experimental line at Wormwood Scrubs	0.5	1 in 120	230	June 1840 to March 1843
Dublin & Kingstown (Kingstown to Dalkey)	1.8	1 in 57	380	March 1844 to April 1854
London & Croydon (Croydon to Forest Hill)	5	1 in 50	380	Jan 1846 to May 1847
South Devon (Exeter to Newton)	20	≈ level	380	Sept 1847 to Sept 1848
Paris & St. Germaine (Bois de Vesinet to St. Germaine)	1.4	1 in 29	630	April 1847 to July 1860

had to be cancelled in cold weather because of frozen valves. The Kingstown–Dalkey section was eventually converted to steam locomotives in 1854 as part of a general programme of improvements.

The first really big application of the atmospheric system was on the South Devon Railway. Authorised by Act of Parliament on 4 July 1844 the line was intended to extend the metals of the Bristol and Exeter Railway as far as Plymouth, a distance of 52 miles. Between Exeter and Newton the line was almost level and ran mainly along the coast. After Newton it swung inland to climb over the foothills of Dartmoor before dropping down to sea level again at Plymouth. The engineer for the new construction was none other than I. K. Brunel. Not a person to do things by halves, Brunel was impressed by the hill-climbing abilities of the atmospheric system and advised that it should be adopted for the whole route. His report to the Directors of the South Devon contains some gems.

> "I have given much consideration to the question referred to me by you at your last meeting—namely, that of the advantage of the application of the Atmospheric system to the South Devon Railway. The question is not new to me, as I have foreseen the possibility of its arising, and have frequently considered it.
>
> I shall assume, and I am not aware that it is disputed by anybody, that stationary power ... must be cheaper, is more under command, and is susceptible of producing much higher speeds than locomotive power; and when it is considered that for high speeds, such as sixty miles per hour, the locomotive engine with its tender cannot weigh much less than half of the gross weight of the train, the advantage and economy of dispensing with the necessity of putting this great weight also in motion will be evident.
>
> I am aware that this opinion is directly opposed to that of Mr. Robert Stephenson, who has written and published an elaborate statement of experiments and calculations founded upon them, the results of which support his opinion.

It does not seem to me that we can obtain the minute data required for the mathematical investigation of such a question, and that such calculations, dependent as they are upon an unattained precision in experiments, are as likely to lead you very far from the truth as not.

Experience has led me to prefer what some may consider a more superficial, but what I should call a more general and broader view, and more capable of embracing all the conditions of the question—a practical view.

Having considered the subject for several years past, I have cautiously, and without any cause for favourable bias, formed an opinion which subsequent experiments at Dalkey have fully proved to be correct; viz. that the mere mechanical difficulties can be overcome, and that the full effect of the partial vacuum produced by an air-pump can be communicated, without any loss or friction worth taking into consideration, to a piston attached to the train."

After comparing the costs of the rival systems Brunel concluded:

"For all the reasons above quoted, I have no hesitation in taking upon myself the full and entire responsibility of recommending the adoption of the Atmospheric system on the South Devon Railway, and of recommending as a consequence that the line and works should be constructed for a single line only."

By September 1847 the system had been installed between Exeter and Teignmouth and test trains had been run with a dynamometer car. Maximum speeds of 52 and 42 mph were reached with trains of 6 coaches (62 tons) and 8 coaches (78 tons). Service trains began at the end of the month and by January 1848 the system had been opened as far as Newton.

Problems with the Valve

The winter from December 1847 to February 1848 was cold and it revealed the first of the problems from which the valve was to suffer. Rainwater collected in the valve channel and soaked into the open-cell structure of the leather; when the temperature fell below 0°C the water froze and the leather became stiff. The frozen valve could be opened well enough but it tended to stay off the seating when the train had passed. This allowed air to leak into the vacuum tube and the stationary engines had to be worked much harder to make up for the losses. These difficulties were temporary and when the weather improved the train service returned to normal. But by May 1848 the leather had begun to deteriorate. Brunel wrote:

"Within the last few months, but more particularly during the dry weather of last May and June, a considerable extent of longitudinal valve failed by the tearing of the leather, at the joints between the plates; the leather first partially cracked at these points, which causes a considerable leakage, particularly in dry weather; after a time it tears completely through, and that part of the valve is destroyed, and requires to be replaced."

This sounds suspiciously like fatigue. The leather would have been subjected to repeated cycles of bending where it passed over the sharp edges of the plates.

When the leather was damp it was flexible and this would have relieved the bending stresses. But in dry conditions the leather was stiff and the bending stresses were much greater. There was also an indication of chemical attack from the iron plates. These had not been treated against corrosion and they were rusting in the salt-laden environment of the seaside. Brunel observed:

> "I have examined carefully portions of the valve that have been removed, and I find that at the part which has given way the texture of the leather seems to be destroyed—it is black, and has evidently been acted upon by the iron of the plates. Upon some parts of the line the injury seems to be more general than others; but it is very difficult to examine the valve in place, so as to form any correct opinion of the extent of the evil.
>
> By painting but, better still, by zincing or galvanising the iron plates, and making them overlap a short distance, both the chemical action and the mechanical action of the plate upon the leather appears to be prevented, and I believe, therefore, that this evil may be remedied at a small extra cost in any new or repaired valve that might be laid down; but of the existing valve I can say no more than I have done. It is not now in good working condition, and I can see no immediate prospect of its being rendered so."

Presumably the iron had reacted with the tannic acid used to cure the leather. It would have cost £25,000 (£2.5 million at today's prices) to replace the valve between Exter and Newton. The risk of further trouble and expense was too great and Brunel courageously had to admit defeat. The Exeter–Newton section went over to steam locomotives in September 1848 and the atmospheric equipment was sold off. Of the £433,000 (£43 million) spent on the system, only £81,000 (£8 million) was recovered. But this was not all: to this day the line from Newton to Plymouth suffers from excessive gradients for a main line, with banks of ≈ 1 in 40 at Dainton, Rattery and Hemerdon. For the 120 years when this scenic switchback route was worked by steam it presented a daunting challenge to the engines and the men on the footplate.

Design Implications

The atmospheric system would probably have been a success if only modern rubbers had been available at the time. A valve made from rubber would have been impervious to air and water, unaffected by temperature or humidity, immune to biological or chemical attack and resistant to sunlight. It should have been possible to eliminate the stiffening plates by moulding textile plies into the rubber itself, as in automotive tyres. This would have smoothed out the structural discontinuities and greatly improved the fatigue life. The case study provides a graphic illustration of how engineering function can depend critically on the availability of adequate materials. It also provides a sobering reminder of how we tend to take modern materials for granted. Without just one limited class of materials—the elastomers—it would not be possible to produce rubber tyres, and road transport as we know it today would simply not exist. We end with an intriguing thought. If the atmospheric system had been invented not in 1840 but in 1880 it might have had its first trials on one of London's early tube railways. In the dry, sunless and temperate environment of the tube tunnels the leather valve would probably have performed well. It would have been the obvious alternative to the condensing steam locomotive (in the brief window of opportunity when electric traction was still a novelty) and might have survived into the twentieth century silently carrying millions of passengers below city streets.

References

I. Brunel, *The Life of Isambard Kingdom Brunel, Civil Engineer*, Longmans Green, 1870.

C. Hadfield, *Atmospheric Railways*, David and Charles, 1967.

J. Pudney, *Brunel and his World*, Thames and Hudson, 1974.

D. J. Thomas, *A Regional History of the Railways of Great Britain*. Vol 1: *The West Country*, Phoenix, 1960.

H. Great engineering disasters

CHAPTER 26

FLIXBOROUGH

Introduction

Flixborough is a small village which sits in a flat and marshy area of land on the east coast of England near the point at which the River Trent flows into the estuary of the River Humber. The immediate surroundings are thinly populated, but 3 miles to the south lies the quite sizeable town of Scunthorpe. In 1938 a company called Nitrogen Fertilizers Ltd bought a plot of land near the village and put up a chemical plant to manufacture ammonium sulphate. In 1964 a company called Nypro acquired the site from Nitrogen Fertilizers Ltd and between 1964 and 1967 they built a plant to produce caprolactam, which is a basic ingredient for making Nylon 6. The first stage in producing the caprolactam was to obtain cyclohexanone and this was done by subjecting phenol to a hydrogenation reaction.

In 1967 Nypro began the construction of an additional plant to increase the production capacity of caprolactam from 20,000 to 70,000 tonnes per year. The cyclohexanone was to be produced by a route involving the oxidation of cyclo-hexane rather than the hydrogenation of phenol. The new plant was finished in 1972. Now cyclohexane is a highly flammable and volatile liquid, similar to petrol. In the process large volumes of the liquid were to be circulated through the plant under conditions of quite high temperature and pressure. A leak in the plant would obviously lead to the escape of flammable vapour and the introduction of the new process therefore had serious safety implications.

At 4.53 pm on Saturday 1 June 1974 the works were virtually demolished by an explosion estimated to have been equivalent to 45 tons of TNT. The explosion was followed by a fierce fire. Immediately to the south of the reactors lay the main office block. This was completely flattened by the blast. Because the explosion occurred on a Saturday there was no one in the offices; if the accident had occurred on a week day it is likely that many of the office staff would have been killed. The staff of the control room were less fortunate. In accordance with normal practice the control room was close to the centre of the plant. The eighteen people in it were all killed instantly, and a further ten people died out in the open. Thirty-six others were injured, some seriously.

The blast wave travelled as far away as Scunthorpe, damaging 786 houses. In the village of Burton-on-Stather, 2 miles north of Flixborough, 644 out of a total number of 756 houses were damaged. Worst affected was the village of Flixborough itself, which was only half a mile from the plant. In all, a total of 1821 houses, 167 shops and several factories were damaged by the blast. Outside the plant no one was killed, but fifty-three people were injured; hundreds more suffered minor injuries which were not logged.

Description of the Plant

Figure 26.1 shows a simplified diagram of the plant. The oxidation reaction was carried out in a train of six reactors. Each reactor contained liquid cyclohexane up to the level of the outlet. In the intermediate reactors the level of the outlet pipe was 356 mm lower than the level of the inlet pipe. This meant that the liquid could flow along the train of reactors by gravity. The oxygen for the oxidation process was supplied by lancing air into the bottom of each reactor from a compressed air main. Catalyst was supplied to each reactor as well to initiate the reaction. The normal working temperature and pressure were 155°C and 0.88 MPa. However, the vapour space at the top of each reactor was connected to a safety valve which was set to blow at a pressure of 1.08 MPa. If any vapour had blown off from the system it would have been led to a flare stack where it would have been burnt off harmlessly. The reactors were quite large (about 3 m in diameter and 5 m high).

Figure 26.2 gives details of the piping connections between the reactors. Flanged nozzles were welded into the side of each reactor at positions corresponding to both inlet and outlet. The outlet on one reactor was coupled to the inlet on the next by means of a convoluted bellows pipe. This construction meant that the connection was able to accept both axial and shear deflections and this allowed for any relative movement between the adjacent reactors due to thermal expansion or settlement. The bellows and nozzles were made from stainless steel and had an internal diameter of 711 mm.

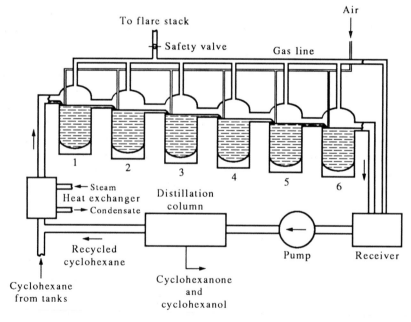

FIG 26.1   Simplified diagram of the cyclohexane oxidation plant.

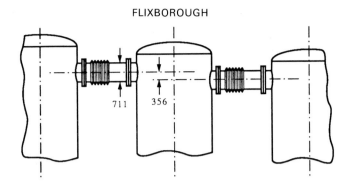

FIG 26.2   Details of the piping connections between the reactors. Dimensions in mm.

Background to the Failure

On 27 March 1974 it was found that cyclohexane was leaking from reactor no. 5. This was traced to a vertical crack about 2 m long running along the axis of the vessel at right angles to the hoop stress. The wall of the reactor was fabricated from 13-mm mild steel plate which was clad on the inside with a layer of stainless steel 3 mm thick to prevent corrosion. The crack in the mild steel had only penetrated the stainless steel cladding over a short distance and this was why the 2-m crack resulted in only a small leak of cyclohexane. It was obvious that there was a serious risk that the crack could go unstable by fast fracture. The plant was therefore shut down and the offending reactor was removed from the reactor train.

It was decided to return the plant to service by putting a temporary connection into the gap which had been left between reactors 4 and 6. The nature of this repair is crucial to the failure but before we discuss it it is worth questioning the basic wisdom of returning the plant to service in the first place. Put simply, if a dangerously long crack had been found in reactor 5, how could anyone suppose that the remaining reactors would be safe? Although this did not in fact affect the final outcome the other reactors were not stripped of their outer covering and were not searched for cracks. Neither were they subjected to a repeat hydraulic test to the pressure specified by the Code of construction. In the event a metallurgical analysis of samples removed from reactor 5 showed that the damage had been caused by stress-corrosion cracking. During the operation of the plant cooling water had been sprayed onto the outside of the reactor to dilute small leakages of cyclohexane. The water had been treated with sodium nitrate, a chemical which is known to cause stress-corrosion cracking in mild steel.

The Temporary Connection

Figure 26.3 shows the basic features of the temporary connection. The nearest size of stainless steel pipe that could be found on site had a bore of only 508 mm. However, calculations showed that a pipe having this bore could handle the flow of liquid along the reactor train and it was decided to use it for the connection. Because of the difference in level between the outlet from reactor 4 and the inlet to

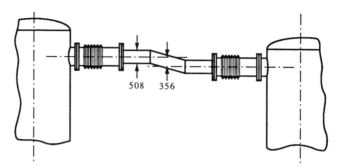

FIG 26.3   Details of the temporary connection. Dimensions in mm.

reactor 6 the pipe had to have an axial offset of 356 mm. This was achieved by welding three lengths of pipe together using mitred joints. A light framework of scaffolding pipes was erected immediately beneath the assembly in order to take the weight off the bellows. Finally the assembly was tested *in situ* by pressurising the whole system to the normal working pressure of 0.88 MPa with nitrogen gas. The plant was back in action by 1 April, having been out of operation for only 3 days.

Once the plant was working again the temporary connection was lagged. It gave no trouble and was looked at on several occasions by various people. One of these reported that when the system was brought on stream the pipe seemed to move upwards slightly, but this was apparently not thought important.

Details of the Failure

On 29 May a leak was found in a small fitting on one of the reactors and the plant was shut down. During 30 and 31 May the leak was repaired and some other maintenance work was done. The start-up process was begun in the early hours of Saturday 1 June by the shift which had come on duty at 11 pm the previous night. At about 4 am the circulation pump was started and steam was put into the heat exchanger to start warming up the liquid. When the new shift came on duty at 7 am the pressure in the reactors was about 0.4 MPa. By 12 noon the pressure had risen to the operating pressure of 0.88 MPa but the temperature of the liquid was still below the operating figure of 155°C. The afternoon shift came on duty at 3 pm and it is presumed that they continued to warm up the circulating cyclohexane. The explosion took place two hours later.

The failure was traced to the temporary connection which had been fitted between reactors 4 and 6. As Fig. 26.4 shows, this had come away from both reactors and had fallen to the ground. The flexible bellows had been torn from their mountings and the pipe itself had "jack-knifed". A number of eyewitnesses reported that shortly before the explosion they had seen and heard unusual events in the region of the reactors. They were probably observing the disintegration of the temporary connection. Then all fell quiet again. But because the connection had failed vapour and liquid were able to pour out of reactors 4 and 6 through the flanged nozzles. A large cloud of explosive potential developed around the reactors and ignition took place minutes later.

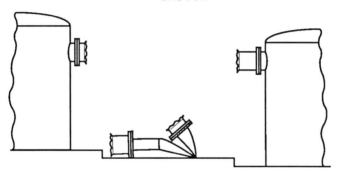

FIG 26.4   Schematic of the failure.

Simulation Test

The failure investigators decided to conduct a full-size simulation test to identify the likely collapse mechanism. Reactor 3 was removed from the devastated plant and a replacement temporary connection was made to fit in between reactors 2 and 4. The assembly of pipe and bellows was heated electrically to 155°C and a length of steel chain was run through it horizontally to simulate the evenly distributed loading from the liquid cyclohexane. The connection was supported on scaffolding pipes as in the original installation. The assembly was pressurised with nitrogen, but to minimise the risk of explosion the reactors themselves were blanked off and were filled with water.

As the pressure was increased the flange at the lower end of the pipe began to move downwards. The deflection was 20 mm at a gauge pressure of just under 0.96 MPa. At 0.96 MPa the lower bellows suddenly went unstable, as shown in Fig. 26.5. This type of instability is called "squirm". It occurs when the shear displacement of the opposite ends of the bellows reaches a critical value at which the internal pressure is sufficient to blow out the convolutions. The pressure was increased past the point of squirm until, at 1.43 MPa, the bellows burst open.

Why should the assembly distort in this way? The situation is shown in Fig. 26.6. The fluid-filled pipe is subjected to an axial force pA, where p is the gauge pressure in the reactors and A is the cross-sectional area of the pipe to which the bellows is attached. This pipe has an internal diameter of 711 mm and an area of 3.97×10^5 mm². At the normal working pressure of 0.88 N mm^{-2} the axial force is

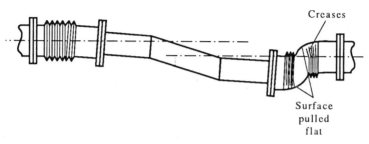

FIG 26.5   The simulation test.

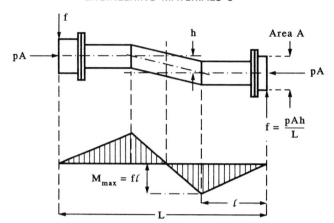

FIG 26.6   Forces and moments acting on the fluid-filled pipe.

nearly 36 tonnef. The axial offset of the connection means that these axial forces are not in line: because of this they exert a turning moment on the assembly. This is resisted by having a shear force on each bellows as shown in Fig. 26.6. The value of the shear force can be found by balancing the moments. Because the assembly is much longer than the axial offset the shear force is much less than 36 tonnef. In fact it turns out to be only 2.8 tonnef at normal operating pressure. At 0.96 MPa the shear force is increased to 3.1 tonnef. However, the bellows is not very stiff and this value of the shear force is enough to cause squirm.

The pressure needed to cause squirm in the simulation test was 9% more than the normal operating pressure, so why should squirm have taken place before the explosion? When a large chemical plant is being brought on stream the operating conditions can oscillate quite widely before the system finally settles down. Unfortunately no records survived from the control room. But the pressure might well have climbed towards the value of 1.08 MPa at which the safety valves would have lifted, and this would have been enough to make the bellows squirm. There is one final catch, however: the maximum pressure of 1.08 MPa is less than the pressure of 1.43 MPa needed to make the bellows burst. In fact the key element in explaining why the assembly blew apart is the jack-knifing of the pipe.

Jack-knifing of the Pipe

As well as trying to rotate in response to the axial load pA the connection also wants to buckle in axial compression. As shown in Fig. 26.6 the maximum bending moment on the pipe occurs at the mitred joints and this is exactly where the pipe jack-knifed in the accident. In order to find out at what pressure the assembly might have buckled a second full-scale test was carried out using the arrangement shown in Fig. 26.7. In this test the replacement connection was blanked off and was filled with water. An axial load F was applied to each end of the assembly as shown in Fig. 26.7. As the load was increased the pressure of the water inside the pipe was increased in proportion so that at any instant p was always equal to F/A. This

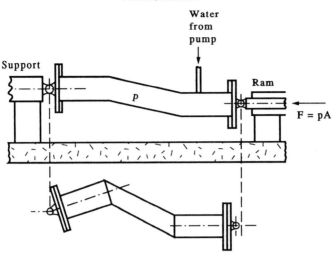

FIG 26.7   Schematic of the axial load test. Above a critical value of F the pipe starts to "jack-knife".

was a direct simulation of the loading experienced by the pipe in the actual installation although the test had to be done at room temperature for safety reasons. As the pressure was increased the mitred joints deflected elastically. Above 1.47 MPa the joints started to deflect plastically. At 1.81 MPa the pipe began to jack-knife. However, the lateral deflection at the joints just before buckling was only 6 mm. The jack-knifing pressure is less at 155°C because both the yield strength and the Young's modulus of stainless steel decrease with temperature. As a result it was estimated that the jack-knifing pressure under operating conditions was about 1.5 MPa.

   This critical pressure for jack-knifing is, of course, greater than the pressure of 1.08 MPa at which the safety valve would have blown. Under static conditions the pipe should not have failed by jack-knifing. However, calculations of the dynamic loads during the failure showed that they were in fact large enough to produce jack-knifing. The reason for this is interesting. When the bellows squirm the volume of the assembly increases because of the very different geometry of the squirmed bellows. As a result the vapour expands and does work. The pipe starts to jack-knife and the flanges at each end of the pipe move towards one another. The flanges pull the bellows out lengthwise: this increases the volume still further and allows the vapour to do more work. Eventually the bellows are pulled away from their mountings and the assembly disintegrates. This chain of events did not occur in the simulation squirming test because the volume of pressurised nitrogen was small: as the bellows squirmed the pressure in the pipe dropped and the situation stabilised. This is a classic example of a displacement–controlled event, like the buckling of the chemical reactor described in Chapter 4. In the disaster itself the volume of vapour was large and the pressure would have been sustained during the squirming process. This is a classic example of a load–controlled event like the flipping of the partition described in Chapter 5.

Lessons to be Learned

As we saw in Chapter 7 a small model steam boiler is typically 130 mm in diameter and 240 mm long. The safety valve operates at about 0.7 MPa. No model engineer would dream of using even a tiny pressure vessel like this without first giving it a hydraulic test to twice the maximum working pressure. By way of comparison the operating pressure of the temporary connection was greater by 50%, the volume was greater by 300 times, the available volume of liquid was greater again and the contents were highly flammable. Yet the assembly was only tested for leaks at the normal operating pressure and was not even taken up to the pressure at which the safety valves would have blown. According to the British Standard for pipework the assembly should have been tested to at least 1.3 × the "design pressure". This term is ambiguous, but even if the pipe had been tested to 1.3 × the normal operating pressure (i.e. to 1.14 MPa) the bellows would probably have squirmed, exposing the weakness of the design. Almost as bad, the pressure testing was carried out using nitrogen. Had the assembly failed during the test there would have been an enormous outrush of high-pressure gas and people might well have been killed. Wherever possible pressure testing is done using water: because liquids are comparatively incompressible the energy released if a vessel fails during hydraulic testing is small.

Whether the assembly had been pressure tested or not, an experienced mechanical design engineer would have spotted the deficiencies of the design on the drawing board. The moral is that modifications should not be made to an existing plant without going through the same rigorous processes of design and testing that were used in the first place. The disaster certainly acts as a reminder of the awesome responsibilities placed on design engineers and the importance of implementing the relevant Codes and Standards. No matter how careful people are they can still make mistakes: Codes and Standards embody the accumulated experience of engineers and designers over many years and are an essential safeguard against getting things wrong.

Reference

Department of Employment, *The Flixborough Disaster: Report of the Court of Inquiry*, HMSO, 1975.

CHAPTER 27

THE TAY BRIDGE—1

A Note on Units

The events described in this case study took place in the United Kingdom between 1849 and 1879. The dimensions of the different parts of the bridge and of the trains that ran on it were given in feet and inches. Weights were given in tons and hundredweights, and wind pressures were quoted in pounds per square foot. We have not converted the data to their SI equivalents. Many of the dimensions of the bridge have been scaled from contemporary photographs: where there is doubt about the exact value of a large dimension it is a pretty safe bet that it will be a whole number of feet. A small dimension, like the thickness of an iron bar, is likely to be a common fractional size such as $\frac{1}{2}$ inch. It is often easier (and safer) to work within a particular system of units than to make frequent conversions. The English units used in this chapter are defined below.

1 ton (long ton) = 20 cwt (hundredweight) = 2240 lb = 1017 kg = 1.017 tonnes (metric tons).
1 ft (foot) = 12 in (inches) = 304.8 mm.
1 mph (mile per hour) = 1.609 kph (km per hour) = 0.447 m s^{-1}.
1 tsi (ton per square inch) = 15.44 MPa.

Background

North of Edinburgh lie two great estuaries, the Firth of Forth and the Firth of Tay. The Firth of Forth begins at the ancient town of Stirling and runs a full 50 miles to the east until it emerges into the North Sea. Edinburgh, the capital of Scotland, is situated on the south shore of the firth 30 miles west of Stirling. Eleven miles east of Edinburgh the Queensferry promontory narrows the firth down to a width of 1 mile but the water is still between 40 and 60 feet deep depending on the tides. The Firth of Tay lies 25 miles to the north of the Forth: it begins at Perth and runs east for 25 miles until it meets the sea. Dundee is situated 18 miles to the east of Perth on the north shore of the firth. At this point the Tay is $1\frac{1}{2}$ miles wide and is up to 80 ft deep at high water.

The firths have always been a major barrier to communication and this was especially so in the 1850s, when the railways were expanding at a frantic pace. A passenger who wanted to go from Edinburgh to Dundee and perhaps on to Aberdeen had either to go the long way round through Stirling and Perth, adding perhaps 60 miles to the journey, or had to endure two ferry crossings, one across the Forth from Granton to Burntisland, and one across the Tay from Tayport to Broughty Ferry. The fastest boat train of the day left Waverley Station, Edinburgh at 6.25 am and was timed to arrive at Dundee at 9.37 am. This meant a journey time of 3 hours and 12 minutes for a distance of 46 miles—an average speed of only 14 miles per hour. In bad weather the ferries might not run at all; if they did the hapless passengers would probably arrive cold and sea-sick. Freight traffic posed special problems because goods had to be off-loaded at the ferry terminals.

The route was operated by the Edinburgh and Northern Railway. In 1849 they appointed a civil engineer, Thomas Bouch, to be their manager. Twenty-six years old at the time, he immediately set about improving the ferry service and by 1850 had built what was the world's first roll-on, roll-off train ferry. Bouch realised, however, that this was only a stop-gap measure: the real answer was to build a railway bridge over the Forth and another over the Tay. In 1854 the E & N was taken over by the rapidly expanding North British Railway. Bouch put his proposal for a pair of bridges to the directors of the NBR but they dismissed it as "the most insane idea that could ever be propounded". In the long run, of course, the case for the bridges was overwhelming; but the chequered progress of the project gives a fascinating insight into the ruthless commercial politics of the railway age. Eventually, on 15 July 1870, a Bill was passed by Parliament which authorised the construction of the bridge over the Tay. Bouch, by then an independent consultant, was appointed engineer to the new bridge.

The Bridge Described

An outline plan of the bridge is shown in Fig. 27.1. In order to allow shipping to pass up the Tay a clearance of 88 ft was required between the girders of the bridge and the high water mark in the middle of the firth. On the south shore of the firth, at Wormit, the land rose steeply to a height of about 200 ft and this provided an ideal jumping-off point for the bridge. After leaving the shore on a short curved alignment the track climbed gradually at 1 in 490 until it reached pier 29. It then ran level to pier 36. After passing pier 37 the track fell rapidly, at 1 in 74, until it reached the north shore at Dundee. At pier 53 the track entered a large, sweeping curve which took it in alongside the shore and down to a height of about 40 ft above high water. The overall length of the bridge was 10,300 ft (nearly 2 miles), which at the time made it the longest iron bridge in the world.

The spans of the bridge were supported on a total of 85 piers. The first 14 were made from brick and looked fairly substantial. The rest, fabricated from iron, looked rather spindly in comparison. Over most of the bridge the track ran on top of the girders. But between piers 28 and 41 the construction was different. This was the place where the navigation channel lay and where the bridge had to have the full headroom of 88 ft. To get the extra height the piers were extended to bring their tops up to the level of the tracks. Thirteen spans of lattice-work box section were then placed end-to-end on top of the piers and the track was carried on the floor of the box. The thirteen spans (eleven of 245 ft and two of 227 ft) were aptly called the High Girders. Made from wrought-iron sections riveted together a single 245-ft

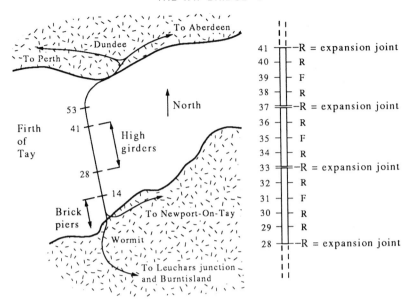

FIG 27.1 Plan of the bridge. The numbers mark the positions of supporting piers. F = fixed joint, R = roller joint.

span weighed about 230 tons complete with track and wooden decking. As shown in Fig. 27.1 the individual spans in the High Girders were joined together into three continuous lengths, one of five spans and two of four spans. The girders were attached to the tops of the piers with either roller bearings or fixed attachments and were terminated in expansion joints.

The track was single, laid to the standard gauge of 4 ft $8\frac{1}{2}$ in and fixed with twin check rails. On the approaches to the high girders the bridge deck was fitted with railings, but these would have been far too flimsy to keep a train on the bridge had it derailed. A passenger would have felt more secure once the train had entered the high girders although some people complained about the drumming noise which the open latticework set up as it passed by the moving carriages.

Figure 27.2 shows the side and end elevations of a 245-ft span from the high girders. The leading dimensions are given in the literature but the depths of the booms and the widths of the diagonal members have been estimated from a close-up photograph (Thomas). Figure 27.3 shows the construction of the iron piers which supported the High Girders. The dimensions have been scaled from outline NBR drawings (Thomas) and from photographs (Barlow, Prebble). A pier consisted of six vertical columns of cast iron tied together by bracing bars made from wrought iron. Each column was built up by bolting together seven flanged cast-iron pipes. The ends of the flanges were skimmed true in a lathe and were then fastened together with eight $1\frac{1}{4}$ in wrought iron bolts. The central four columns had an outside diameter of 17 in and a wall thickness of 1 in. The two outermost columns had an outside diameter of 20 in. The bracing bars were located in "snugs" which were cast in one with the pipes. The horizontal bars were made from T-section iron and were secured at each end with a single wrought iron bolt. The diagonal ties were made

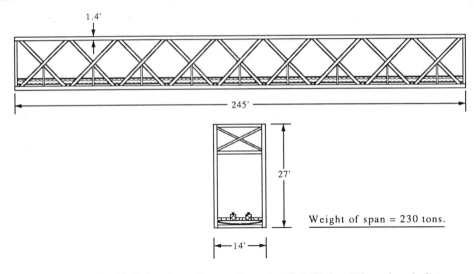

1.4'

245'

27'

14'

Weight of span = 230 tons.

FIG 27.2   Simplified elevations of a span from the High Girders. Dimensions in feet.

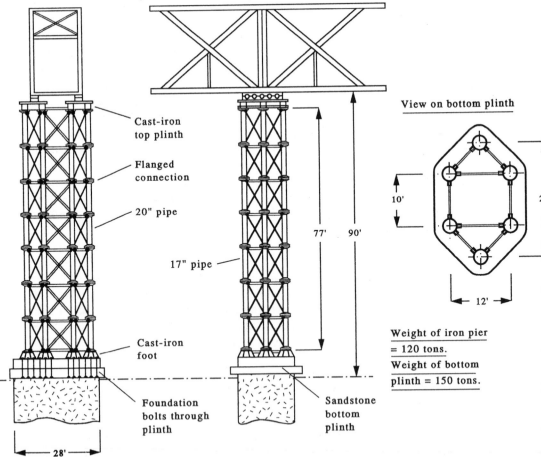

View on bottom plinth

Cast-iron
top plinth

Flanged
connection

20" pipe

17" pipe

77'   90'

10'

12'

Cast-iron
foot

Weight of iron pier
= 120 tons.
Weight of bottom
plinth = 150 tons.

Foundation
bolts through
plinth

Sandstone
bottom
plinth

28'

FIG 27.3   Simplified plan and elevations of an iron pier. Dimensions in feet.

from iron flats with a cross-section measuring 1 in × 4 in. Each tie was held by a single $1\frac{1}{2}$ in bolt at one end and a tapered cotter at the other. In theory this arrangement meant that each tie could be pulled up tight after assembly to make it take its share of the load. As the pier was erected the inside of each column was filled with portland cement. The weight of a pier complete with cement, bracing bars and top plinths was about 120 tons.

Figure 27.4 shows a contemporary extract from the working drawings (published in the once famous but long-defunct journal *The Engineer*) which gives detailed dimensions for the flanged connections and the bracing. There is, of course, no guarantee that the bridge was built exactly to the drawings but they are the best guide that we have.

The Disaster

The contract for the construction was won by the firm of Charles de Bergue. The contract was signed on 8 May 1871 and in it the contractor undertook to have the bridge ready for traffic in three years at a price of £217,000. In the event the bridge was opened on 31 May 1878 by which time it had cost £300,000.

The morning of Sunday 28 December 1879 was quiet. When Captain Wright took his ferry boat, the *Dundee* across the firth at 1.15 pm he noted that the weather was good and the water was calm. The 4.15 pm crossing was just as uneventful but the Captain noted that the wind had freshened. By 5.15 pm a gale was moving in from the west and the river, in the words of the Captain, "was getting up very fast". The local shuttle train left Newport at 5.50 pm and arrived in Dundee station shortly after 6 pm. The passengers had had a worrying crossing. Their carriages were buffeted by the growing force of the storm and lines of sparks flew from the flanges of the wheels under the sideways force of the wind.

The mail train from Edinburgh had left Burntisland at 5.20 pm and by the time that the local had arrived in Dundee the mail had reached Thornton Junction, 27 miles south of the bridge. The last station was St Fort, which lay in a small depression 2 miles short of the bridge. The station staff collected the tickets of the passengers who were going on to Dundee. In addition to three men on the footplate of the locomotive there were seventy-two passengers. By 7.13 pm the train had reached the signal cabin at Wormit. The driver slowed the locomotive down to walking pace so his fireman could take the staff for the single line from the signalman. Then he opened up the regulator and took the train out onto the bridge and into the teeth of the westerly gale. The signalman returned to the shelter of his cabin and sent the "train entering section" signal to his opposite number in the signal box at Dundee.

The train receded into the darkness and the light of the three red tail lamps grew dimmer. Sparks flew from the wheels and merged into a continuous sheet that was dragged to the lee of the bridge by the wind. Eyewitnesses saw a bright glow of light from the direction of the train just after it must have entered the High Girders and then all went dark. The train was timed to pass the Dundee box at 7.19 pm. When it failed to arrive the signalman tried to telegraph the Wormit box but to no avail. The obvious conclusion was that the telegraph wires had been severed where

(A)

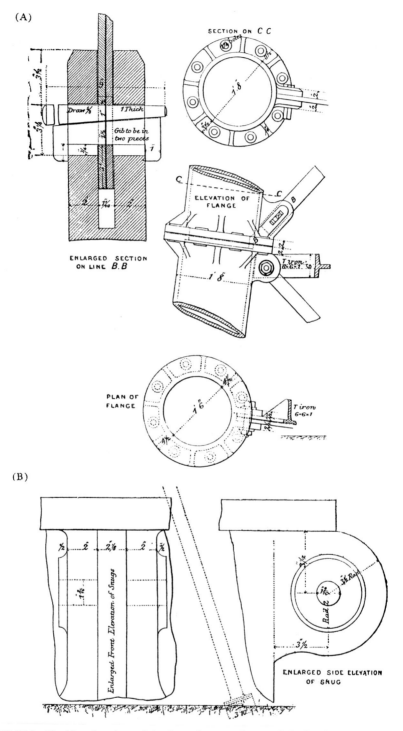

(B)

FIG 27.4   Working drawings of the flanged connections and the bracing arrangements.
Dimensions in feet and inches.

they passed over the bridge. James Roberts, the locomotive foreman at the Dundee engine sheds, walked out along the bridge to investigate. Although at times he was forced to crawl on all fours by the force of the gale he eventually made his way to the end of the low-level girders. Further progress was impossible: the whole length of the High Girders had disappeared into the river taking the train with it.

Photographs taken after the disaster (see Thomas, Prebble) show that the iron piers which had supported the High Girders had fractured. Pier 29 had broken off at the flanges 22 ft above the base of the columns. The upper part of the pier had fallen into the river. In the lower part of the pier, which was still anchored to its masonry base, four of the bracing bars were left hanging free. Pier 31 had broken away at the flanges 11 ft above the base. The remaining piers (30 and 32 to 40) had broken off where the columns were bolted to the base and had fallen into the river. Bracing bars were also hanging loose from piers 28 and 41. Divers found that the High Girders were lying on their sides on the river bed to the east of the bridge. As shown in Fig. 27.5, most of the train was lying inside the fifth span counting from the south end of the High Girders. The horizontal distance between the girders and the edges of the piers varied between 16 and 51 ft; the fifth span was between 21 and 39 ft from the piers. When the locomotive was raised from the river bed it was found that the regulator was fully open, the reversing lever was in the third notch from mid-gear and the brakes were fully off. This is the normal position for the controls when an engine is up to speed and the driver could have had no warning of the disaster. The engine was not badly damaged (it was later returned to service

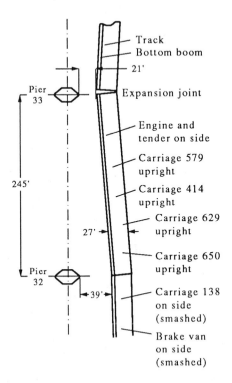

FIG 27.5 Location of the fallen train. Dimensions in feet.

and was nicknamed "The Diver"); most of the damage was confined to the left-hand side. The carriages and the brake van were made from wood. The leading carriages were damaged most badly on their left-hand sides, just like the engine. The last two vehicles in the train had, however, been reduced to matchwood. The axles of the brake van and the carriages were all bent.

The Court of Inquiry

The Board of Trade appointed a formal Court of Inquiry to investigate the causes of the disaster. Their main conclusions, which are summarised below, make interesting reading.

(a) "There is no absolute knowledge of the mode in which the structure broke down; the evidence of persons who happened to be looking at the bridge at the time agrees in describing lights falling into the river, and that these appearances lasted only a few seconds, but the evidence is not sufficiently clear and definite to determine by it which portion of the bridge fell first."

(b) "The storm which occurred at Dundee on the night of the disaster was recorded on board the *Mars* training ship, lying near Newport, as being of the force of 10 to 11 of the Beaufort scale, and was especially characterised by strong gusts at intervals ... from the configuration of the land, the main force of the gusts would probably take the line of the river."

(c) "The first indication of weakness in the bridge itself was the loosening of a number of the ties of the cross bracing." This refers to the bracing bars between the cast-iron columns. Maintenance staff reported hearing the ties chattering when trains passed overhead; the tapered cotters tended to work loose, allowing the ends of the ties to come free. "The loosening ... must have ensued from lateral action, and was most probably due, as was suggested, to strains on the cross-bracing produced by storms of wind."

(d) "The storm of the 28th December 1879, would necessarily produce great tension on the ties, varying as the heavy gusts bore upon different parts of the bridge." The fractures in the columns might have arisen "by the yielding of the cross-bracing, and the consequent distortion of the form of the piers, which would throw unequal strains on the flanges and connecting bolts."

(e) "It is observable in the ruins of the bridge, that the columns have for the most part separated where they had been bolted to the base pieces; in two piers the separation has taken place higher up the pier, one being at the first and the other at the second tier of columns."

(f) "The wrought iron employed was of fair strength, though not of high quality as regards toughness; and the cast iron was also fairly good in strength, but sluggish when melted, and presented difficulty in obtaining sound castings." Sometimes columns at the foundry were found with their "snugs" broken off. They were re-used by having new lugs "burned on": a special mould was placed over the column and a new snug was cast *in-situ* on to the cold surface of the pipe. This was a highly questionable procedure and the junction between the snug and the pipe often cracked when the snug cooled down.

(g) "That the first or southern set of continuous girders, covering five spans, was the first that fell after the engine and part of the train had passed over the fourth pier, and that the two consecutive sets of continuous girders, each covering four spans, were in succession pulled off the piers on which their northern ends rested, by the action of the first set of continuous girders falling over, and probably breaking some of the supporting columns." This finding seems pretty speculative in view of conclusion (a).

(h) Finally, the most revealing finding was that "we have to state that there is no requirement issued by the Board of Trade respecting wind-pressure, and there does not appear to be any understood rule in the engineering profession regarding wind-pressure in railway structures; and we therefore recommend that the Board of Trade should take such steps as may be necessary for the establishment of rules for that purpose". Some bridge designers were still using figures given by Smeaton (of lighthouse fame) to a meeting of the Royal Society *in 1759*—120 years before the Tay bridge was built! These varied from 6 lb/sq ft for "high winds" to 12 lb/sq ft for "storm or tempest". Bouch sought advice on this very point from Colonel William Yolland, the chief inspector of railways. He asked "and is it necessary to take the pressure of wind into account for spans not exceeding 200 feet span, the girders being open lattice work?" The reply was "and we do not take the force of the wind into account when open lattice girders are used for spans not exceeding 200 feet". Bouch's design was also examined and approved by two leading engineering consultants, T. E. Harrison and J. M. Hepper.

In this case study we look at the failure using two techniques which were not available at the time of the disaster: we assess the wind loading on the bridge using the British Standard Code for the design of bridges; and we investigate the collapse of a pier using plastic theory.

Wind Pressure

Conclusion (b) of the Court of Inquiry was that the wind speed in the firth near the bridge was between 10 and 11 on the Beaufort scale. The Beaufort scale is summarised in Table 27.1. From it one can see that the wind speed at the time of the disaster might have been between 55 and 75 mph.

What pressure should a wind of this speed exert on the bridge? BS 5400 defines a maximum wind gust speed given by

$$v_c = vS, \tag{27.1}$$

where v is the mean hourly wind speed and S is the gust factor. S increases as the size of the wind-loaded component decreases and increases with height above ground (or sea) level. For a component 100 m wide situated 30 m above ground level $S = 1.63$. If the width of the component is ≤ 20 m, then $S = 1.73$. A reasonable value of S for both the High Girders and the train is ≈ 1.65. The wind load in $N\,m^{-2}$ is given by

$$P = 0.613 v_c^2 C_D, \tag{27.2}$$

TABLE 27.1   *The Beaufort Scale of Wind Force*

Beaufort number	Description of wind	Wind speed (mph)	Effect of wind
0	Calm	<1	Smokes rises straight
1	Light air	1 to 3	Smoke moves in wind
2	Light breeze	4 to 7	Wind felt on face
3	Gentle breeze	8 to 12	Leaves move
4	Moderate breeze	13 to 18	Small branches move
5	Fresh breeze	19 to 24	Small trees sway
6	Strong breeze	25 to 31	Umbrellas difficult
7	Moderate gale	32 to 38	Whole trees move
8	Fresh gale	39 to 46	Hard to walk into gale
9	Strong gale	47 to 54	Roofing tiles taken off
10	Whole gale	55 to 63	Trees uprooted
11	Storm	64 to 75	Major damage
12	Hurricane	>75	Tropical storms

where v_c is in m s^{-1}. C_D is the drag coefficient, which depends on the shape of the wind-loaded component: for a train $C_D = 1.45$; for a lattice work truss of the High Girders $C_D = 1.75$. If we apply Eqns. (27.1) and (27.2) to the train at a mean wind speed of 55 mph (24.6 m s^{-1}) we get a maximum gust pressure of 1.463 kN m^{-2} (31 lb/sq ft). However, since the wind force increases with the *square* of the mean wind speed, the maximum gust pressure at 75 mph goes up to 58 lb/sq ft. It is interesting to see that BS 5400 would require the designer of a new Tay bridge to use a mean wind speed of 38.7 m s^{-1} (87 mph) in his or her calculations.

Another clue to the wind force can be gleaned from the train, which got as far as the High Girders without being blown off. Light railway carriages can be capsized by high winds, and by the time of the disaster there had been at least three reported cases of this in the UK and four in France. The wind pressure needed to capsize a carriage in the train can be estimated by taking moments as shown in Fig. 27.6. Table 27.2 gives the weights and dimensions of the vehicles and Table 27.3 gives the results of the calculations.

Length of train = 223 ft.
Net weight of train = 114 tons 14 cwt.
Payload: 46 mail bags in brake van (≈ 1 ton).
Passengers: 72 (≈ 5 tons; assume 1 ton in each carriage).
Gross weight of train $\approx$ 120 tons.
Height from top of rails to top of carriages $\approx$ 11 ft.
Height from top of rails to bottom of carriage body $\approx$ 3 ft.
Track gauge $\approx$ 56 in.
Length of uncompressed buffer $\approx$ 1.5 ft.

The last carriage in the train was obviously the least stable, and it should have capsized when the wind pressure on it was about 28 lb/sq ft. This is essentially identical with the wind pressure of 31 lb/sq ft which we calculated for the bottom end of the Beaufort 10 range. In the following section we find what the load on the bridge would have been when the pressure on the train was 28 lb/sq ft. Although the calculations are rather tedious they are straightforward and need no explaining.

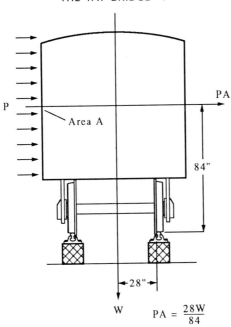

FIG 27.6   Finding the wind pressure needed to capsize a carriage. Dimensions in inches.

TABLE 27.2   *Details of the Train*

Item of rolling stock	Weight	Length over buffers (ft)
4-4-0 Locomotive no. 224	34 tons 12 cwt	28
Tender (6 wheels)	≈ 24 tons 17 cwt	18
Carriage no. 579 (4 wheels)	8 tons 8 cwt	28
Carriage no. 414 (6 wheels)	14 tons 5 cwt	38
Carriage no. 629 (4 wheels)	8 tons 8 cwt	28
Carriage no. 650 (4 wheels)	9 tons 16 cwt	31
Carriage no. 138 (4 wheels)	5 tons 19 cwt	26
Brake van (4 wheels)	8 tons 9 cwt	26

TABLE 27.3   *Wind Pressures to Capsize Rolling Stock*

Item of rolling stock	Wind area (sq ft)	Capsize pressure (lb/sq ft)
Locomotive	≈ 7 × 25 = 175	≈ 170
Tender	≈ 6 × 15 = 90	≈ 360
Carriage no. 579	8 × 25 = 200	35
Carriage no. 414	8 × 35 = 280	41
Carriage no. 629	8 × 25 = 200	35
Carriage no. 650	8 × 28 = 224	36
Carriage no. 138	8 × 23 = 184	28
Brake van	8 × 23 = 184	38

Wind Loads

Single Span of High Girders with Train Inside

(a) *Train*

Area of train $\approx$ 1540 sq ft.
Length of cross-bracing shielding train $\approx$ 170 ft.
Width of cross-bracing $\approx$ 1.4 ft.
Area of cross-bracing shielding train $\approx$ 170 $\times$ 1.4 $\approx$ 240 sq ft.
Unshielded area of train $\approx$ 1540 $-$ 240 $\approx$ 1300 sq ft.
Wind force $\approx$ 28 $\times$ 1300 lb $\approx$ 16 tons acting $\approx$ 7 ft above rail level ($\approx$ 10 ft above bottom of girder).
Drag coefficient C_D = 1.45 (BS 5400: unshielded live load).

(b) *Windward Truss*

Length of top boom = 245 ft.
Length of bottom boom = 245 ft.
Combined length of end beams $\approx$ 48 ft.
Combined length of cross-bracing $\approx$ 595 ft.
Combined length of all sections $\approx$ 1130 ft.
Width of all sections $\approx$ 1.4 ft.
Area $\approx$ 1130 $\times$ 1.4 $\approx$ 1580 sq ft.
Gross area of truss = 245 $\times$ 27 = 6615 sq ft.
Solidity ratio of truss = 1580/6615 = 0.24.
Drag coefficient = 1.75 (BS 5400: solidity ratio = 0.24; flatsided members).
Wind drag relative to train = 1.75/1.45 = 1.21.
Wind force $\approx$ 1.21 $\times$ 28 $\times$ 1580 lb $\approx$ 24 tons acting 13.5 ft above bottom of girder.

(c) *Bridge Deck*

Depth of deck $\approx$ 1 ft.
Area of deck $\approx$ 1 $\times$ 245 $\approx$ 245 sq ft.
Area shielded by windward truss $\approx$ 30 sq ft.
Unshielded area $\approx$ 215 sq ft.
Drag coefficient = 1.1 (BS 5400: truss girder superstructures).
Wind drag relative to train = 1.1/1.45 = 0.76.
Wind force $\approx$ 0.76 $\times$ 28 $\times$ 215 lb $\approx$ 2 tons acting $\approx$ 3 ft above bottom of girder.

(d) *Leeward Truss*

Area $\approx$ 1580 sq ft.
Area shielded by deck and train $\approx$ 30 + 240 $\approx$ 270 sq ft.
Unshielded area $\approx$ 1310 sq ft.
Spacing ratio of trusses = 14/27 = 0.52.
Shielding factor = 0.85 (BS 5400: solidity ratio = 0.24; spacing ratio <1).
Drag coefficient = 0.85 $\times$ 1.75 = 1.49.
Wind drag relative to train = 1.49/1.45 = 1.03.
Wind force $\approx$ 1.03 $\times$ 28 $\times$ 1310 lb $\approx$ 17 tons acting $\approx$ 14 ft above bottom of girder.

Combined Load
   59 tons acting 12.5 ft above bottom of girder.

Single Pier

   Height of one column $\approx$ 77 ft.
   Diameter of column $\approx$ 1.5 ft.
   Area of one column $\approx$ 116 sq ft.

(a) *Windward Truss*
   Area of windward truss $\approx$ 3 × 116 = 348 sq ft.
   Solidity ratio $\approx$ 0.35.
   Drag coefficient = 0.8 (BS 5400: solidity ratio = 0.35; round members at high
      wind speed).
   Wind drag relative to train = 0.8/1.45 = 0.55.
   Wind force = 0.55 × 28 × 348 lb $\approx$ 2.4 tons acting $\approx$ 39 ft above base of pier.

(b) *Leeward Truss*
   Spacing ratio of trusses $\approx$ 1.
   Shielding factor = 0.70 (BS 5400: solidity ratio = 0.35; spacing ratio < 1).
   Drag coefficient = 0.70 × 0.8 = 0.56.
   Wind drag relative to train = 0.56/1.45 = 0.39.
   Wind force = 0.39 × 28 × 348 lb $\approx$ 1.7 tons acting $\approx$ 39 ft above base of pier.

Combined Load
   $\approx$ 4 tons acting $\approx$ 39 ft above base of pier.

CHAPTER 28

THE TAY BRIDGE—2

The Wind Loading on a Pier

We saw in Chapter 27 that the lateral wind force on a 245-ft length of the High Girders containing the train was ≈ 59 tons. The lateral force on an empty 245-ft girder would obviously have been significantly less: the wind resistance of the superstructure would have been greatest when the train was running through it. The lateral force on the superstructure in turn applies a lateral force to the top of each pier. We can put an upper bound of ≈ 59 tons on this force by considering a piece of superstructure 245 ft long containing the train and balanced on a single pier, as shown in Fig. 28.1. We can get a *lower* bound on the wind force by working out the wind loading on the *continuous* length of superstructure between piers 28 and 33 and dividing this by the number of piers. In practice the maximum lateral force on a pier will be somewhat less than 59 tons: away from the expansion joints the superstructure has a finite bending stiffness and this spreads the additional wind loading generated by the train over more than one pier.

Stability of the Bridge

We can see from Fig. 27.3 that the iron columns in each pair were bolted down to a massive sandstone plinth weighing about 150 tons. It seems that this plinth simply rested on the foundation tower which in turn rested on the bed of the firth. If the piers had been sufficiently strong and the wind force had been great enough the whole bridge would have rolled off the foundation towers. As we saw in Chapter 27, this did not happen—the plinths were left standing on the foundations. In fact this observation can be used to provide an upper bound for the lateral force on the bridge.

Before looking at the stability of the bridge we must first check out the stability of the superstructure. Figure 28.2 shows the forces acting on a 245-ft girder containing the train. The total weight of the unit is 350 tons. The wind force needed to make it topple over is found by taking moments about the leeward support. The answer is 196 tons: this is much more than our estimate of 59 tons and there seems little chance that the wind could have rolled the girders off the piers. Could the wind have *pushed* the girders off? Again, this seems unlikely: even if we neglect the lateral shear strength of the attachments a coefficient of friction of 0.17 would have been enough to resist the estimated lateral force of 59 tons. Since the superstructure

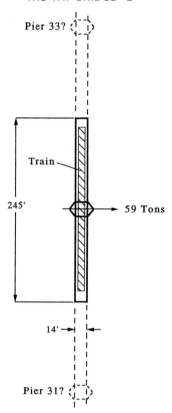

Pier 33?

Train

245'

59 Tons

14'

Pier 31?

FIG 28.1   The worst-case loading on a pier. Dimensions in feet.

was stable we can treat the bridge as a single free body for the purposes of calculating its stability.

The stability calculation is summarised in Fig. 28.3. The estimated weight of a 245-ft girder plus train, iron pier and sandstone plinth is 620 tons. In addition to the lateral wind force on the superstructure there is the lateral wind force on the pier itself: as we saw in Chapter 27 this acts about half-way up the pier and has a value of 4 tons for every 59 tons delivered by the superstructure. The bridge starts to topple when the wind is sufficiently strong to apply a lateral force of 83 tons to the girder plus train. This is 24 tons more than the force of 59 tons which the loaded superstructure would have delivered when the least stable vehicle (carriage no. 138) was about to topple.

Table 27.3 shows that the most stable carriage in the train (no. 414) would have turned over when the wind pressure on it was 41 lb/sq ft. By comparison carriage no. 138 would have capsized at 28 lb/sq ft. If the wind pressure on the train had been 41 lb/sq ft the lateral force delivered by the girder plus train would have been $(41/28) \times 59 = 86$ tons. This means that, if the train had been composed entirely of carriages like no. 414, a strong enough wind would have toppled the whole bridge over *before* the train would have become unstable. As we can see from Chapter 27, to get a pressure of 41 lb/sq ft on the train we need a mean wind speed of 63 mph.

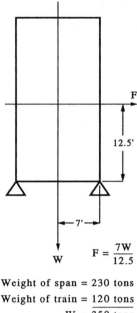

$$F = \frac{7W}{12.5}$$

Weight of span = 230 tons
Weight of train = 120 tons
W = 350 tons

FIG 28.2   Finding the wind force needed to capsize the girder. Dimensions in feet.

Table 27.1 shows that this wind speed is achieved at the top end of the force 10 band. The conclusion is that, even if the piers had been strengthened, a force 10 gale could well have rolled the bridge off its foundations without first derailing a train composed of the large six-wheeled carriages.

Plastic Collapse of the Piers

How did the piers fail? It is difficult to be certain in our diagnosis without making a thorough examination of the wreckage—and this is scarcely possible in the present case. But much progress can be made by assuming a reasonable collapse mechanism which is consistent with the overall mode of failure and the likely deformation behaviour of the various parts of the structure. Now the pier is a statically indeterminate structure with a large degree of redundancy. In order to carry out an elastic analysis we would need to know the precise fits between all the components and even then we would probably have to run the problem on a computer. As it is, one of the main features of the structure is that the fits were *not* reproducible or quantifiable: there were no specifications for loading up the bracing bars; the jibs worked loose in service; some were re-tightened but others were not; some of the snugs were cracked and some had come loose; the flange bolts were not tightened up evenly; and so on. In such a case the obvious way to proceed is to use *plastic* analysis: we set up a plausible *collapse mechanism* in which the active parts of the structure deform by *plastic deformation*. The attraction of this method is that the structure *shakes down*. If part of the structure starts off by carrying more

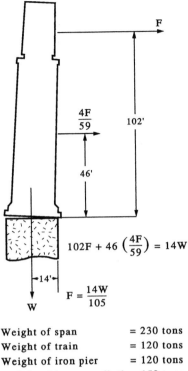

than its fair share of the load (an over-tightened bracing bar for example) then it will deform by plastic deformation, shedding load to other parts of the structure. Eventually, when all the active components undergo a big enough plastic strain, the loads in the various elements of the structure will be evenly distributed and can be related to the overall collapse load through the equations of static equilibrium.

In order to analyse the collapse we divide the pier into two independent structures. The first is shown in Fig. 28.4. In this, each vertical column is assumed to behave as a rigid bar fixed to a hinge at top and bottom. The diagonal bracing bars have been removed. If the hinges were free the structure would shear as a pin-jointed assembly: the top plinth would move horizontally relative to the bottom plinth. In practice the hinges are *not* free: they are the bolted joints at the ends of the columns. As shown in Example 28.1 the plastic moment M needed to make the hinge rotate is ≈ 126 ton ft. The shear force F_M which must be applied to the top plinth to make all six columns rotate is easily found from statics. The appropriate equation is given in Fig. 28.4; since $h = 77$ ft, $F_M \approx 20$ tons.

The second structure is shown in Fig. 28.5. As before, each vertical column is assumed to behave as a rigid bar. However, the structure is pin-jointed throughout and the resistance to shear is provided entirely by the diagonal bracing bars

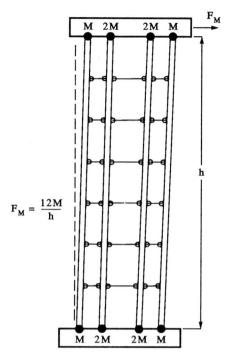

FIG 28.4   Finding the shear force needed to create fully-plastic moments at the ends of
the columns.

acting in tension. We have ignored the resistance offered by the bracing bars in compression. As shown in Example 27.1, they buckle under an axial compressive load of ≈ 0.5 ton which is small compared to their strength in tension. The tension T in the tie is assumed to be the same in all parts of the structure and is taken to be equal to the yield load of a bracing bar in tension. Figure 28.6 shows how the tensions in the ties are resolved into the direction of the applied shear force F_T. As shown in Fig. 28.5, $F_T = 2.7T$. As we will see later, the main length of a bracing bar yields in tension at about 44 tons. Thus $F_T \approx 2.7 \times 44 \approx 120$ tons.

The lateral force needed to deform the whole structure in accordance with our fully plastic collapse mechanism is given by

$$F = F_M + F_T \approx 20 \text{ tons} + 120 \text{ tons} = 140 \text{ tons}. \tag{28.1}$$

The lateral force which is *actually available* has two components. The main one is the force of ≈ 59 tons from the girder plus train. The second comes from the wind loading on the pier itself. As we have already seen this is 4 tons acting half-way up the pier: it is equivalent to 2 tons acting at the level of the top plinth. The total force is therefore ≈ 61 tons. Actually this figure is deceptively precise; it would be more sensible to say that the lateral force was probably ≈ 60 tons with a chance that it might have been as high as ≈ 80 tons if the wind had suddenly become strong enough to capsize the lighter parts of the train inside the High Girders. These figures are a good deal less than the calculated value of 140 tons, so what has gone wrong?

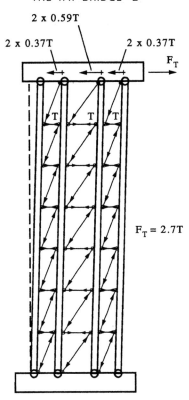

FIG 28.5   Finding the shear force needed to generate plastic flow in the diagonal bracing bars.

Deformation and Fracture of a Bracing Bar

In fact the deformation behaviour of a bracing bar is not nearly as simple as we have assumed. Although the main section of the bar should give a large plastic extension at a load of ≈ 44 tons the behaviour of the end attachments is actually likely to be the determining factor. Figure 28.7 shows the probable ways in which the components of the bracing system deform and fracture. The corresponding loads are listed in Table 28.1.

Mechanical Properties of Wrought Iron

Tensile strength $\sigma_{TS} \approx 20$ tsi (Thomas).
Yield strength $\sigma_y \approx 0.56\sigma_{TS}$ (Pugsley) ≈ 11 tsi.
Shear yield stress $k = \sigma_y/1.732$ (Appendix 1C, Von Mises) ≈ 6 tsi.
Shear failure stress $k_u \approx \sigma_{TS}/1.6$ (Appendix 1C) ≈ 12.5 tsi.
Unconstrained indentation pressure $\approx \sigma_y \approx 11$ tsi.

TABLE 28.1 *Failure of a Diagonal Bracing Bar under Axial Tension*

Failure mechanism	Area of fracture surface (sq in)	Force to give yield (tons)	Force to give fracture (tons)
Tear out from slot	$2(\sqrt{2} \times 1.5 \times 1) = 4.2$	25	53
Tensile from slot	$2(1.5 \times 1) = 3.0$	33	60
Tensile in bar	$1 \times 4 = 4.0$	44	80
Indentation of gib and end of slot	$\approx 1 \times 1 = 1.0$	≈ 11	–
Indentation of bolt and hole	$\approx 1.5 \times 1 = 1.5$	≈ 17	–
Tensile from hole	$2(1.5 \times 1) = 3.0$	33	60
Tear out from hole	$2(\sqrt{2} \times 1.5 \times 1) = 4.2$	25	53
Double shear of gib	$2(2.5 \times 1) = 5.0$	30	63
Double shear of bolt	$2(\pi 0.75^2) = 3.5$	21	44

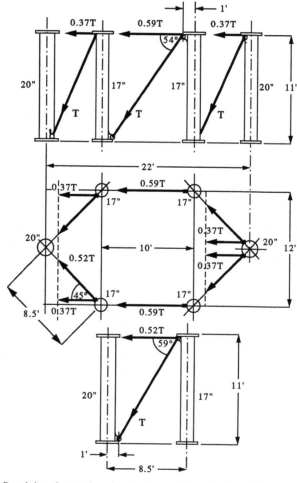

FIG 28.6 Resolving the tensions in the diagonal bracing bars. Dimensions in feet.

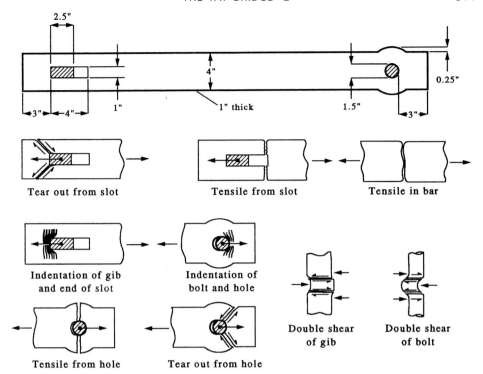

FIG 28.7   Mechanisms of deformation and fracture in a bracing bar under axial tension.
Dimensions in inches.

The table shows some interesting features. The load at which the main length of the bar yields in tension is comparable to the load at which the attachment bolt *fractures* in double shear. Our values for σ_y and k_u are only estimates and it is quite possible that the bolts can fracture when the main length of the bar is still elastic. The shear offset needed to shear a bolt is probably no more than ≈ 0.5 in. This displacement is only 0.3% of the length of a bracing bar and the ductility of the whole system is very small. Actually, deformation starts at a load of only ≈ 11 tons where the head of the gib bears on the end of the slot. At ≈ 17 tons flow starts where the bolt contacts its hole and at ≈ 21 tons the bolt itself starts to deform in double shear. Both slot and hole deform by tear-out shear at ≈ 25 tons, and yield in tension at ≈ 33 tons. These mechanisms all contribute to the plastic extension of the bar. However, the lengths of the deforming elements are small and their contribution to the overall ductility of the bar is not great.

Deformation and Fracture of a Snug

The cast-iron snugs to which the bracing bars were secured are shown in Fig. 27.4. In order to pull a single snug away from the column we have to produce a tensile fracture surface ≈ 20 sq in. in area. If the strength of the casting is ≈ 10 tsi (as in a Grade 10 cast iron) then the snug should withstand a tensile load of ≈ 200

tons. The Court of Inquiry commissioned a series of tests in which unbroken columns were recovered from the bridge and individual snugs were loaded in tension exactly as they would have been in service. A total of 14 separate tests was carried out: the fracture loads ranged from 20.5 tons to 27.9 tons; the average strength was 24.1 tons. The snugs were far weaker than they should have been, confirming that the castings were seriously defective.

If the snugs could all be loaded simultaneously to their breaking load then $F_T \approx 2.7 \times 24 = 65$ tons. The corresponding shear strength of the pier is $F \approx 20$ tons + 65 tons = 85 tons. However, the bracing bars will exhibit a negligible plastic extension at 24 tons and there will be no shakedown. If we assume that in practice only half of the ties carried the load then $F \approx 20$ tons $+ \frac{1}{2} \times 65$ tons = 53 tons. This value is comfortably below the likely wind loading of 60 tons.

Failure Mechanism

The failure was almost certainly caused by the poor tensile strength of the cast-iron snugs combined with the absence of gross yielding in the wrought-iron bracing bars. To begin with, a number of bars would have been exposed to a tensile load which was enough to break their snugs. When these ties failed they would have transferred their load to other ties in the structure and a "domino" effect of sequential failures would have followed, leading to total collapse.

It is interesting to note that the failure could probably have been avoided by making the bracing bars *weaker*: if we halve the cross-sectional area of the main length of each bar then general yield will take place at ≈ 20 tons. Shakedown occurs before the snugs can snap and the pier should take a force of about 20 tons + 2.7 × 20 tons = 74 tons.

Increasing the strength of the snugs to the "theoretical" value of ≈ 200 tons does little to strengthen the pier. As we have seen, the securing bolts are likely to fail before the main length of the bar can yield; shakedown is unlikely and the structure will be weakened as a result. If, as before, we assume that only half of the ties carry the load then $F \approx 20$ tons $+ \frac{1}{2} \times 120$ tons = 80 tons. The maximum strength of 140 tons can only be developed if the ties are re-designed to make sure that the main length of the bar will yield well before the end fittings fail.

Implications for Bridge Design

The collapse of the Tay Bridge sent shock waves through the engineering community. The immediate reaction was to cancel Bouch's next major project for a bridge across the Forth and to build a replacement bridge across the Tay. The new bridge carried a double railway track and needed much wider girders and piers to support the greater width of permanent way. The main piers were fabricated from iron plates in the form of a massive tubular arch and had an ample resistance to lateral wind forces. The new bridge runs parallel to the course of the old one and the stumps of Bouch's piers can still be seen from a passing train. Many of the girders from the old bridge were incorporated into the structure of its successor and they survive as a fitting memorial to a great but ill-fated enterprise.

When the Forth was eventually bridged in 1890 it marked a new dimension in bridge construction. The main crossing is 5330 ft long and has a headroom above high water of fully 157 ft. It consists of three huge double cantilevers fabricated from steel with a maximum height above high water of 361 ft. The bridge contains 58,000 tons of steel of which 4200 tons are just rivets! The steelwork has an external area of 145 *acres* and it is a full-time job for a gang of twenty-nine painters to protect the structure against corrosion. The Forth bridge was as massive as the old Tay bridge was slender; its designer—Sir Benjamin Baker—was taking no chances.

Such over-design could not be sustained for long and bridge designers gradually pared back their margins of safety. There is elegance and economy in having the lightest structure compatible with function. But history has a habit of repeating itself. In 1940 a new suspension bridge with a central span of 2800 ft was built over the Tacoma Narrows in the United States. It was soon noticed that the bridge deck was prone to oscillate in certain winds: the vertical amplitude of the oscillations was as much as 5 ft, but the bridge was never closed to traffic. Four months after the opening the deck went out of control in a 40 mph gale and literally shook itself to pieces. The bridge had been a victim of "flutter" which was caused by the inadequate torsional stiffness of the bridge deck. The designers might have avoided disaster if they had realised that the Old Chain Pier in Brighton, England had collapsed for the very same reason way back in 1836.

Moving on to the 1960s the new craze was the box-girder deck. For short spans this was simply supported on piers although longer spans could be suspended from central towers using cable stays. Problems started in June 1970 when a bridge over the river Cleddau in South Wales collapsed during erection, killing five people. As a span was being cantilevered out from a pier the bottom of the deck buckled and the whole span collapsed into the river. A similar incident occurred in November 1971 with a new bridge which was being built across the Rhine near Koblenz. Twelve people were killed in this mishap. Meanwhile, in Australia, a major disaster had occurred at the West Gate bridge in Melbourne. In this incident a span had just been erected on its piers. The deck had been fabricated with an incorrect camber and "kentledge" (large blocks of concrete) had been lowered onto the span to deform the bridge into the correct shape. Not surprisingly this had raised a buckle in the top of the deck. In an attempt to remove the buckle the bolts were removed from one of the transverse splice joints in the top of the deck. This lunatic course of action weakened the bridge to the extent that the deck could no longer support the bending moment generated by the self-weight at mid span and the whole structure descended to the ground, killing thirty-five people. Again, full-scale experiments had established the limits of a new form of construction with disastrous results.

References

C. Barlow, *The New Tay Bridge: A Course of Lectures Delivered at the Royal School of Military Engineering at Chatham, November 1888*, Spon, 1889.

British Standards Institution, BS 5400: 1978: "Steel, Concrete and Composite Bridges": Part 2: "Specification for Loads".

Department of the Environment, *Inquiry into the Basis of Design and Method of Erection of Steel Box Girder Bridges*, Interim Report, HMSO, 1971; Final Report, HMSO, 1973.

Government of Victoria, *Report of the Royal Commission into the Failure of the West Gate Bridge*, 1971.
R. Hammond, *The Forth Bridge and its Builders*, Eyre and Spottiswoode, 1964.
A. C. Palmer, *Structural Mechanics*, Oxford, 1976.
J. Prebble, *The High Girders*, Pan Books, 1959.
A. Pugsley, "Clifton Suspension Bridge", in *The Works of Isambard Kingdom Brunel*, edited by A. Pugsley, Cambridge University Press, 1980, p. 51.
C. J. Smithells, *Metals Reference Book*, 6th edition, Butterworth, 1984.
The Engineer, "The Tay Bridge", 1873, p. 197.
J. Thomas, *The Tay Bridge Disaster: New Light on the 1879 Tragedy*, David and Charles, 1972.

APPENDICES

ESSENTIAL DATA FOR PROBLEM SOLVING

In order to analyse a materials failure we almost always need extra information. For example, if we want to analyse the fast fracture of a pressure vessel we would hopefully know what the operating pressure of the vessel was, and the location and dimensions of the crack that triggered the failure. But we would need to know the stress equations that allow us to link the operating pressure to the tensile stress acting on the crack. And we would need to know how to go about finding a valid figure for the fracture toughness of the material.

This Appendix gives a collection of formulae and data which keep on cropping up in failure analysis. They can be regarded as a basic "toolkit" which can be supplemented when required by information from the specialised data sources to be found in most engineering libraries.

The Appendix is divided up into a number of separate sections. So there is a section on Stress and Strain, a section on Elastic Deformation, a section on Plastic Deformation, and so on. The presentation is deliberately concise and factual—this is not a book on stress analysis or plastic theory. The sections contain worked examples to illustrate how the formulae and data are handled. Finally, each section has a list of References giving suggestions for both background reading and for sources of more detailed information. Several sourcebooks require separate mention, however. These are:

R. J. Roark and W. C. Young, *Formulas for Stress and Strain*, 5th edition, McGraw-Hill, 1975.

This book, which runs to over 600 pages, is a mine of information and an essential tool for the materials, mechanical or civil engineer.

Metals Handbook, 9th edition, American Society for Metals, 1980.

This comprises sixteen substantial volumes, covering processing and fabrication; properties and selection; mechanical testing; metallography and fractography; and corrosion and failure analysis. This is an exceptionally informative collection of information, and is essential reference material for anyone involved with the application of engineering materials.

C. J. Smithells, *Metals Reference Book*, 6th edition, Butterworth, 1984.

This has much information on the properties of metals and alloys including a comprehensive set of phase diagrams.

G. W. C. Kaye and T. H. Laby, *Tables of Physical and Chemical Constants*, 14th edition, Longmans, 1973.

This has a lot of useful data for solids, liquids and gases.

M. F. Ashby and D. R. H. Jones, *Engineering Materials 1*, Pergamon, 1980.
M. F. Ashby and D. R. H. Jones, *Engineering Materials 2*, Pergamon, 1986.

These two books contain data for metals, ceramics, polymers and composites. They are also assumed background reading for the present book.

Summary of Contents

Section A: Stress and Strain

General Three-dimensional Stress System: stress tensor; principal stresses; examples of simple stress states; plane stress.

Finding the Principal Stresses: diagonalisation; example—pure shear.

Finding the Principal Directions: direction cosines; example—pure shear.

General Three-dimensional Strain System: strain tensor; principal strains; examples of simple strain states; plane strain.

Elastic Stress–Strain Relations for Isotropic Materials: definitions of elastic moduli; relations between principal stresses and strains; relations between moduli; plane strain.

Thin Pressure Vessels (elastic or plastic): equations for principal stresses at operating pressure.

Thick Pressure Vessels (elastic): equations for principal stresses and changes in diameter.

Thermal Stress (elastic): thermal strains; thermal stresses.

Elastic Stress Concentration Factors (SCFs): round hole in infinite plate; shouldered shaft in tension; shouldered shaft in bending.

References.

Section B: Elastic Deformation

Bending of Beams: general equations for curvature and stress; deflections of beams (six cases).

Second Moments of Area: definition; common cases; parallel-axis theorem; examples.

Critical Whirling Speeds/Mode 1 Natural Vibration Frequencies: six cases for shafts or beams.

Buckling of Struts: equations for three cases; example.

Buckling Under External Pressure: tubes; spheres; spherical caps.

Trunnion Loading: peripherally supported circular disc with central boss subjected to twisting moment.

References.

Section C: Plastic Deformation

Plastic Stress and Strain: Von Mises and Tresca yield criteria; equivalent stress and strain; Levy–Mises equations; examples.

Bending of Beams: stress distribution; fully plastic moments: definition and common cases.

Shearing Torques: common cases.

Plastic Buckling: substituting the tangent modulus for Young's modulus in the elastic buckling equations.

Mechanical Property Correlations: relations between: tensile strength, Vickers hardness, Brinell hardness and Rockwell hardness; yield strength, Vickers hardness and Brinell hardness; shear strength and tensile strength.

Plastic Constraint Factors: flat indenter, double edge notch, centre notch; effect of notch depth and included angle.

Ductile Fracture Surfaces: in tension, shear and torsion.

References.

Section D: Creep Deformation

Creep Strain: equivalent tensile and shear stresses; Levy–Mises equations for strain rate; equivalent tensile and shear strain rates; constitutive equations for power-law and diffusion creep; examples.

Deformation-Mechanism Maps: description of maps; identification of creep mechanisms; effects of metallurgical variables (grain size, cold work, solid solution strengthening, precipitation hardening); applications and limitations.

Fracture-Mechanism Maps: description of maps; identification of fracture mechanisms; effect of solid solutions and precipitates; applications and limitations.

References.

Section E: Fracture and Fatigue

Linear Elastic Fracture Mechanics: fast fracture equation; values of stress intensity factors (nine cases); crack-tip elastic stress field; geometry of crack-tip plastic zone; validity of plane-strain LEFM; effect of material thickness; geometry of fast fracture surfaces.

Charpy V-Notch Impact Test: details of test; ductile-brittle transition in ferritic steels; correlation between Charpy energy and fracture toughness.

Fatigue Strengths of Welds: fatigue curve for 97.5% survival; fatigue curve for 50% survival; classes of welds on surface of member; classes of welds at end connections of member.

References.

Section F: Corrosion Data

Standard Electrode Potentials: table of potentials for common anodic and cathodic reactions.

Rates of Uniform Corrosion: definition of mils per year; formulae for penetration rates from weight loss and current density; table of atomic weights and densities for metals.

Electrochemical Equilibrium Diagrams: description and limitations of diagrams; diagrams for pure copper, lead, iron, zinc, titanium, aluminium and magnesium; cathodic reactions involving hydrogen and oxygen.

Reference.

Section G: Heat Treatment of Steels

Hardness of Martensite: diagram for hardness as a function of carbon content; hardening effect of other alloying elements.

Hardenability: table of diameter of an oil-quenched bar to give 100% martensite for typical carbon, carbon–manganese and low-alloy steels.

Phases in Stainless Steels: Schaeffler diagram.

References.

Section H: Thermodynamics Data

Steam Table: data from the triple point to the critical point.

Saturation Tables: refrigerant-12; methyl chloride; ammonia; propane; butane.

References.

SECTION A: STRESS AND STRAIN

General Three-dimensional Stress System

Right-handed set of orthogonal axes labelled 1, 2, 3 (see top of page 321).
Stress components σ_{ij} where i = 1, 2, 3 and j = 1, 2, 3.
i represents the direction of the normal to the plane on which a stress component acts.
j represents the direction in which the stress component acts.
A component with i = j is a normal stress.
A component with i ≠ j is a shear stress.

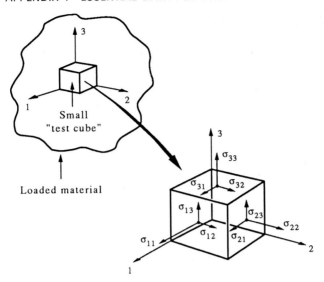

The stress tensor is

$$\sigma_{ij} = \begin{pmatrix} \sigma_{11} & \sigma_{12} & \sigma_{13} \\ \sigma_{21} & \sigma_{22} & \sigma_{23} \\ \sigma_{31} & \sigma_{32} & \sigma_{33} \end{pmatrix}.$$

For static equilibrium $\sigma_{21} = \sigma_{12}$, $\sigma_{31} = \sigma_{13}$, $\sigma_{32} = \sigma_{23}$. Thus tensor is symmetrical about leading diagonal.

The test cube can always be rotated into one particular orientation where all the shear components vanish.

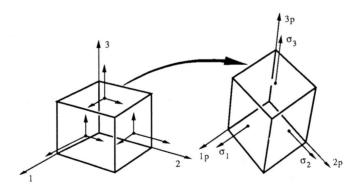

Then:

Tensile components σ_{11}, σ_{22}, σ_{33} become principal stresses σ_1, σ_2, σ_3.
Axes 1, 2, 3 become principal directions 1_p, 2_p, 3_p.
Cube planes become principal planes.

The principal stress tensor is

$$\sigma_p = \begin{pmatrix} \sigma_1 & 0 & 0 \\ 0 & \sigma_2 & 0 \\ 0 & 0 & \sigma_3 \end{pmatrix}.$$

Property of transformation: $\sigma_{11} + \sigma_{22} + \sigma_{33} = \sigma_1 + \sigma_2 + \sigma_3$.

EXAMPLES

UNIAXIAL TENSION

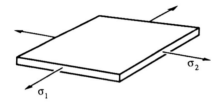

$$\sigma_p = \begin{pmatrix} \sigma_1 & 0 & 0 \\ 0 & 0 & 0 \\ 0 & 0 & 0 \end{pmatrix}.$$

BIAXIAL TENSION

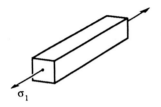

$$\sigma_p = \begin{pmatrix} \sigma_1 & 0 & 0 \\ 0 & \sigma_2 & 0 \\ 0 & 0 & 0 \end{pmatrix}.$$

HYDROSTATIC PRESSURE

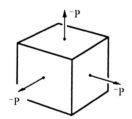

$$\sigma_p = \begin{pmatrix} -p & 0 & 0 \\ 0 & -p & 0 \\ 0 & 0 & -p \end{pmatrix}.$$

TRIAXIAL TENSION

$$\sigma_p = \begin{pmatrix} \sigma_1 & 0 & 0 \\ 0 & \sigma_2 & 0 \\ 0 & 0 & \sigma_3 \end{pmatrix}.$$

When any one of $\sigma_1, \sigma_2, \sigma_3$ is zero we have a condition called *plane stress*.

Finding the Principal Stresses

Initial stress state

$$\sigma_{ij} = \begin{pmatrix} \sigma_{11} & \sigma_{12} & \sigma_{13} \\ \sigma_{21} & \sigma_{22} & \sigma_{23} \\ \sigma_{31} & \sigma_{32} & \sigma_{33} \end{pmatrix}.$$

Form the determinant

$$\begin{vmatrix} (\sigma_{11} - \lambda) & \sigma_{12} & \sigma_{13} \\ \sigma_{21} & (\sigma_{22} - \lambda) & \sigma_{23} \\ \sigma_{31} & \sigma_{32} & (\sigma_{33} - \lambda) \end{vmatrix} = 0.$$

Then the three values of λ given by this equation give the values of $\sigma_1, \sigma_2, \sigma_3$, with

$$\sigma_p = \begin{pmatrix} \sigma_1 & 0 & 0 \\ 0 & \sigma_2 & 0 \\ 0 & 0 & \sigma_3 \end{pmatrix}.$$

EXAMPLE

PURE SHEAR

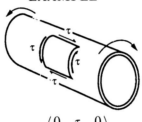

$$\sigma_{ij} = \begin{pmatrix} 0 & \tau & 0 \\ \tau & 0 & 0 \\ 0 & 0 & 0 \end{pmatrix}.$$

Then

$$\begin{vmatrix} (0 - \lambda) & \tau & 0 \\ \tau & (0 - \lambda) & 0 \\ 0 & 0 & (0 - \lambda) \end{vmatrix} = 0,$$

which gives the characteristic equation

$$\lambda^3 = \tau^2 \lambda.$$

The roots of this equation are

$$\lambda = \tau, -\tau, 0.$$

This gives

$$\sigma_p = \begin{pmatrix} \tau & 0 & 0 \\ 0 & -\tau & 0 \\ 0 & 0 & 0 \end{pmatrix}.$$

Finding the Principal Directions

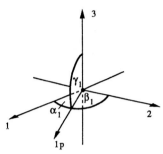

Direction cosines:

$$l_1 = \cos \alpha_1,$$

$$m_1 = \cos \beta_1,$$

$$n_1 = \cos \gamma_1.$$

Property:

$$l_1^2 + m_1^2 + n_1^2 = 1.$$

Solve the equations:

$$l_1(\sigma_{11} - \sigma_1) + m_1\sigma_{12} + n_1\sigma_{13} = 0,$$

$$l_1\sigma_{21} + m_1(\sigma_{22} - \sigma_1) + n_1\sigma_{23} = 0,$$

$$l_1\sigma_{31} + m_1\sigma_{32} + n_1(\sigma_{33} - \sigma_1) = 0,$$

$$l_1^2 + m_1^2 + n_1^2 = 1,$$

to give l_1, m_1, n_1 and hence $\alpha_1, \beta_1, \gamma_1$.

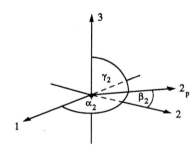

Solve:

$$l_2(\sigma_{11} - \sigma_2) + m_2\sigma_{12} + n_2\sigma_{13} = 0,$$

$$l_2\sigma_{21} + m_2(\sigma_{22} - \sigma_2) + n_2\sigma_{23} = 0,$$

$$l_2\sigma_{31} + m_2\sigma_{32} + n_2(\sigma_{33} - \sigma_2) = 0,$$

$$l_2^2 + m_2^2 + n_2^2 = 1,$$

to give l_2, m_2, n_2 and hence $\alpha_2, \beta_2, \gamma_2$.

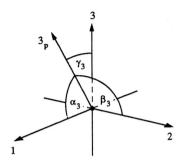

Solve:

$$l_3(\sigma_{11} - \sigma_3) + m_3\sigma_{12} + n_3\sigma_{13} = 0,$$
$$l_3\sigma_{21} + m_3(\sigma_{22} - \sigma_3) + n_3\sigma_{23} = 0,$$
$$l_3\sigma_{31} + m_3\sigma_{32} + n_3(\sigma_{33} - \sigma_3) = 0,$$
$$l_3^2 + m_3^2 + n_3^2 = 1.$$

to give l_3, m_3, n_3 and hence $\alpha_3, \beta_3, \gamma_3$.

EXAMPLE

PURE SHEAR

$$l_1(0 - \tau) + m_1\tau + n_10 = 0,$$
$$l_1\tau + m_1(0 - \tau) + n_10 = 0,$$
$$l_10 + m_10 + n_1(0 - \tau) = 0,$$
$$l_1^2 + m_1^2 + n_1^2 = 1.$$

The solutions of these equations are

$$l_1 = \frac{1}{\sqrt{2}}, \qquad m_1 = \frac{1}{\sqrt{2}}, \qquad n_1 = 0,$$

which give

$$\alpha_1 = 45°, \qquad \beta_1 = 45°, \qquad \gamma_1 = 90°.$$

This appears as follows:

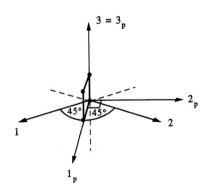

Because the transformation is a simple 45° rotation about axis 3 there is no need to work out $\alpha_2, \beta_2, \gamma_2$ or $\alpha_3, \beta_3, \gamma_3$ in this case.

SUMMARY OF TRANSFORMATION FOR PURE SHEAR

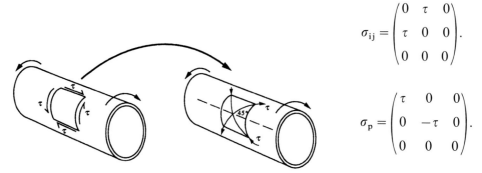

$$\sigma_{ij} = \begin{pmatrix} 0 & \tau & 0 \\ \tau & 0 & 0 \\ 0 & 0 & 0 \end{pmatrix}.$$

$$\sigma_p = \begin{pmatrix} \tau & 0 & 0 \\ 0 & -\tau & 0 \\ 0 & 0 & 0 \end{pmatrix}.$$

This transformation is a good example of the rule that planes of maximum shear stress always lie at 45° to principal planes.

General Three-dimensional Strain System

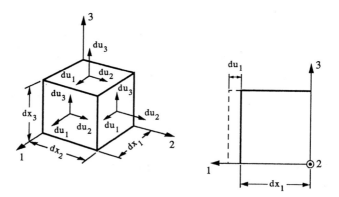

Strain components ε_{ij} where $i = 1, 2, 3$ and $j = 1, 2, 3$.
A component with $i = j$ is an axial strain.
The three axial strains are:

$$\varepsilon_{11} = \frac{du_1}{dx_1}, \qquad \varepsilon_{22} = \frac{du_2}{dx_2}, \qquad \varepsilon_{33} = \frac{du_3}{dx_3}.$$

A component with $i \neq j$ is a shear strain.
Example of how a component of shear strain is defined:

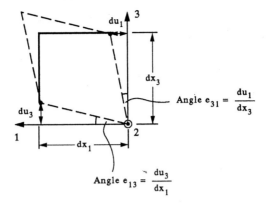

$$\varepsilon_{13} = \varepsilon_{31} = \tfrac{1}{2}(e_{13} + e_{31}) = \frac{1}{2}\left(\frac{du_3}{dx_1} + \frac{du_1}{dx_3}\right).$$

Thus the strain tensor is:

$$\varepsilon_{ij} = \begin{pmatrix} \varepsilon_{11} & \varepsilon_{12} & \varepsilon_{13} \\ \varepsilon_{21} & \varepsilon_{22} & \varepsilon_{23} \\ \varepsilon_{31} & \varepsilon_{32} & \varepsilon_{33} \end{pmatrix}$$

$$= \begin{pmatrix} \dfrac{du_1}{dx_1} & \dfrac{1}{2}\left(\dfrac{du_1}{dx_2} + \dfrac{du_2}{dx_1}\right) & \dfrac{1}{2}\left(\dfrac{du_1}{dx_3} + \dfrac{du_3}{dx_1}\right) \\ \dfrac{1}{2}\left(\dfrac{du_2}{dx_1} + \dfrac{du_1}{dx_2}\right) & \dfrac{du_2}{dx_2} & \dfrac{1}{2}\left(\dfrac{du_2}{dx_3} + \dfrac{du_3}{dx_2}\right) \\ \dfrac{1}{2}\left(\dfrac{du_3}{dx_1} + \dfrac{du_1}{dx_3}\right) & \dfrac{1}{2}\left(\dfrac{du_3}{dx_2} + \dfrac{du_2}{dx_3}\right) & \dfrac{du_3}{dx_3} \end{pmatrix}.$$

Note that the tensor is symmetrical about the leading diagonal.

The test cube can always be rotated into one particular orientation where all the shear components vanish. Then:

Axial components $\varepsilon_{11}, \varepsilon_{22}, \varepsilon_{33}$ become principal strains $\varepsilon_1, \varepsilon_2, \varepsilon_3$.
Axes 1, 2, 3 become principal directions $1_p, 2_p, 3_p$.
In an isotropic material the principal strain directions are the same as the principal stress directions.
The principal strain tensor is

$$\varepsilon_p = \begin{pmatrix} \varepsilon_1 & 0 & 0 \\ 0 & \varepsilon_2 & 0 \\ 0 & 0 & \varepsilon_3 \end{pmatrix}.$$

Property of transformation: $\varepsilon_{11} + \varepsilon_{22} + \varepsilon_{33} = \varepsilon_1 + \varepsilon_2 + \varepsilon_3$.
When any one of $\varepsilon_1, \varepsilon_2, \varepsilon_3$ is zero we have a condition called *plane strain*.
The *dilatation*, $\Delta = \varepsilon_{11} + \varepsilon_{22} + \varepsilon_{33} = \varepsilon_1 + \varepsilon_2 + \varepsilon_3$.

EXAMPLES

UNIAXIAL TENSION (ELASTIC)

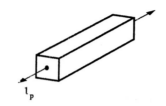

$$\varepsilon_p = \begin{pmatrix} \varepsilon_1 & 0 & 0 \\ 0 & -\nu\varepsilon_1 & 0 \\ 0 & 0 & -\nu\varepsilon_1 \end{pmatrix} \quad \text{and} \quad \varDelta = \varepsilon_1(1 - 2\nu).$$

HYDROSTATIC TENSION

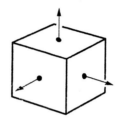

$$\varepsilon_p = \begin{pmatrix} \varepsilon & 0 & 0 \\ 0 & \varepsilon & 0 \\ 0 & 0 & \varepsilon \end{pmatrix} \quad \text{and} \quad \varDelta = 3\varepsilon.$$

ENGINEERING SHEAR

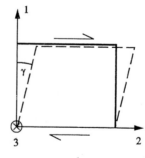

$$e_{12} = \gamma \quad \text{and} \quad e_{21} = 0.$$

Thus

$$\varepsilon_{12} = \varepsilon_{21} = \tfrac{1}{2}(e_{12} + e_{21}) = \frac{\gamma}{2}.$$

$$\varepsilon_{ij} = \begin{pmatrix} 0 & \dfrac{\gamma}{2} & 0 \\[2mm] \dfrac{\gamma}{2} & 0 & 0 \\[2mm] 0 & 0 & 0 \end{pmatrix} \qquad \text{and} \qquad \Delta = 0.$$

Elastic Stress–Strain Relations for Isotropic Materials

$$\varepsilon = \frac{\sigma}{E}, \qquad \gamma = \frac{\tau}{G}, \qquad \Delta = -\frac{p}{K}.$$

$$\varepsilon_1 = \frac{\sigma_1}{E} - v\,\frac{\sigma_2}{E} - v\,\frac{\sigma_3}{E}.$$

$$\varepsilon_2 = \frac{\sigma_2}{E} - v\,\frac{\sigma_1}{E} - v\,\frac{\sigma_3}{E}.$$

$$\varepsilon_3 = \frac{\sigma_3}{E} - v\,\frac{\sigma_1}{E} - v\,\frac{\sigma_2}{E}.$$

EXAMPLE

FIND AN EQUATION RELATING E AND K.

Under hydrostatic pressure,

$$\varepsilon_1 = \varepsilon_2 = \varepsilon_3 = \varepsilon \qquad \text{and} \qquad \sigma_1 = \sigma_2 = \sigma_3 = -p.$$

$$\varepsilon_1 = \frac{\sigma_1}{E} - v\,\frac{\sigma_2}{E} - v\,\frac{\sigma_3}{E}$$

becomes

$$\varepsilon = -\frac{p}{E}(1 - 2v).$$

$$K = -\frac{p}{\Delta} = -\frac{p}{3\varepsilon} = \frac{1}{3\varepsilon} \times \frac{\varepsilon E}{(1 - 2v)} = \frac{E}{3(1 - 2v)}.$$

Thus

$$K = \frac{E}{3(1 - 2v)}.$$

Note:

$$\text{when } v = \tfrac{1}{3}, \qquad K = E.$$

EXAMPLE

FIND AN EQUATION RELATING E AND G.

The strain tensor for pure shear is:

$$\varepsilon_{ij} = \begin{pmatrix} 0 & \dfrac{\gamma}{2} & 0 \\ \dfrac{\gamma}{2} & 0 & 0 \\ 0 & 0 & 0 \end{pmatrix}.$$

To find the principal strains, form the determinant:

$$\begin{vmatrix} (0 - \lambda) & \dfrac{\gamma}{2} & 0 \\ \dfrac{\gamma}{2} & (0 - \lambda) & 0 \\ 0 & 0 & (0 - \lambda) \end{vmatrix} = 0.$$

This gives the characteristic equation

$$\lambda^3 = \left(\dfrac{\gamma}{2}\right)^2 \lambda$$

which has roots

$$\lambda = \dfrac{\gamma}{2}, \; -\dfrac{\gamma}{2}, \; 0.$$

This gives

$$\varepsilon_p = \begin{pmatrix} \dfrac{\gamma}{2} & 0 & 0 \\ 0 & -\dfrac{\gamma}{2} & 0 \\ 0 & 0 & 0 \end{pmatrix}.$$

Remember that, for pure shear,

$$\sigma_p = \begin{pmatrix} \tau & 0 & 0 \\ 0 & -\tau & 0 \\ 0 & 0 & 0 \end{pmatrix}.$$

$$\varepsilon_1 = \dfrac{\sigma_1}{E} - v\dfrac{\sigma_2}{E} - v\dfrac{\sigma_3}{E}$$

then becomes

$$\dfrac{\gamma}{2} = \dfrac{\tau}{E} + v\dfrac{\tau}{E}.$$

This gives

$$\gamma = \dfrac{2\tau(1 + v)}{E}.$$

Finally,

$$G = \frac{\tau}{\gamma} = \frac{E}{2(1 + v)}.$$

Thus

$$G = \frac{\tau}{\gamma} = \frac{E}{2(1 + v)}.$$

Note:

$$\text{when } v = \tfrac{1}{3}, \qquad G = \frac{3E}{8}.$$

EXAMPLE

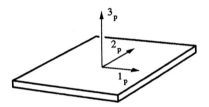

A plate is loaded in tension along direction 1_p but is prevented from contracting along direction 2_p. Find the effective elastic modulus in direction 1_p.

Now $\varepsilon_2 = 0$ (as specified) and $\sigma_3 = 0$ (free surface).

$$\varepsilon_2 = \frac{\sigma_2}{E} - v \frac{\sigma_1}{E} = 0, \qquad \text{so } \sigma_2 = v\sigma_1.$$

$$\varepsilon_1 = \frac{\sigma_1}{E} - v \frac{\sigma_2}{E} = \frac{\sigma_1}{E}(1 - v^2).$$

The required modulus is

$$E_1 = \frac{\sigma_1}{\varepsilon_1} = \frac{E}{(1 - v^2)}$$

which relates to a condition of *plane strain*.
Note: if $v = 0.33$, $E_1 = 1.12E$.

Thin Pressure Vessels (Elastic or Plastic)
(Internal pressure p)

TUBES ($r/t > 10$)

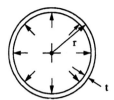
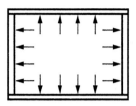

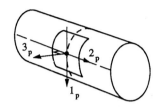

$$\sigma_1 = \frac{pr}{t}.$$

$$\sigma_2 = \frac{pr}{2t} \qquad \text{if end caps fitted; otherwise } \sigma_2 = 0.$$

$$\sigma_3 = 0.$$

SPHERES $(r/t > 10)$

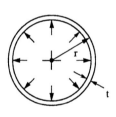
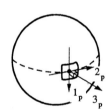

$$\sigma_1 = \sigma_2 = \frac{pr}{2t}.$$

$$\sigma_3 = 0.$$

(Equi-biaxial stress state).

CONES $(r/t > 10)$ (see top of page 333)

$$\sigma_1 = \left(\frac{pr}{t}\right) \times \frac{1}{(\cos \alpha)}.$$

$$\sigma_2 = \left(\frac{pr}{2t}\right) \times \frac{1}{(\cos \alpha)}.$$

$$\sigma_3 = 0.$$

For $\alpha > 0$, either an end cap or an equivalent force is required for equilibrium.

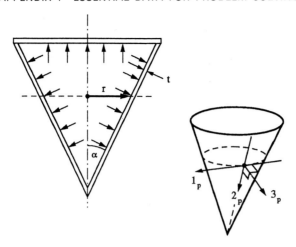

Note: The above results can be applied when the pressure is external provided the signs of the stresses are reversed. However, under external pressure, buckling will intervene if the pressure is high enough.

Thick Pressure Vessel (Elastic)
(Internal pressure p)

No End Caps

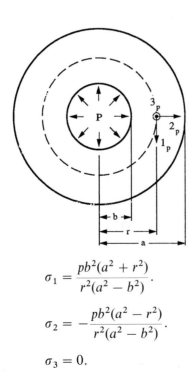

$$\sigma_1 = \frac{pb^2(a^2 + r^2)}{r^2(a^2 - b^2)}.$$

$$\sigma_2 = -\frac{pb^2(a^2 - r^2)}{r^2(a^2 - b^2)}.$$

$$\sigma_3 = 0.$$

Increase in external radius:

$$\Delta a = \left(\frac{p}{E}\right)\left(\frac{2ab^2}{a^2 - b^2}\right).$$

Increase in internal radius:

$$\Delta b = \left(\frac{p}{E}\right)b\left\{\left(\frac{a^2 + b^2}{a^2 - b^2}\right) + v\right\}.$$

When pressure is external, we have:

$$\sigma_1 = -\frac{pa^2(b^2 + r^2)}{r^2(a^2 - b^2)}.$$

$$\sigma_2 = -\frac{pa^2(r^2 - b^2)}{r^2(a^2 - b^2)}.$$

$$\sigma_3 = 0.$$

And:

$$\Delta a = -\left(\frac{p}{E}\right)a\left\{\left(\frac{a^2 + b^2}{a^2 - b^2}\right) - v\right\}.$$

$$\Delta b = -\left(\frac{p}{E}\right)\left(\frac{2a^2b}{a^2 - b^2}\right).$$

Thermal Stress (Elastic)

Thermal strain $= \alpha \Delta T$.
Thus,

$$\varepsilon_1 = \frac{\sigma_1}{E} - v\frac{\sigma_2}{E} - v\frac{\sigma_3}{E} + \alpha \Delta T.$$

$$\varepsilon_2 = \frac{\sigma_2}{E} - v\frac{\sigma_1}{E} - v\frac{\sigma_3}{E} + \alpha \Delta T.$$

$$\varepsilon_3 = \frac{\sigma_3}{E} - v\frac{\sigma_1}{E} - v\frac{\sigma_2}{E} + \alpha \Delta T.$$

EXAMPLE

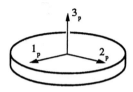

Derive an equation for calculating the surface stresses that are produced when a hot disc of a poor thermal conductor is dropped into cold water. Assume that the stresses are not relieved by cracking.

Since the surface layers are constrained by the bulk material, $\varepsilon_1 = 0$, $\varepsilon_2 = 0$. In addition, $\sigma_3 = 0$ (free surface). Thus

$$0 = \frac{\sigma_1}{E} - v\frac{\sigma_2}{E} + \alpha\Delta T,$$

$$0 = \frac{\sigma_2}{E} - v\frac{\sigma_1}{E} + \alpha\Delta T,$$

which gives $\sigma_1 = \sigma_2 = \sigma$ (equi-biaxial stress state). Then

$$0 = \frac{\sigma}{E} - v\frac{\sigma}{E} + \alpha\Delta T$$

which gives

$$\sigma = \frac{\alpha\Delta TE}{(1 - v)}.$$

Since ΔT in a cooling situation is negative, the stress is a tensile one.

EXAMPLE

Calculate the thermal stresses in a long tube subjected to a uniform radial flow of heat through the tube wall. The temperature of the outside of the tube is greater than the temperature of the inside by amount ΔT.

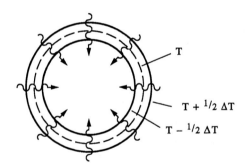

The stress state is equi-biaxial, as in the previous example. The situation is as follows:

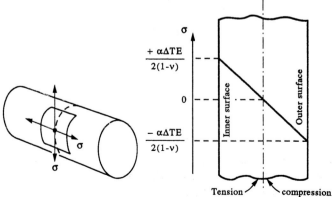

$$\sigma_{max} = \pm \frac{\alpha \Delta TE}{2(1 - v)}.$$

Elastic Stress Concentration Factors (SCFs)

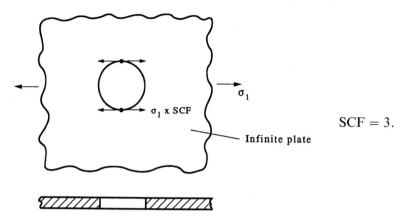

SCF $= 3$.

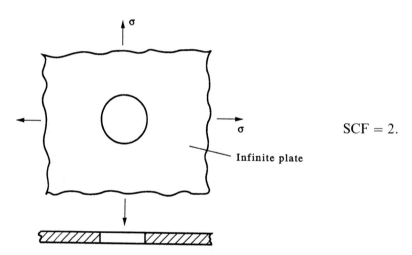

SCF $= 2$.

Stress all around hole circumference $= 2\sigma$.

TABLE OF SCF VALUES

$\dfrac{r}{d}$ / $\dfrac{D}{d}$	0.025	0.05	0.10	0.15	0.20	0.30
1.02	1.80	1.50	1.35	1.25	1.22	1.19
1.05	2.10	1.72	1.47	1.37	1.32	1.25
1.10	2.25	1.88	1.58	1.44	1.36	1.29
1.50		2.40	1.90	1.68	1.55	1.44

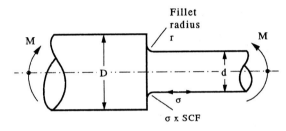

TABLE OF SCF VALUES

$\dfrac{r}{d}$ / $\dfrac{D}{d}$	0.025	0.05	0.10	0.15	0.20	0.30
1.02	1.90	1.64	1.43	1.34	1.24	1.20
1.05	2.13	1.79	1.54	1.40	1.31	1.23
1.10	2.25	1.86	1.59	1.43	1.37	1.26
1.50	2.59	2.06	1.67	1.50	1.40	1.29
3.00	2.85	2.30	1.80	1.58	1.43	1.32

References

M. F. Ashby and D. R. H. Jones, *Engineering Materials 1*, Pergamon, 1980.
G. E. Dieter, *Mechanical Metallurgy*, 2nd edition, McGraw-Hill Kogakusha, 1976.
J. F. Knott, *Fundamentals of Fracture Mechanics*, Butterworth, 1973.
R. J. Roark and W. C. Young, *Formulas for Stress and Strain*, 5th edition, McGraw-Hill Kogakusha, 1975.
F. R. Shanley, *Mechanics of Materials*, McGraw-Hill Kogakusha, 1967.
J. E. Shigley, *Mechanical Engineering Design*, 3rd edition, McGraw-Hill Kogakusha, 1977.

SECTION B: ELASTIC DEFORMATION

Bending of Beams

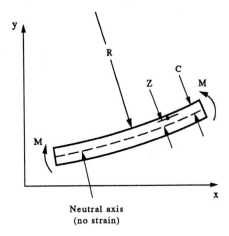

$$\sigma = \frac{Mz}{I}.$$

$$\sigma_{max} = \frac{Mc}{I}.$$

$$\frac{d^2y}{dx^2} = \frac{1}{R} = \frac{M}{EI}.$$

Neutral axis
(no strain)

CASE 1

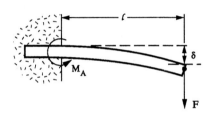

$$\delta = \frac{1}{3}\left(\frac{Fl^3}{EI}\right).$$

$$M_A = Fl.$$

CASE 2

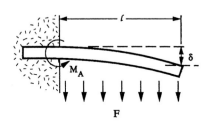

$$\delta = \frac{1}{8}\left(\frac{Fl^3}{EI}\right).$$

$$M_A = \frac{Fl}{2}.$$

CASE 3

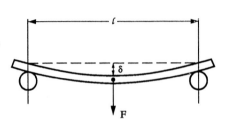

$$\delta = \frac{1}{48}\left(\frac{Fl^3}{EI}\right).$$

Maximum moment is

$\dfrac{Fl}{4}$ at mid span.

CASE 4

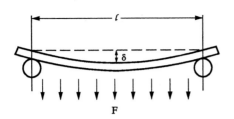

$$\delta = \frac{5}{384}\left(\frac{Fl^3}{EI}\right).$$

Maximum moment is

$\dfrac{Fl}{8}$ at mid span.

CASE 5

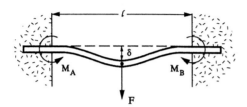

$$\delta = \frac{1}{192}\left(\frac{Fl^3}{EI}\right).$$

$$M_A = M_B = \frac{Fl}{8}.$$

CASE 6

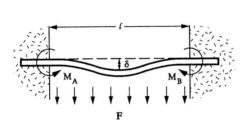

$$\delta = \frac{1}{384}\left(\frac{Fl^3}{EI}\right).$$

$$M_A = M_B = \frac{Fl}{12}.$$

Second Moments of Area

$$I = \int_A dA\, q^2.$$

$$I = \frac{a^4}{12}, \qquad A = a^2.$$

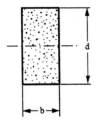

$$I = \frac{bd^3}{12}, \qquad A = bd.$$

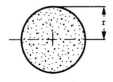

$$I = \frac{\pi r^4}{4}, \qquad A = \pi r^2.$$

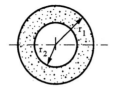

$$I = \frac{\pi}{4}(r_1^4 - r_2^4),$$

$$A = \pi(r_1^2 - r_2^2).$$

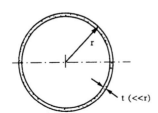

$$I = \pi r^3 t, \qquad A = 2\pi r t.$$

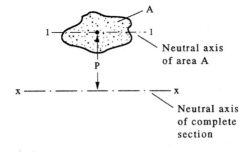

Neutral axis of area A

Neutral axis of complete section

$$I_x = I_1 + Ap^2$$

(Parallel axis theorem).

EXAMPLE

Calculate I_x for the following section (dimensions in mm).

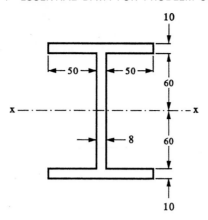

Divide section up as follows:

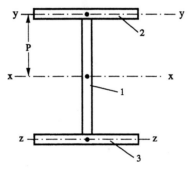

Then

$$I_x = I_x(1) + I_x(2) + I_x(3)$$

$$= I_x(1) + 2I_x(2)$$

$$= I_x(1) + 2\{I_y(2) + A(2)p^2\}$$

$$= \frac{b_1 d_1^3}{12} + 2\frac{b_2 d_2^3}{12} + 2b_2 d_2 p^2$$

$$= \frac{8 \times 120^3}{12} + \frac{2 \times 108 \times 10^3}{12} + (2 \times 108 \times 10 \times 65^2)$$

$$= (1.152 + 0.018 + 9.126) \times 10^6 \text{ mm}^4$$

$$= \underline{\underline{1.03 \times 10^7 \text{ mm}^4}}.$$

EXAMPLE

Siân wants to keep her large collection of history books in an alcove in her sitting room. The alcove is 1600 mm wide and you offer to install a set of pine shelves, each 1600 mm long, supported at their ends on small brackets screwed into the wall. You guess that shelves measuring 20 mm deep by 250 mm wide will work. However, professional integrity demands that you calculate the maximum stress and deflection first. For pine, $E \approx 10$ GPa and $\sigma_y \approx 60$ MPa. A book 40 mm thick weighs about 1 kg.

The appropriate equations are:

$$\delta = \frac{5}{384}\left(\frac{Fl^3}{EI}\right),$$

$$\sigma_{max} = \frac{Mc}{I},$$

$$I = \frac{bd^3}{12},$$

$$M = \frac{Fl}{8}.$$

$$I = \frac{250 \times 20^3}{12} \text{ mm}^4 = 1.67 \times 10^5 \text{ mm}^4.$$

$$F = \frac{1600 \text{ mm}}{40 \text{ mm}} \text{ kgf} = 40 \text{ kgf}.$$

$$M = \frac{40 \times 9.81 \text{ N} \times 1600 \text{ mm}}{8} = 7.85 \times 10^4 \text{ N mm}.$$

$$\sigma_{max} = \frac{7.85 \times 10^4 \text{ N mm} \times 10 \text{ mm}}{1.67 \times 10^5 \text{ mm}^4} = 4.7 \text{ N mm}^{-2} = \underline{\underline{4.7 \text{ MPa.}}}$$

$$\delta = \frac{5 \times 40 \times 9.81 \text{ N} \times 1600^3 \text{ mm}^3}{384 \times 10^4 \text{ N mm}^{-2} \times 1.67 \times 10^5 \text{ mm}^4} = \underline{\underline{12.5 \text{ mm.}}}$$

The shelves will not break but they will deflect noticeably. The deflection is just about acceptable aesthetically, but it may increase with time due to creep. One solution would be to turn the shelf over every six months. This will not be popular, however, and you are better advised to increase the thickness instead.

Critical Whirling Speeds/Mode 1 Natural Vibration Frequencies

$$f_{cr} = \frac{\omega}{2\pi}, \text{ where } f_{cr} \text{ is in revolutions/cycles per second.}$$

CASE 1

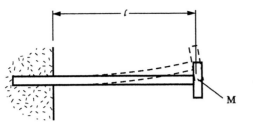

$$f_{cr} = 0.276 \sqrt{\frac{EI}{Ml^3}}.$$

CASE 2

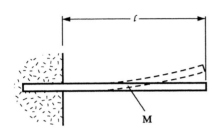

$$f_{cr} = 0.560 \sqrt{\frac{EI}{Ml^3}}.$$

CASE 3

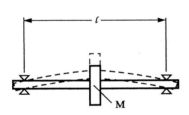

$$f_{cr} = 1.103 \sqrt{\frac{EI}{Ml^3}}.$$

CASE 4

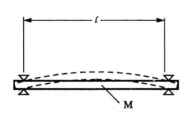

$$f_{cr} = 1.571 \sqrt{\frac{EI}{Ml^3}}.$$

CASE 5

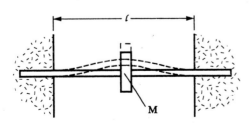

$$f_{cr} = 2.206 \sqrt{\frac{EI}{Ml^3}}.$$

CASE 6

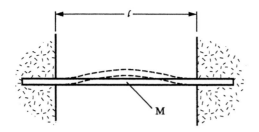

$$f_{cr} = 3.565 \sqrt{\frac{EI}{Ml^3}}.$$

Buckling of Struts

CASE 1

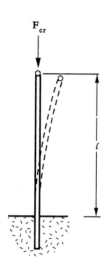

$$F_{cr} = 2.47\left(\frac{EI}{l^2}\right).$$

CASE 2

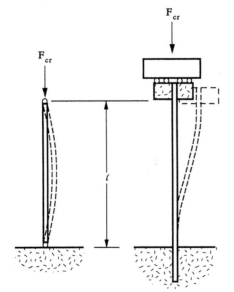

$$F_{cr} = 9.87\left(\frac{EI}{l^2}\right).$$

CASE 3

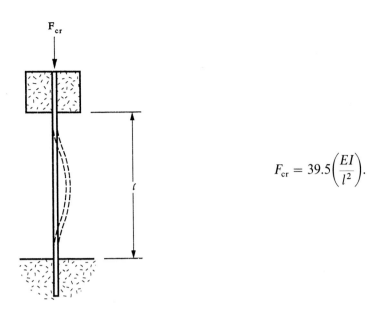

$$F_{cr} = 39.5\left(\frac{EI}{l^2}\right).$$

EXAMPLE

Uncle Albert has been given an elegant hickory walking stick for his 80th birthday. It is 750 mm long, 17 mm in diameter (being an engineer you always carry a pair of vernier calipers, of course) and has a curved handle which can be gripped quite firmly if necessary. He is very pleased with his stick, but you have your doubts that it can support his weight (90 kg on a good day). Do the calculations to see whether you are right. Assume that $E = 20$ GPa (2×10^4 N mm^{-2}).

Appropriate situation is probably Case 1, with

$$F_{cr} = 2.47\left(\frac{EI}{l^2}\right) \quad \text{and} \quad I = \frac{\pi r^4}{4},$$

which gives

$$F_{cr} = \frac{2.47\pi E}{4}\left(\frac{r}{l}\right)^2 r^2$$

$$= \frac{2.47\pi \times 2 \times 10^4 \text{ N mm}^{-2}}{4}\left(\frac{8.5 \text{ mm}}{750 \text{ mm}}\right)^2 8.5^2 \text{ mm}^2$$

$$= 360 \text{ N} = 37 \text{ kgf.}$$

Your doubts seem confirmed by this result. However, if he does not grip the handle very firmly, but just leans on it close to the axis of the stick, Case 2(a) is probably more appropriate, giving $F_{cr} = 148$ kgf. In reality, the best estimate is probably to halve this figure, giving $F_{cr} \approx 74$ kgf. So he should be reasonably safe.

Buckling Under External Pressure

TUBES $(r/t > 10)$

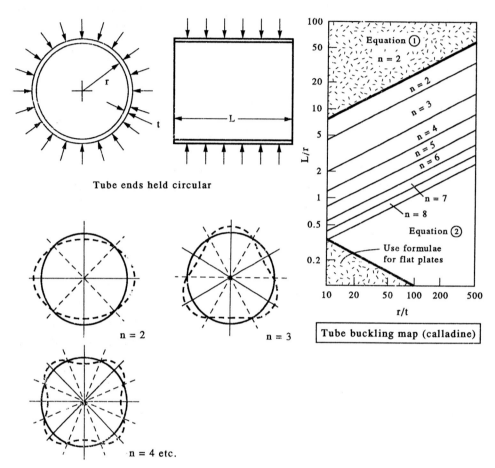

Tube ends held circular

n = 2

n = 3

n = 4 etc.

Tube buckling map (calladine)

Equation (1): $p_{cr}\text{(theoretical)} = \dfrac{E}{4(1 - v^2)}\left(\dfrac{t}{r}\right)^3.$

Equation (2): $p_{cr}\text{(theoretical)} = \dfrac{0.86E}{(1 - v^2)^{3/4}}\left(\dfrac{t}{L}\right)\left(\dfrac{t}{r}\right)^{3/2}.$

Actual p_{cr} may be as much as 20% less than the theoretical values given above because real tubes are never exactly round due to manufacturing tolerances.
    Generally,

$$p_{cr}\text{(actual minimum)} \approx 0.8 p_{cr}\text{(theoretical).}$$

EXAMPLE

A plastic bottle that once contained Caledonian Spring Water has a cylindrical portion that is 200 mm long and 102 mm in diameter. The thickness of the plastic is 0.25 mm. Best estimates of E and v are 3 GPa and 0.5 respectively. Your colleague Christopher, who is an expert on shell structures, offers to bet you a bottle of wine that you will not be able to induce a buckling instability by putting the top of the plastic bottle into your mouth and sucking in. He gives you five minutes to decide whether to accept, while he retains the bottle! You have no option but to do the calculations.

$$\frac{L}{r} = \frac{200 \text{ mm}}{51 \text{ mm}} = 3.9.$$

$$\frac{r}{t} = \frac{51 \text{ mm}}{0.25 \text{ mm}} = 204.$$

Plotted on the Tube Buckling Map these values put the failure in the middle of the $n = 5$ field. The relevant buckling equation is then Eqn. (2), from which

$$p_{cr}(\text{theoretical}) = \frac{0.86 \times 3 \times 10^3 \text{ MPa}}{(1 - 0.5^2)^{3/4}} \left(\frac{0.25 \text{ mm}}{200 \text{ mm}}\right)\left(\frac{1}{204}\right)^{3/2}.$$

$$= 1.37 \times 10^{-3} \text{ MPa gauge} = \underline{\underline{13.7 \text{ mbar gauge}}}.$$

This is a very small pressure differential (1 atmosphere $= 1013$ mbar) and should readily be achievable. You can look forward to claiming your reward when the trial has been carried out!

SPHERES ($r/t > 10$)

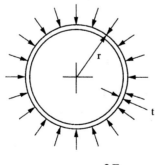

$$p_{cr}(\text{theoretical}) = \frac{2E}{\{3(1 - v^2)\}^{1/2}} \left(\frac{t}{r}\right)^2.$$

$$p_{cr}(\text{actual minimum}) \approx 0.25 p_{cr}(\text{theoretical}).$$

SPHERICAL CAPS

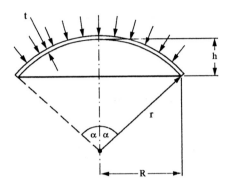

Periphery simply supported:

$$p_{cr}(\text{actual}) \approx (0.2 \text{ to } 0.6)p_{cr}(\text{theoretical}).$$

Periphery clamped:

$$p_{cr}(\text{actual}) \approx (0.3 \text{ to } 0.8)p_{cr}(\text{theoretical}).$$

Above equations are valid for $5 \leq \lambda \leq 50$, where

$$\lambda = \{12(1 - v^2)\}^{1/4}\left(\frac{r}{t}\right)^{1/2} \alpha.$$

For $\alpha \ll 1$,

$$\lambda \approx 2\{3(1 - v^2)\}^{1/4}\left(\frac{h}{t}\right)^{1/2}.$$

Note that $r = (R^2 + h^2)/2h$. $r \approx R^2/2h$ when $h/2r \ll 1$.

Trunnion Loading

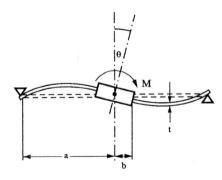

Thin circular plate of outside diameter $2a$ and thickness t. Twisting moment M applied to central circular boss of diameter $2b$. At $r = b$ the maximum (surface)

stress in the plate is given by:

$$\sigma = \pm \frac{\beta M}{at^2}.$$

The angle of twist is given by:

$$\vartheta = \frac{\alpha M}{Et^3}.$$

If the perimeter of the plate is simply supported, values of α and β are as follows.

$b/a =$	0.10	0.15	0.20	0.25	0.30	0.40	0.50	0.60	0.70	0.80
$\beta =$	9.48	6.25	4.62	3.63	2.95	2.06	1.49	1.07	0.731	0.449
$\alpha =$	1.40	1.06	0.82	0.641	0.500	0.301	0.169	0.084	0.035	0.010

If the perimeter is built in, values for α and β are as follows.

$b/a =$	0.10	0.15	0.20	0.25	0.30	0.40	0.50	0.60	0.70	0.80
$\beta =$	9.36	6.08	4.41	3.37	2.66	1.73	1.146	0.749	0.467	0.262
$\alpha =$	1.15	0.813	0.595	0.439	0.320	0.167	0.081	0.035	0.013	0.003

References

C. R. Calladine, *Theory of Shell Structures*, Cambridge University Press, 1983.
A. Kaplan, "Buckling of Spherical Shells", in *Thin-Shell Structures*, edited by Y. C. Fung and E. E. Sechler, Prentice-Hall, 1974, p. 247.
R. J. Roark and W. C. Young, *Formulas for Stress and Strain*, 5th edition, McGraw-Hill Kogakusha, 1975.
J. E. Shigley, *Mechanical Engineering Design*, 3rd edition, McGraw-Hill Kogakusha, 1977.
S. P. Timoshenko and J. M. Gere, *Theory of Elastic Stability*, 2nd edition, McGraw-Hill, 1961.

SECTION C: PLASTIC DEFORMATION

Plastic Stress and Strain

VON MISES YIELD CRITERION

For yield require

$$(\sigma_1 - \sigma_2)^2 + (\sigma_2 - \sigma_3)^2 + (\sigma_3 - \sigma_1)^2 \geq 2\sigma_y^2 = 6k^2.$$

$$k = \frac{\sigma_y}{\sqrt{3}} = \frac{\sigma_y}{1.732}.$$

EXAMPLES

UNIAXIAL TENSION

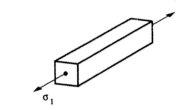

$$(\sigma_1 - 0)^2 + (0 - 0)^2 + (0 - \sigma_1)^2 \geq 2\sigma_y^2,$$

$$2\sigma_1^2 \geq 2\sigma_y^2,$$

$$\sigma_1 \geq \sigma_y.$$

EQUI-BIAXIAL TENSION

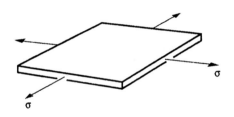

$$(\sigma - \sigma)^2 + (\sigma - 0)^2 + (0 - \sigma)^2 \geq 2\sigma_y^2,$$

$$\sigma \geq \sigma_y.$$

HYDROSTATIC PRESSURE

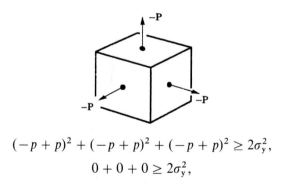

$$(-p + p)^2 + (-p + p)^2 + (-p + p)^2 \geq 2\sigma_y^2,$$

$$0 + 0 + 0 \geq 2\sigma_y^2,$$

i.e. yield can never occur in a purely hydrostatic stress field.

THIN PRESSURISED TUBE WITH END CAPS ($r/t > 10$)

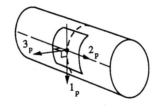

$$\sigma_1 = \sigma,$$
$$\sigma_2 = \sigma/2,$$
$$\sigma_3 = 0.$$

$$\left(\sigma - \frac{\sigma}{2}\right)^2 + \left(\frac{\sigma}{2} - 0\right)^2 + (0 - \sigma)^2 \geq 2\sigma_y^2,$$

$$\frac{\sigma^2}{4} + \frac{\sigma^2}{4} + \sigma^2 \geq 2\sigma_y^2,$$

$$\sigma \geq 1.16\sigma_y.$$

PURE SHEAR

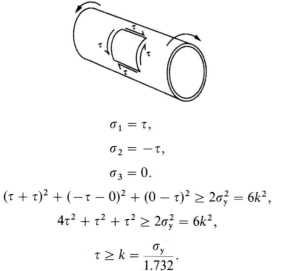

$$\sigma_1 = \tau,$$
$$\sigma_2 = -\tau,$$
$$\sigma_3 = 0.$$

$$(\tau + \tau)^2 + (-\tau - 0)^2 + (0 - \tau)^2 \geq 2\sigma_y^2 = 6k^2,$$
$$4\tau^2 + \tau^2 + \tau^2 \geq 2\sigma_y^2 = 6k^2,$$

$$\tau \geq k = \frac{\sigma_y}{1.732}.$$

VON MISES EQUIVALENT STRESS

$$\sigma_e = [\tfrac{1}{2}\{(\sigma_1 - \sigma_2)^2 + (\sigma_2 - \sigma_3)^2 + (\sigma_3 - \sigma_1)^2\}]^{1/2}.$$

Yield when $\sigma_e \geq \sigma_y$.

$$k_e = [\tfrac{1}{6}\{(\sigma_1 - \sigma_2)^2 + (\sigma_2 - \sigma_3)^2 + (\sigma_3 - \sigma_1)^2\}]^{1/2}.$$

Yield when $k_e \geq k$.

EXAMPLE

A metal with a yield stress of 280 MPa is subjected to a stress state with principal stresses of 300 MPa, 200 MPa and 50 MPa. Will the metal yield?

$$\sigma_e \text{ (MPa)} = [\tfrac{1}{2}\{(300 - 200)^2 + (200 - 50)^2 + (50 - 300)^2\}]^{1/2}$$

$$= \underline{\underline{218 \text{ MPa}}}.$$

This is less than the yield stress so the metal will not yield.

TRESCA YIELD CRITERION
Yield when

$$|\sigma_1 - \sigma_2| \text{ or } |\sigma_2 - \sigma_3| \text{ or } |\sigma_3 - \sigma_1| \geq 2k = \sigma_y.$$

This is an approximate result, often used for convenience. Depending on the stress state it can be in error by up to 16% on the conservative side.

EXAMPLES

UNIAXIAL TENSION

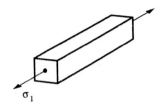

$$|\sigma_1 - 0| \text{ or } |0 - 0| \text{ or } |0 - \sigma_1| \geq \sigma_y,$$

$$\sigma_1 \text{ or } 0 \text{ or } \sigma_1 \geq \sigma_y,$$

$$\sigma_1 \geq \sigma_y.$$

Note that this is the same as the Von Mises result.

EQUI-BIAXIAL TENSION

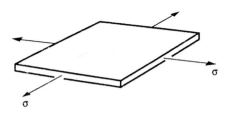

$$|\sigma - \sigma| \text{ or } |\sigma - 0| \text{ or } |0 - \sigma| \geq \sigma_y,$$

$$\sigma \geq \sigma_y.$$

This is the same as the Von Mises result.

HYDROSTATIC PRESSURE

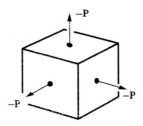

$$|-p + p| \text{ or } |-p + p| \text{ or } |-p + p| \geq \sigma_y,$$

$$0 \text{ or } 0 \text{ or } 0 \geq \sigma_y.$$

As with the Von Mises result, yield can never occur in a purely hydrostatic stress field.

THIN PRESSURISED TUBE WITH END CAPS $(r/t > 10)$

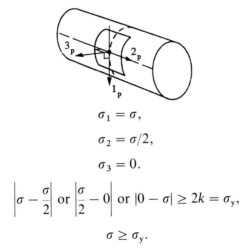

$$\sigma_1 = \sigma,$$

$$\sigma_2 = \sigma/2,$$

$$\sigma_3 = 0.$$

$$\left|\sigma - \frac{\sigma}{2}\right| \text{ or } \left|\frac{\sigma}{2} - 0\right| \text{ or } |0 - \sigma| \geq 2k = \sigma_y,$$

$$\sigma \geq \sigma_y.$$

The Von Mises result is 16% more than this.

PURE SHEAR

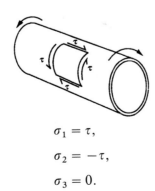

$$\sigma_1 = \tau,$$

$$\sigma_2 = -\tau,$$

$$\sigma_3 = 0.$$

$$|\tau + \tau| \text{ or } |-\tau - 0| \text{ or } |0 - \tau| \geq 2k = \sigma_y,$$

$$2\tau \geq 2k,$$

$$\tau \geq k = \frac{\sigma_y}{2}.$$

The Von Mises result is 16% greater than this.

EXAMPLE

A metal with a yield stress of 280 MPa is subjected to a stress state with principal stresses of 300 MPa, 200 MPa and 50 MPa. Will the metal yield?
Tresca requires

$$|300 - 200| \text{ or } |200 - 50| \text{ or } |50 - 300| \geq \sigma_y,$$

$$100 \text{ or } 150 \text{ or } 250 \geq \sigma_y.$$

The inequality is not satisfied, so the metal will not yield. The metal would just yield if it had a yield stress of 250 MPa. As we saw earlier, the Von Mises criterion requires a yield stress of 218 MPa in an identical stress state. In the present situation the Tresca yield stress is therefore 15% higher than the Von Mises yield stress.

LEVY–MISES EQUATIONS
Applicable when plastic strains are large compared to elastic strains.

$$\delta\varepsilon_i \propto \{\sigma_i - \sigma_h\}$$

where $\delta\varepsilon$ is a small increment of plastic strain. $i = 1, 2, 3$. σ_h is the hydrostatic component of the stress field and is given by

$$\sigma_h = \frac{\sigma_1 + \sigma_2 + \sigma_3}{3}.$$

Thus

$$\delta\varepsilon_1 = \lambda\left\{\sigma_1 - \frac{\sigma_2 + \sigma_3}{2}\right\},$$

$$\delta\varepsilon_2 = \lambda\left\{\sigma_2 - \frac{\sigma_1 + \sigma_3}{2}\right\},$$

$$\delta\varepsilon_3 = \lambda\left\{\sigma_3 - \frac{\sigma_1 + \sigma_2}{2}\right\}.$$

λ is an arbitrary constant with units of stress^{-1}.
Note that adding the three Levy–Mises equations together gives

$$\delta\varepsilon_1 + \delta\varepsilon_2 + \delta\varepsilon_3 = 0$$

which shows that there is no volume change as a result of large-strain plasticity.

EXAMPLES

UNIAXIAL TENSION

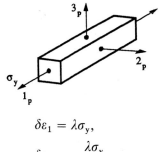

$$\delta\varepsilon_1 = \lambda\sigma_y,$$

$$\delta\varepsilon_2 = -\frac{\lambda\sigma_y}{2},$$

$$\delta\varepsilon_3 = -\frac{\lambda\sigma_y}{2}.$$

HYDROSTATIC PRESSURE

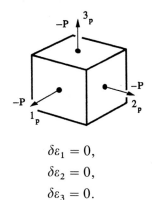

$$\delta\varepsilon_1 = 0,$$

$$\delta\varepsilon_2 = 0,$$

$$\delta\varepsilon_3 = 0.$$

THIN PRESSURISED TUBE WITH END CAPS $(r/t > 10)$

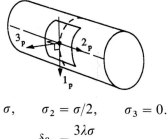

$$\sigma_1 = \sigma, \qquad \sigma_2 = \sigma/2, \qquad \sigma_3 = 0.$$

$$\delta\varepsilon_1 = \frac{3\lambda\sigma}{4},$$

$$\delta\varepsilon_2 = 0,$$

$$\delta\varepsilon_3 = -\frac{3\lambda\sigma}{4}.$$

EQUIVALENT PLASTIC STRAIN

$$\varepsilon_e = [\tfrac{2}{9}\{(\varepsilon_1 - \varepsilon_2)^2 + (\varepsilon_2 - \varepsilon_3)^2 + (\varepsilon_3 - \varepsilon_1)^2\}]^{1/2}.$$

EXAMPLES

UNIAXIAL TENSION

$$\varepsilon_1 = \varepsilon_1, \qquad \varepsilon_2 = -\varepsilon_1/2, \qquad \varepsilon_3 = -\varepsilon_1/2.$$

$$\varepsilon_e = \left[\frac{2}{9}\left\{\left(\frac{3\varepsilon_1}{2}\right)^2 + (0)^2 + \left(-\frac{3\varepsilon_1}{2}\right)^2\right\}\right]^{1/2} = \varepsilon_1.$$

PURE SHEAR

$$\varepsilon_1 = \frac{\gamma}{2}, \qquad \varepsilon_2 = -\frac{\gamma}{2}, \qquad \varepsilon_3 = 0.$$

$$\varepsilon_e = \left[\frac{2}{9}\left\{\gamma^2 + \frac{\gamma^2}{4} + \frac{\gamma^2}{4}\right\}\right]^{1/2} = \frac{\gamma}{\sqrt{3}} = \frac{\gamma}{1.732}.$$

Bending of Beams

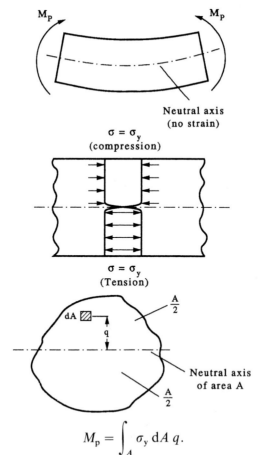

$$M_p = \int_A \sigma_y \, dA \, q.$$

PLASTIC MOMENTS

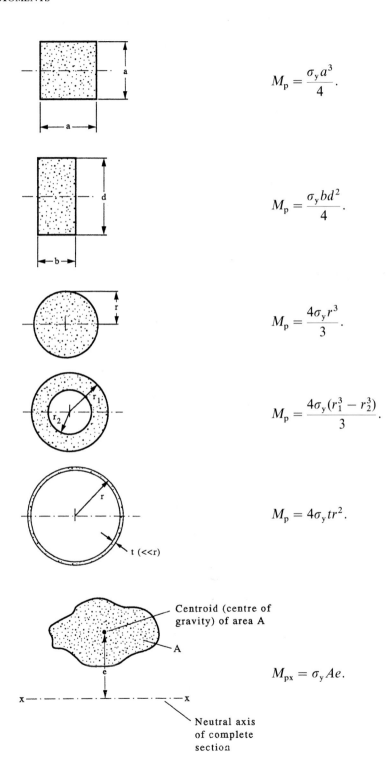

$$M_p = \frac{\sigma_y a^3}{4}.$$

$$M_p = \frac{\sigma_y b d^2}{4}.$$

$$M_p = \frac{4\sigma_y r^3}{3}.$$

$$M_p = \frac{4\sigma_y (r_1^3 - r_2^3)}{3}.$$

$$M_p = 4\sigma_y t r^2.$$

Centroid (centre of gravity) of area A

A

$$M_{px} = \sigma_y A e.$$

Neutral axis of complete section

Shearing Torques

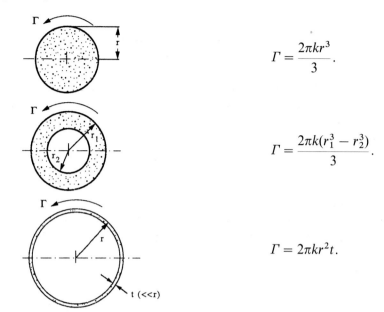

$$\Gamma = \frac{2\pi k r^3}{3}.$$

$$\Gamma = \frac{2\pi k (r_1^3 - r_2^3)}{3}.$$

$$\Gamma = 2\pi k r^2 t.$$

Plastic Buckling

The results given in Section B for the elastic buckling of struts, tubes, spheres and spherical caps can be applied to situations where the buckling is plastic provided Young's modulus E is replaced in the equations by the *tangen modulus* E_t which is defined in the following diagram.

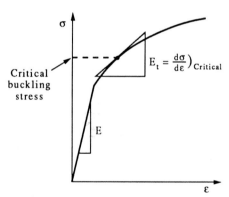

Mechanical Property Correlations

For steels (carbon, low-alloy, high-alloy, stainless, cast) experimental data for hardness and tensile strength give the following correlations:

$$\sigma_{TS} \text{ (MPa)} \approx 3.2 \times HV$$

where HV is the Vickers hardness (expressed in kgf mm^{-2}).

$$HB \approx 0.94 \, HV$$

where HB is the Brinell hardness (expressed in kgf mm^{-2}).

$$HRC \approx 0.09 \, HV$$

where HRC is the Rockwell C-scale hardness (arbitrary units).

Smithells gives data, conversion tables and details of the standard methods of hardness testing.

For most metals the following correlations hold.

$$\sigma_y \, (\text{MPa}) \approx \frac{H}{3}$$

where H is the indentation hardness (Vickers or Brinell) converted to MPa. σ_y is the yield stress of the metal after an additional plastic strain of 8%, which represents the average strain produced by the indenter during the hardness test.

$$k_u \approx \frac{\sigma_{TS}}{1.6}$$

where k_u is the failure stress in shear (see experimental data from ASTME and Smithells).

Plastic Constraint Factors

Plastic constraint factor $= L$.
Dimension perpendicular to diagrams $= l$. $l \gg w$ (plane strain).

FLAT INDENTER

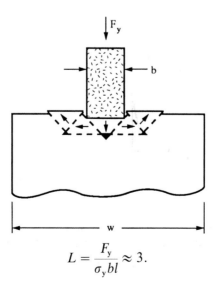

$$L = \frac{F_y}{\sigma_y bl} \approx 3.$$

DOUBLE EDGE NOTCH

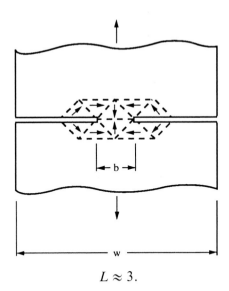

$L \approx 3.$

Using a more accurate slip-line-field model with the Tresca criterion we get the following data for both the flat indenter and the double edge notch.

$w/b =$	8.6	7.1	5.8	4.7	4.3
$L =$	2.57	2.39	2.22	2.05	1.98

$L = 2.57$ for $w/b > 8.6$.

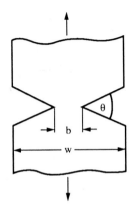

θ	Maximum L	w/b for maximum L
0	2.57	8.6
20	2.40	7.1
40	2.22	5.8
60	2.05	4.7
80	1.87	3.8
100	1.70	3.0
120	1.52	2.4
140	1.35	1.8
160	1.17	1.4
180	1	1

CENTRE NOTCH

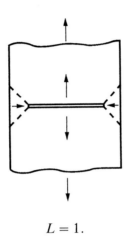

$$L = 1.$$

Ductile Fracture Surfaces

IN TENSION

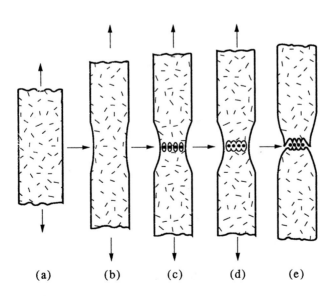

(a)   (b)   (c)   (d)   (e)

(a) Metals generally contain inclusions.
(b) A neck forms because of plastic instability.
(c) The metal in the neck is plastically constrained and is subjected to triaxial tension as a result. This pulls the metal away from the inclusions, forming voids.
(d) The voids link by microvoid coalescence to give a fibrous fracture surface.
(e) The remaining ligaments shear on 45° planes (like the centre notch situation) to give shear lips.

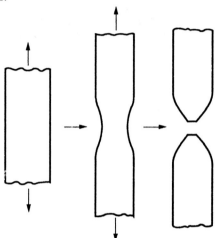

With metals that are inclusion-free or intrinsically very ductile, necking continues, giving a large reduction in area. There are no voids or shear lips.

IN SHEAR

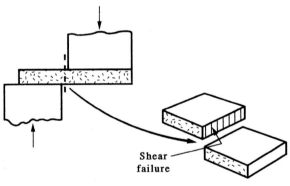

Shear failure

IN TORSION

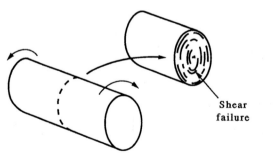

Shear failure

References

M. F. Ashby and D. R. H. Jones, *Engineering Materials 1*, Pergamon, 1980.
ASTME, *Tool Engineers' Handbook*, McGraw-Hill, 1959.
C. R. Calladine, *Plasticity for Engineers*, Ellis Horwood, 1985.
G. E. Dieter, *Mechanical Metallurgy*, 2nd edition, McGraw-Hill Kogakusha, 1976.
R. W. Hertzberg, *Deformation and Fracture Mechanics of Engineering Materials*, 3rd edition, Wiley, 1989.
W. Johnson and P. B. Mellor, *Plasticity for Mechanical Engineers*, Van Nostrand, 1962.
J. F. Knott, *Fundamentals of Fracture Mechanics*, Butterworth, 1973.
C. J. Smithells, *Metals Reference Book*, 6th edition, Butterworth, 1984.

SECTION D: CREEP DEFORMATION

Creep Strain

EQUIVALENT TENSILE STRESS

$$\sigma_e = [\tfrac{1}{2}\{(\sigma_1 - \sigma_2)^2 + (\sigma_2 - \sigma_3)^2 + (\sigma_3 - \sigma_1)^2\}]^{1/2}.$$

EQUIVALENT SHEAR STRESS

$$\tau_e = [\tfrac{1}{6}\{(\sigma_1 - \sigma_2)^2 + (\sigma_2 - \sigma_3)^2 + (\sigma_3 - \sigma_1)^2\}]^{1/2}.$$

$$\tau_e = \frac{\sigma_e}{\sqrt{3}} = \frac{\sigma_e}{1.732}.$$

LEVY–MISES EQUATIONS FOR STRAIN RATE

$$\dot{\varepsilon}_i \propto \{\sigma_i - \sigma_h\}.$$

$$\dot{\varepsilon}_1 = C\left\{\sigma_1 - \frac{\sigma_2 + \sigma_3}{2}\right\},$$

$$\dot{\varepsilon}_2 = C\left\{\sigma_2 - \frac{\sigma_1 + \sigma_3}{2}\right\},$$

$$\dot{\varepsilon}_3 = C\left\{\sigma_3 - \frac{\sigma_1 + \sigma_2}{2}\right\}.$$

C is an arbitrary constant with units of stress^{-1} time^{-1}.

EQUIVALENT TENSILE STRAIN RATE

$$\dot{\varepsilon}_e = [\tfrac{2}{9}\{(\dot{\varepsilon}_1 - \dot{\varepsilon}_2)^2 + (\dot{\varepsilon}_2 - \dot{\varepsilon}_3)^2 + (\dot{\varepsilon}_3 - \dot{\varepsilon}_1)^2\}]^{1/2}.$$

EQUIVALENT SHEAR STRAIN RATE

$$\dot{\gamma}_e = [\tfrac{2}{3}\{(\dot{\varepsilon}_1 - \dot{\varepsilon}_2)^2 + (\dot{\varepsilon}_2 - \dot{\varepsilon}_3)^2 + (\dot{\varepsilon}_3 - \dot{\varepsilon}_1)^2\}]^{1/2}.$$

$$\dot{\gamma}_e = \sqrt{3}\,\dot{\varepsilon}_e = 1.732\dot{\varepsilon}_e.$$

CONSTITUTIVE EQUATION

$$\dot{\varepsilon}_e = A\sigma_e^n e^{-(Q/RT)}.$$

For power-law creep $n \approx 3$ to 8. For diffusion creep $n \approx 1$.

EXAMPLES

UNIAXIAL TENSION

$$\sigma_2 = 0 \text{ and } \sigma_3 = 0, \text{ giving } \sigma_e = \sigma_1.$$

$$\dot{\varepsilon}_2 = \dot{\varepsilon}_3 = -\frac{\dot{\varepsilon}_1}{2}, \text{ which gives } \dot{\varepsilon}_e = \dot{\varepsilon}_1.$$

The constitutive equation then becomes

$$\dot{\varepsilon}_1 = A\sigma_1^n e^{-(Q/RT)}.$$

THIN PRESSURISED TUBE WITH END CAPS $(r/t > 10)$

$$\sigma_1 = \sigma, \qquad \sigma_2 = \sigma/2, \qquad \sigma_3 = 0, \text{ giving}$$

$$\tau_e = \left[\frac{1}{6}\left\{\frac{3\sigma^2}{2}\right\}\right]^{1/2} = \frac{\sigma}{2}.$$

VISCOUS LIQUID

$$\dot{\gamma} = \left(\frac{1}{10\eta}\right)\tau$$

where η is the viscosity in units of poise (P) or 10^{-1} N m^{-2} s. If a viscous solid is loaded in uniaxial tension along direction 1_p, then

$$\sqrt{3}\,\dot{\varepsilon}_1 = \left(\frac{1}{10\eta}\right) \times \frac{\sigma_1}{\sqrt{3}},$$

and

$$\dot{\varepsilon}_1 = \frac{1}{3}\left(\frac{1}{10\eta}\right)\sigma_1.$$

Deformation-Mechanism Maps

EXAMPLE OF MAP (see top of page 365)

The map is a plot of experimental data obtained from creep tests. The vertical axis is the equivalent shear stress normalised by the shear modulus, G. The contours are the equivalent shear strain rate in steady-state creep. Because the shear modulus of metals and alloys decreases with temperature the lines of constant shear stress rise (typically by a factor of 2) from the left to the right of the map.

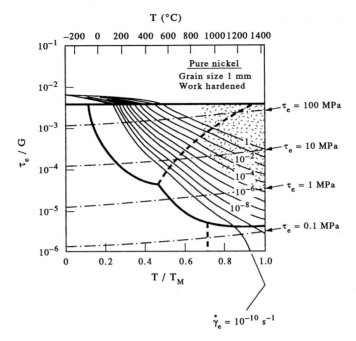

Although the map shown is for pure nickel its general features are typical of most metals and alloys and also inorganic compounds such as the alkali halides, oxides, minerals and even ice. However, the map for a material which undergoes a polymorphic phase change (e.g. ferritic steels and titanium alloys) is more complicated because it is like two separate overlapping maps. Further examples of deformation-mechanism maps can be found elsewhere in the text.

IDENTIFICATION OF CREEP MECHANISMS

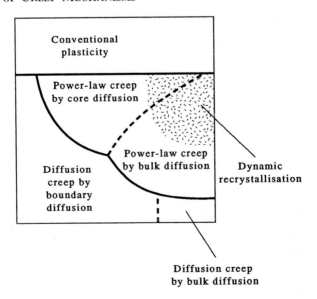

Further details of the mechanisms are given in Ashby and Jones, and Frost and Ashby. Note that although the boundaries between the various fields on the map have been drawn as distinct lines, there is in practice a smooth transition between the mechanisms. At high temperatures and strain rates recrystallisation can occur during the creep process, giving rise to "dynamic recrystallisation".

EFFECT OF GRAIN SIZE

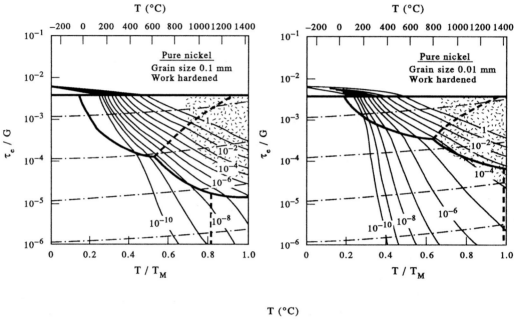

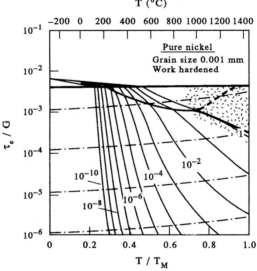

The sequence of maps given for pure nickel shows the effect of reducing the grain size in steps from 1 mm to 0.001 mm. As the grain size is reduced, creep by diffusion along the grain boundaries becomes more important. This field expands, and the fields for power-law creep and bulk diffusion creep contract. Eventually the field for bulk diffusion creep vanishes altogether. At any point in the boundary diffusion field the creep rate increases rapidly as the grain size falls.

EFFECT OF ANNEALING

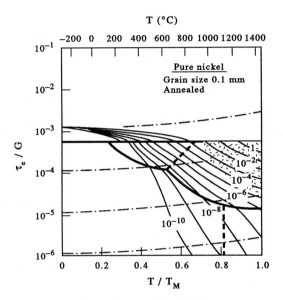

The only effect of annealing is to lower the yield stress, expanding the field for conventional plastic flow at the expense of the creep fields.

EFFECT OF SOLID SOLUTION

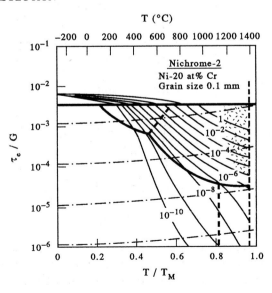

The pinning effect of the dissolved atoms reduces the creep rate in the power-law fields and suppresses dynamic recrystallisation. The power-law fields contract in order to ensure that the strain-rate contours remain continuous where they emerge from the diffusion-creep fields.

EFFECT OF SOLID SOLUTION PLUS PRECIPITATES

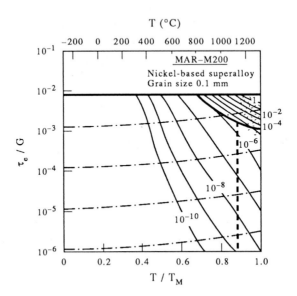

The combination of dissolved atoms and stable precipitates gives more effective pinning and leads to a further contraction of the power-law fields. Dynamic recrystallisation is almost prevented by the pinning of the grain boundaries.

COMMENTS

Deformation-mechanism maps are useful for deciding whether a given combination of stress and temperature is likely to cause significant creep. They can also tell us whether we are likely to cross a mechanism boundary if we extrapolate laboratory creep data. The above examples show that creep rates and mechanisms are very structure-sensitive. For this reason design and accurate analysis must always use data obtained from a batch of the material actually specified.

Fracture-Mechanism Maps

EXAMPLE OF MAP (see top of page 369)

On the vertical axis σ_n is the nominal tensile stress in a uniaxial tensile test. The contours are times to failure in seconds. Because the Young's modulus of metals and alloys decreases with temperature the lines of constant tensile stress rise (typically by a factor of 2) from the left to the right of the map.

Although the map shown is for pure nickel its general features are typical of metals and alloys with a face-centred-cubic structure. Ferritic steels have an additional cleavage field at low temperature. In other, more brittle, materials

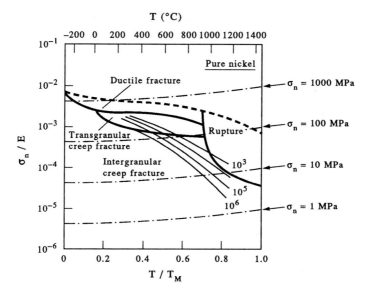

(e.g. inorganic compounds) the fields of ductile fracture and rupture are usually absent. As before, the map for a material which undergoes a polymorphic phase change is more complicated because it is like two separate overlapping maps.

DESCRIPTION OF FRACTURE MECHANISMS

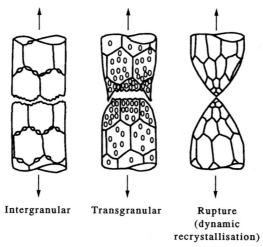

Intergranular   Transgranular   Rupture
(dynamic
recrystallisation)

In intergranular creep fracture wedge cracks or voids nucleate at grain boundaries under the action of the applied tensile stress. The defects grow and eventually the remaining ligaments fail. The tensile ductility and the reduction in area at the break are small.

In transgranular creep fracture voids nucleate and grow throughout the grains. Failure occurs by microvoid coalescence in a way that is similar to ductile failure at ordinary temperatures.

In rupture the process of dynamic recrystallisation inhibits the nucleation of voids. Eventually the specimen necks down because of plastic instability. Rupture is typified by a large ductility and reduction in area at break.

Further details of the fracture mechanisms are given in the references.

EFFECT OF SOLID SOLUTION

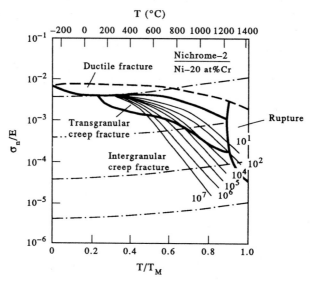

The dissolved atoms slow down the processes of creep and the times to failure are increased. The boundaries between the mechanisms are shifted upwards.

EFFECT OF PRECIPITATES

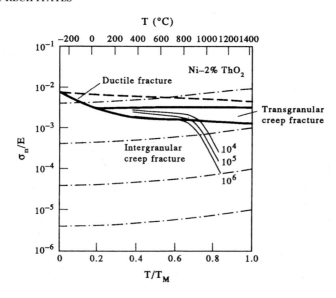

The thoria particles are stable up to the melting point. The pinning effect prevents any recrystallisation so rupture is absent. The transgranular field contracts and the times to failure increase further.

COMMENTS

Fracture-mechanism maps are useful for deciding how creep fracture is likely to take place under given conditions of stress and temperature. They also give us a rough idea of the times to failure, although the data in the maps are usually restricted and are often highly variable. The boundaries between the fields are not distinct, and mixed failure modes are possible. There are problems in applying the data (which are obtained from tensile tests) to situations involving triaxial stresses. The maps can be very structure sensitive: design and accurate analysis must use data from a batch of the material specified.

References

M. F. Ashby, C. Gandhi and D. M. R. Taplin, "Fracture-Mechanism Maps and their Construction for FCC Metals and Alloys", *Acta metall.*, **27**, 699 (1979).

M. F. Ashby and D. R. H. Jones, *Engineering Materials 1*, Pergamon, 1980.

R. J. Fields, T. Weerasooriya and M. F. Ashby, "Fracture Mechanisms in Pure Iron, Two Austenitic Steels, and One Ferritic Steel", *Metall. Trans. A*, **11A**, 333 (1980).

H. J. Frost and M. F. Ashby, *Deformation-Mechanism Maps*, Pergamon, 1982.

C. Gandhi and M. F. Ashby, "Fracture-Mechanism Maps for Materials which Cleave: FCC, BCC and HCP Metals and Ceramics", *Acta metall.*, **27**, 1565 (1979).

SECTION E: FRACTURE AND FATIGUE

Linear Elastic Fracture Mechanics (LEFM)

FAST FRACTURE EQUATION

$$K_{1c} = K_1 \equiv Y\sigma\sqrt{\pi a}$$

where K_{1c} is the plane strain fracture toughness, K_1 is the mode 1 stress intensity factor and Y is a dimensionless factor which is governed by the geometry of the crack and the surrounding material.

Values of fracture toughness are commonly given in MPa $\sqrt{m}$, N mm$^{-3/2}$ or ksi $\sqrt{in}$ (kilopounds per square inch $\sqrt{inch}$). The conversion factors between these units are as follows.

$$1 \text{ N mm}^{-3/2} = 0.032 \text{ MPa } \sqrt{m} = 0.029 \text{ ksi } \sqrt{in}.$$

VALUES OF Y

CASE 1

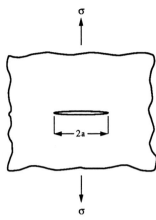

$$Y = 1.$$

CASE 2

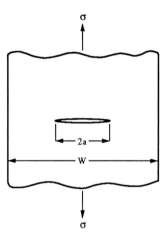

$$Y = \left\{ \cos\left(\frac{\pi a}{W}\right) \right\}^{-1/2}.$$

Note: when $W \gg a$, $Y = 1$ (Case 1).

    Examples: when $W = 4a$, $Y = 1.20$; when $W = 3a$, $Y = 1.43$.

CASE 3

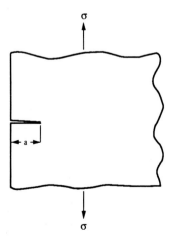

$$Y = 1.12.$$

This situation is like one half of Case 1. The factor of 1.12 is added to compensate for introducing a free surface.

CASE 4

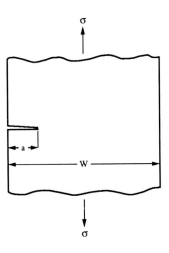

Y	a/W
1.12	0 (Case 3)
1.37	0.2
2.11	0.4
2.83	0.5

CASE 5

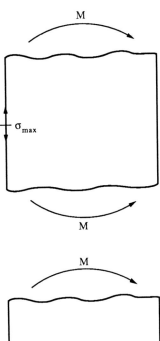

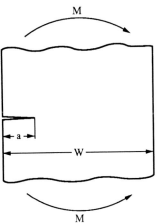

$$K_1 = Y\sigma_{max}\sqrt{\pi a}.$$

Y	a/W
1.00	0
1.06	0.2
1.32	0.4
1.62	0.5
2.10	0.6

CASE 6

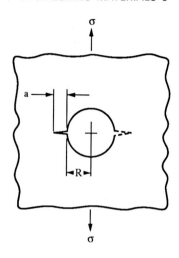

Y (one crack)	Y (two cracks)	a/R
3.36	3.36	0
2.73	2.73	0.1
2.30	2.41	0.2
1.86	1.96	0.4
1.64	1.71	0.6
1.47	1.58	0.8
1.37	1.45	1.0
1.18	1.29	1.5
0.71	1.00 (Case 1)	∞

Note that, for a round hole in uniaxial tension, the stress concentration factor is 3 (see Appendix 1A). For $a/R = 0$ we have a Case 3 crack embedded in a local stress field of 3σ. Thus $Y = 3 \times 1.12 = 3.36$ as shown in the table.

CASE 7

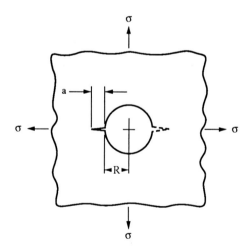

Y (one crack)	Y (two cracks)	a/R
2.24	2.24	0
1.98	1.98	0.1
1.82	1.83	0.2
1.58	1.61	0.4
1.42	1.52	0.6
1.32	1.43	0.8
1.22	1.38	1.0
1.06	1.26	1.5
0.71	1.00	∞

Note that, for a round hole in equi-biaxial tension, the stress concentration factor is 2 (see Appendix 1A). For $a/R = 0$ we have a Case 3 crack embedded in a local stress field of 2σ. Thus $Y = 2 \times 1.12 = 2.24$.

CASE 8

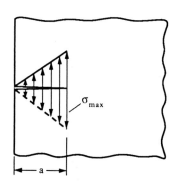

$$K_1 = 0.683\sigma_{max}\sqrt{\pi a}.$$

CASE 9

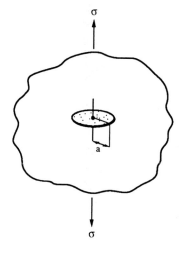

$$Y = 0.64.$$

CASE 10

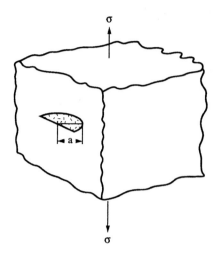

$$Y = 1.12 \times 0.64.$$

CRACK-TIP ELASTIC STRESS FIELD

$$\sigma_1 = \sigma_2 = \frac{K_1}{\sqrt{2\pi r}}.$$

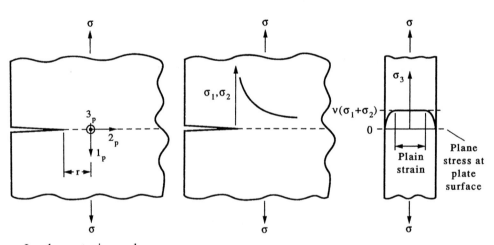

In plane strain we have

$$\sigma_1 = \sigma_2 = \sigma,$$

$$\sigma_3 = \nu(\sigma + \sigma) \approx 0.6\sigma.$$

The Von Mises equivalent stress is then given by

$$\sigma_e = [\tfrac{1}{2}\{(0)^2 + (0.4\sigma)^2 + (-0.4\sigma)^2\}]^{1/2} = 0.4\sigma.$$

In plane stress the corresponding expressions are

$$\sigma_1 = \sigma_2 = \sigma,$$
$$\sigma_3 = 0,$$
$$\sigma_e = [\tfrac{1}{2}\{(0)^2 + \sigma^2 + \sigma^2\}]^{1/2} = \sigma.$$

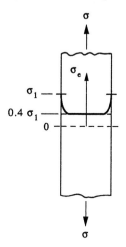

CRACK-TIP PLASTIC ZONE

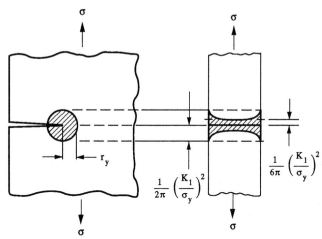

For plane stress we have

$$\sigma_e = \frac{K_1}{\sqrt{2\pi r_y}} = \sigma_y$$

which gives

$$r_y = \frac{1}{2\pi}\left(\frac{K_1}{\sigma_y}\right)^2.$$

For plane strain

$$\sigma_e = \frac{0.4 K_1}{\sqrt{2\pi r_y}} = \sigma_y$$

and

$$r_y = \frac{1}{12\pi}\left(\frac{K_1}{\sigma_y}\right)^2.$$

As the plastic zone grows the triaxial stresses relax and the radius of the plastic zone increases to

$$r_y = \frac{1}{6\pi}\left(\frac{K_1}{\sigma_y}\right)^2.$$

VALIDITY OF PLANE STRAIN LEFM

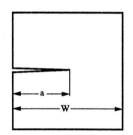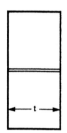

Experiments show that an analysis using plane strain LEFM is only valid if

$$16r_y(\text{plane stress}) \leq a, t, (W - a),$$

which leads to the condition

$$2.5\left(\frac{K_1}{\sigma_y}\right)^2 \leq a, t, (W - a).$$

EFFECT OF MATERIAL THICKNESS

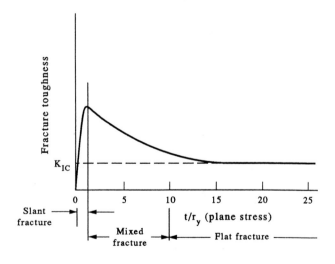

GEOMETRY OF FAST FRACTURE

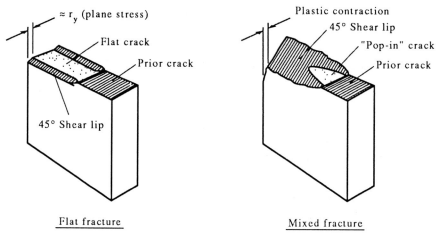

Flat fracture   Mixed fracture

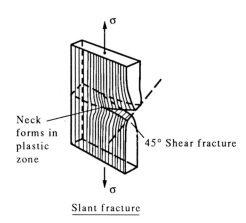

Slant fracture

Charpy V-notch Impact Test

DETAILS OF TEST

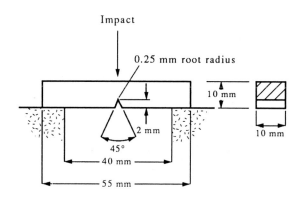

The test gives the energy in joules required to break the specimen in half at the notch. Older data are often given in lbf ft. The conversion is

$$1\,J = 0.737\,\text{lbf ft}.$$

DUCTILE-BRITTLE TRANSITION IN FERRITIC STEELS

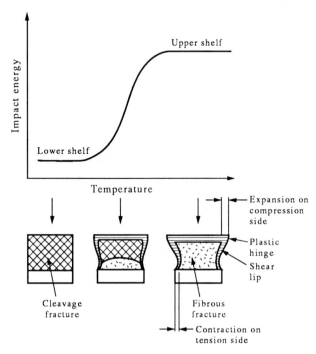

Below a measured impact energy of about 1 to 2 J machine effects (e.g. kinetic energy of broken halves of specimen) are dominant and the true fracture energy is essentially zero.

CORRELATIONS BETWEEN CHARPY ENERGY AND FRACTURE TOUGHNESS

A number of researchers have correlated experimental values of Charpy energy and fracture toughness for steels (see Pisarski). The correlations are approximate and must not be used for design purposes. However, they are useful for analysing failures when direct measurements of K_{1c} are not available. The more reliable results are for impact energies on or close to the lower shelf, where the differences between impact and fracture toughness test methods are not as marked as they are on the upper shelf. A reasonably reliable result, due to Sailors and Corten, is

$$K_{1c} = 15.5\,(\text{CVN})^{1/2}.$$

K_{1c} is in ksi $\sqrt{\text{in}}$ and CVN is the Charpy V-notch impact energy in lbf ft. The expression is valid for CVN values in the range 5 to 50 lbf ft (7 to 70 J). The British Standards Institution considers that K_{1c} for ferritic steels cannot fall below 30 MPa $\sqrt{\text{m}}$ in the fully embrittled state.

Fatigue Strengths of Welds

CURVES FOR 97.7% SURVIVAL

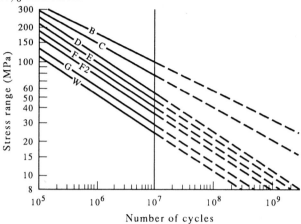

Number of cycles

CURVES FOR 50% SURVIVAL

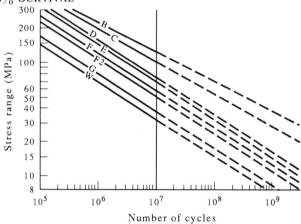

Number of cycles

CLASSES OF WELDS
Details on Surface of Member

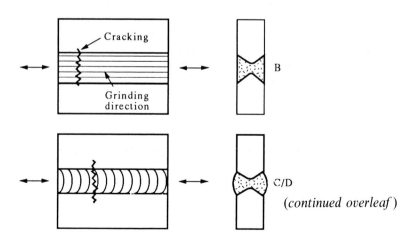

(continued overleaf)

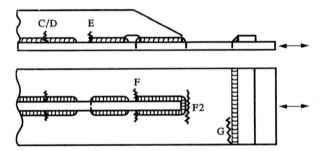

Weld Details on End Connections of Member

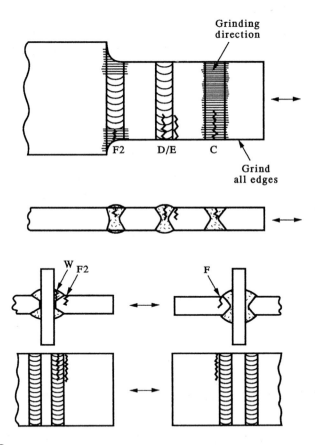

References

M. F. Ashby and D. R. H. Jones, *Engineering Materials 1*, Pergamon, 1980.

British Standards Institution, BS 5400: 1980: "Steel, Concrete and Composite Bridges": Part 10: "Code of Practice for Fatigue".

British Standards Institution, PD 6493: 1991: "Guidance on Methods for Assessing the Acceptability of Flaws in Fusion Welded Structures".

R. W. Hertzberg, *Deformation and Fracture Mechanics of Engineering Materials*, 3rd edition, Wiley, 1989.

J. F. Knott, *Fundamentals of Fracture Mechanics*, Butterworth, 1973.
Y. Murakami, *Stress Intensity Factors Handbook*, Pergamon, 1987.
P. C. Paris and G. C. Sih, "Stress Analysis of Cracks", in *Fracture Toughness Testing and its Applications*, ASTM, 1965, p. 30.
H. G. Pisarski, "A Review of Correlations Relating Charpy Energy to K_{1c}", *Welding Institute Research Bulletin*, **19**, 351 (1978).
D. P. Rooke and D. J. Cartwright, *Compendium of Stress Intensity Factors*, HMSO, 1976.
C. J. Smithells, *Metals Reference Book*, 6th edition, Butterworth, 1984.

SECTION F: CORROSION DATA

Standard Electrode Potentials

Volts at 25°C relative to the standard hydrogen electrode. All solutions have a concentration of dissolved ions of 1 mol/litre. All gases have partial pressures of 1 standard atmosphere.

Reaction	Volts	Reaction	Volts
$Au = Au^{3+} + 3e$	+1.50	$Ni = Ni^{2+} + 2e$	−0.25
$O_2 + 4H^+ + 4e = 2H_2O$	+1.23	$Co = Co^{2+} + 2e$	−0.28
$Pt = Pt^{2+} + 2e$	+1.2	$Cd = Cd^{2+} + 2e$	−0.40
$Pd = Pd^2 + 2e$	+0.99	$Fe = Fe^{2+} + 2e$	−0.44
$Ag = Ag^+ + e$	+0.80	$Cr = Cr^{3+} + 3e$	−0.74
$Fe^{3+} + e = Fe^{2+}$	+0.77	$Zn = Zn^{2+} + 2e$	−0.76
$O_2 + 2H_2O + 4e = 4OH^-$	+0.40	$Ti = Ti^{2+} + 2e$	−1.63
$Cu = Cu^{2+} + 2e$	+0.34	$Al = Al^{3+} + 3e$	−1.66
$Sn^{4+} + 2e = Sn^{2+}$	+0.15	$Mg = Mg^{2+} + 2e$	−2.36
$2H^+ + 2e = H_2$	0	$Na = Na^+ + e$	−2.71
$Pb = Pb^{2+} + 2e$	−0.13	$K = K^+ + e$	−2.93
$Sn = Sn^{2+} + 2e$	−0.14		

Rates of Uniform Penetration

Mils per year (mpy): 1 mpy = 0.0254 mm/year = 25.4 µm/year.

Penetration rate in µm/year = 87,600 $(w/\rho At)$, where w is the weight of metal lost in mg, ρ is the density of the metal in $Mg\ m^{-3}$ or $g\ cm^{-3}$, A is the corroding area in cm^2 and t is the time of exposure in hours.

Penetration rate in µm/year = 3.27 $(ai/z\rho)$, where a is the atomic weight of the corroding metal, i is the current density at the corroding surface in $\mu A\ cm^{-2}$, z is the number of electrons lost by each metal atom in the corrosion reaction and ρ is the density of the metal in $Mg\ m^{-3}$ or $g\ cm^{-3}$.

Metal	Atomic weight	Density (Mg m^{-3})
Ag	107.9	10.5
Al	26.98	2.70
Au	197.0	19.3
Cd	112.4	8.6
Co	58.93	8.9
Cr	52.00	7.1
Cu	63.54	8.96
Fe	55.85	7.87
K	39.10	0.86
Mg	24.31	1.74
Na	22.99	0.97
Ni	58.71	8.9
Pb	207.2	11.7
Pd	106.4	12.0
Pt	195.9	21.5
Sn	118.7	7.3
Ti	47.90	4.5
Zn	65.37	7.14

Electrochemical Equilibrium Diagrams

Electrochemical equilibrium diagrams for seven metals of industrial importance are given below. The vertical axis of the diagram is the electrochemical potential of the metal in volts measured relative to the standard hydrogen electrode. The horizontal axis is the pH of the water-based solution in which the metal is immersed. The diagram is divided into a number of fields as follows.

A field of immunity shows the range of potential and pH where corrosion of the metal is thermodynamically impossible.

A field of corrosion shows the range of conditions where there is a thermodynamic driving force trying to make the metal dissolve as ions in the solution.

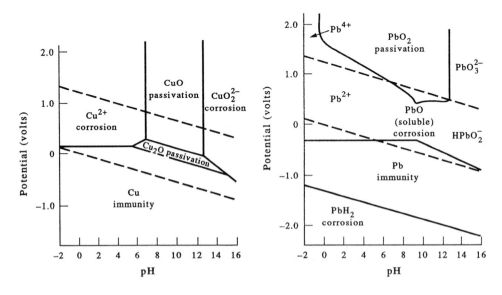

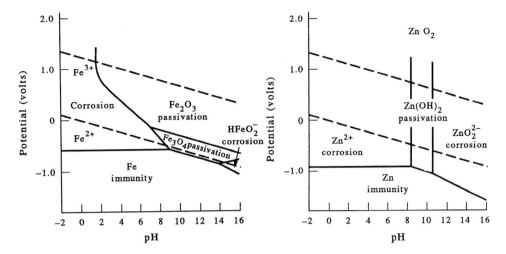

A field of passivation shows where there is a thermodynamic driving force trying to make a stable film (such as an oxide or a hydroxide) form on the surface of the metal. It is important to note that the film may or may not be an effective barrier to corrosion. If it is, then the passivation field is also a field of no corrosion. If it is not, the passivation field will be a field in which corrosion takes place.

The diagrams give no indication of the rates of corrosion: these can only be found by doing suitable experiments. The diagrams do show (a) where corrosion cannot occur, and (b) where a surface film should form which might prevent or slow down corrosion.

The edges of the corrosion fields are defined by a concentration of metal ion in solution of 10^{-6} mol/litre. This is an arbitrary value which is taken to represent such a low tendency of the solid to dissolve that the rate of attack is essentially zero.

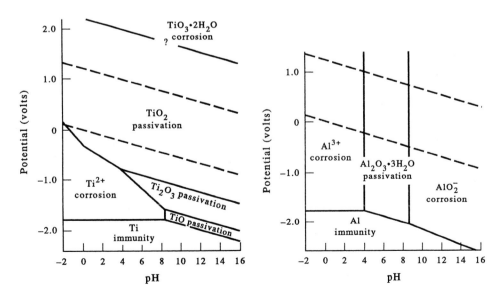

It should be noted that diagrams for alloys will be different from those of the parent metals mainly because of the effect of the alloying elements on the nature of the surface film. The stability of the film will also be affected by the presence in the solution of ions which have a chemical interaction with the film. Specific examples are given elsewhere in the text.

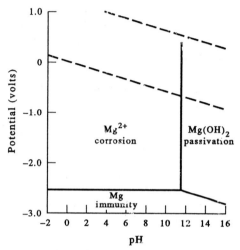

Corrosion cannot occur in practice unless an electron-accepting (cathodic) reaction also takes place. The two most important cathodic reactions (which involve hydrogen and oxygen respectively) are marked on the diagrams as dashed lines. They are identified in the following diagram.

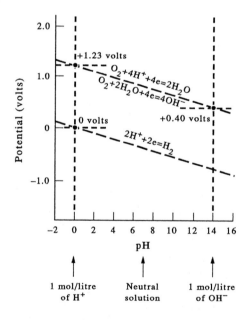

Reference

M. Pourbaix, *Atlas of Electrochemical Equilibria in Aqueous Solutions*, Pergamon, 1966.

SECTION G: HEAT TREATMENT OF STEELS

Hardness of Martensite

The following diagram shows the Vickers hardness of plain carbon martensite as a function of carbon content. The diagram also gives a good guide to the hardness of martensitic carbon–manganese and low-alloy steels. However, for low-alloy steels with lean carbon contents (about 0.1 wt % C or less) the graph generally predicts a hardness that is less than that achieved in practice. This is because of the strengthening effect of the alloying elements. The hardness of the martensite also decreases with decreasing cooling rate. It is therefore common to find that the centre of a quenched bar is softer than the outside even though the critical cooling rate has been achieved throughout the whole cross section. At any given cooling rate the relative strengthening effects of the principal alloying elements are given approximately by the empirical equation

$$\text{HV (martensite)} = \text{constant} + 949\text{C} + 27\text{Si} + 11\text{Mn} + 16\text{Cr} + 8\text{Ni}$$

where the chemical symbols represent the concentration in weight % of the element concerned.

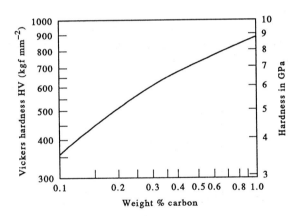

Hardenability

The addition of the alloying elements C, Mn, Cr, Mo and Ni decreases the critical cooling rate for the formation of martensite and increases the size of section that can be quenched to give a full martensitic structure. The following table gives approximate values for the maximum diameter of round bar which can be quenched in oil to give 100% martensite. The table lists typical ranges of compositions for carbon, carbon-manganese and low-alloy steels as used in engineering applications. The concentrations of the alloying elements are given as weight %.

The data have been obtained from continuous cooling transformation diagrams. The diagrams show significant variability even between steels of very similar composition. This is because the critical cooling rate is affected by the prior thermal and mechanical history of the steel. The data should be used only as a rough guide to tell us whether a quenched component of a particular size is likely to be fully martensitic.

Steel type	C	Si	Mn	Cr	Mo	Ni	Diameter (mm)
C	0.13	–	0.60	–	–	–	–
	0.25	0.20	0.70	–	–	–	5
	0.40	0.20	0.70	–	–	–	10
	0.86	0.20	0.60	–	–	–	15
C–Mn	0.19	0.20	1.20	–	–	–	4
	0.28	0.20	1.20	–	–	–	8
	0.28	0.20	1.50	–	–	–	11
	0.36	0.20	1.20	–	–	–	13
	0.38	0.25	1.80	–	–	–	30
Ni	0.10	0.26	0.53	–	–	3.65	7
	0.16	0.25	0.60	0.20	–	1.50	15
	0.10	0.20	0.40	–	–	4.8	20
	0.09	0.25	0.45	0.10	–	9.00	60
	0.40	0.26	0.62	0.23	0.10	3.45	60
Ni–Cr	0.16	0.20	0.80	0.85	–	1.15	8
	0.16	0.31	0.50	1.95	–	2.02	25
	0.40	0.23	0.75	0.65	–	1.30	25
	0.15	0.15	0.40	1.15	–	4.10	40
	0.30	0.20	0.50	1.25	–	4.10	300
Cr	0.20	0.30	0.75	0.95	–	–	8
	0.38	0.25	0.70	0.50	–	–	20
	0.39	0.20	0.70	1.05	–	–	30
	0.59	0.25	0.60	0.65	–	0.20	40
	0.24	0.37	0.27	13.3	–	0.32	500
Cr–Mo	0.14	0.25	0.55	0.60	0.55	–	5
	0.12	0.30	0.45	0.85	0.60	0.16	8
	0.27	0.13	0.60	0.74	0.55	0.19	20
	0.40	0.20	0.85	1.05	0.30	–	50
	0.32	0.25	0.55	3.05	0.40	0.30	110

Phases in Stainless Steels

The structure of a stainless steel at room temperature depends to a large extent on its composition. The following diagram, called a Schaeffler diagram, is a map showing the fields of dominance of the phases as a function of composition. The labellings A, F and M stand for austenite, ferrite and martensite. Also shown is the region where the sigma phase can form. This is an intermetallic with the approximate formula FeCr. Because the sigma phase is brittle, stainless steels must have compositions which avoid it.

The vertical axis of the diagram is the nickel equivalent, which is given approximately by

$$\text{Nickel Equivalent} = \text{Ni} + 30\text{C} + 0.5\text{Mn}.$$

The horizontal axis is the chromium equivalent, given approximately by

$$\text{Chromium Equivalent} = \text{Cr} + \text{Mo} + 1.5\text{Si} + 0.5\text{Nb}.$$

The chemical symbols represent the concentration in weight % of the element concerned.

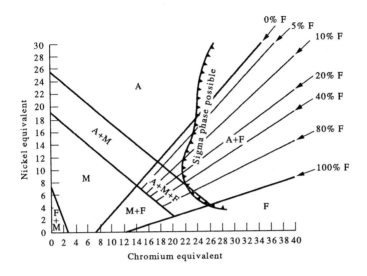

Chromium equivalent

References

M. Atkins, *Atlas of Continuous Cooling Transformation Diagrams for Engineering Steels*, British Steel Corporation.
R. W. K. Honeycombe, *Steels: Microstructure and Properties*, Arnold, 1981.
C. J. Smithells, *Metals Reference Book*, 6th edition, Butterworth, 1984.

SECTION H: THERMODYNAMICS DATA

Steam Table

Equilibrium Mixture of Water and Steam from the Triple Point (0.01°C)
to the Critical Point (374.15°C).
100 kPa = 1 bar. 101.3 kPa = 1 atmosphere.

Absolute pressure (kPa)	Temperature (°C)	Absolute pressure (kPa)	Temperature (°C)	Absolute pressure (kPa)	Temperature (°C)
0.611	0.01	80	93.5	1500	198.3
1.0	7.0	90	96.7	2000	212.4
2.0	17.5	100	99.6	2500	223.9
3.0	24.1	101.3	100.0	3000	233.8
4.0	29.0	120	104.8	3500	242.6
5.0	32.9	140	109.3	4000	250.3
6	36.2	160	113.3	4500	257.4
					(continued)

Steam Table (*continued*)

Absolute pressure (kPa)	Temperature (°C)	Absolute pressure (kPa)	Temperature (°C)	Absolute pressure (kPa)	Temperature (°C)
7	39.0	180	116.9	5000	263.9
8	41.5	200	120.2	5500	270.0
9	43.8	250	127.4	6000	275.6
10	45.8	300	133.5	6500	280.8
12	49.4	350	138.9	7000	285.8
14	52.6	400	143.6	7500	290.5
16	55.3	450	147.9	8000	295.0
18	57.8	500	151.8	8500	299.3
20	60.1	600	158.8	9000	303.3
25	65.0	700	165.0	10000	311.0
30	69.1	800	170.4	11000	318.0
35	72.7	900	175.4	12000	324.6
40	75.9	1000	179.9	13000	330.8
45	78.7	1100	184.1	14000	336.6
50	81.3	1200	188.0	15000	342.1
60	86.0	1300	191.6	20000	365.7
70	90.0	1400	195.0	22120	374.15

Saturation Tables

REFRIGERANT-12 (CCl_2F_2)
Equilibrium Mixture of Liquid and Vapour
100 kPa = 1 bar.

Absolute pressure (kPa)	Temperature (°C)	Absolute pressure (kPa)	Temperature (°C)	Absolute pressure (kPa)	Temperature (°C)
64.1	−40	261	−5	745	30
80.6	−35	308	0	847	35
100.3	−30	362	5	960	40
123.6	−25	423	10	1084	45
150.8	−20	491	15	1219	50
182.5	−15	567	20		
219	−10	651	25		

METHYL CHLORIDE (CH_3Cl)
Equilibrium Mixture of Liquid and Vapour
100 kPa = 1 bar.

Absolute pressure (kPa)	Temperature (°C)	Absolute pressure (kPa)	Temperature (°C)	Absolute pressure (kPa)	Temperature (°C)
47.4	−40	214	−5	653	30
60.7	−35	256	0	748	35
76.7	−30	304	5	852	40
96.0	−25	358	10	967	45
119	−20	420	15	1092	50
146	−15	490	20		
177	−10	567	25		

AMMONIA (NH$_3$)
Equilibrium Mixture of Liquid and Vapour
100 kPa = 1 bar.

Absolute pressure (kPa)	Temperature (°C)	Absolute pressure (kPa)	Temperature (°C)	Absolute pressure (kPa)	Temperature (°C)
71.8	−40	355	−5	1167	30
93.2	−35	429	0	1350	35
119.6	−30	516	5	1554	40
151.6	−25	615	10	1782	45
190	−20	728	15	2033	50
236	−15	857	20		
291	−10	1001	25		

PROPANE (C$_3$H$_8$)
Equilibrium Mixture of Liquid and Vapour
100 kPa = 1 bar.

Absolute pressure (kPa)	Temperature (°C)	Absolute pressure (kPa)	Temperature (°C)	Absolute pressure (kPa)	Temperature (°C)
44	−60	467	0	2131	60
71	−50	630	10	2605	70
110	−40	842	20	3137	80
172	−30	1074	30	3760	90
238	−20	1358	40		
338	−10	1734	50		

BUTANE (C$_4$H$_{10}$)
Equilibrium Mixture of Liquid and Vapour
100 kPa = 1 bar.

Absolute pressure (kPa)	Temperature (°C)	Absolute pressure (kPa)	Temperature (°C)	Absolute pressure (kPa)	Temperature (°C)
60	−20	292	30	1018	80
80	−10	381	40	1265	90
103	0	496	50	1543	100
156	10	642	60	1876	110
211	20	816	70		

References

R. E. Bolz and G. L. Tuve (editors), *Handbook of Tables for Applied Engineering Science*, Chemical Rubber Co., 1970.

R. W. Haywood, *Thermodynamics Tables in SI (metric) Units*, 2nd edition, Cambridge University Press, 1972.

G. W. C. Kaye and T. H. Laby, *Tables of Physical and Chemical Constants*, 14th edition, Longmans, 1973.

R. H. Perry and C. H. Chilton (editors), *Chemical Engineers' Handbook*, 5th edition, McGraw-Hill, 1974.

APPENDIX 2

EXAMPLES

1.1   An elastic material is loaded in tension along direction 3_p but is prevented from straining along directions 1_p and 2_p. Calculate the value of the tensile elastic modulus in direction 3_p in terms of E and v. (The required modulus is called the axial modulus, E_{ax}.)

Answer: $E_{ax} = \dfrac{E(1 - v)}{(1 - 2v)(1 + v)}$.

1.2   The sole of a shoe is to be surfaced with soft synthetic rubber having a Poisson's ratio of 0.49. The cheapest solution is to use a solid rubber slab of uniform thickness. However, a colleague suggests that the sole would give better cushioning if it were moulded as shown in the diagram. One way of comparing the cushioning effect of

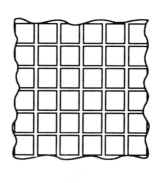

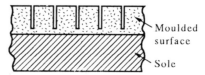

Moulded
surface

Sole

two surfaces is to apply a compressive stress normal to the sole of the shoe and to increase the stress until a given amount of elastic strain energy has been put into the rubber. The peak stress required for each surface is then compared. Following this procedure calculate the ratio of the two peak stress values. Use your answer to decide whether your colleague was right and if so why. You can use the result from Example 1.1.

Answer: σ(solid) = 4.1 × σ(moulded). Your colleague was right because the solid surface transmits a much bigger shock during an impact loading.

1.3 Having been convinced by your colleague that the change gives a better performance, you are still worried by the additional cost of the mouldings. A cheaper option is to mould the surface into a unidirectional pattern as shown in the diagram.

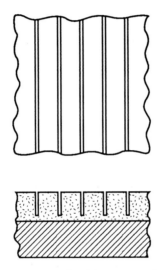

Calculate the ratio of the peak stresses in this case and comment on your findings. You can use the appropriate result from Appendix 1A.

Answer: σ(solid) = 3.6 × σ(moulded). The unidirectional pattern is nearly as good as the more complex two-dimensional one.

2.1 The diagram at the top of page 394 shows a rig for demonstrating whirling instability. A solid circular steel rod 2.34 mm in diameter is held in the chuck of a hand-held electric drill with an operating speed of 2400 rpm. The other end of the rod passes freely through a hole in the top of the bench. A round wheel weighing 25 g is glued to the rod at a distance of 110 mm from the chuck. The drill is started up when the wheel is only a few mm above the top of the bench. Gradually the drill is lifted until the weighted rod fails by whirling. Estimate the distance between the bench top and the drill chuck when whirling takes place. *Note:* if you try this experiment you will find that the whirling instability will take you by surprise and may cause injury unless the rig is surrounded by a safety screen. Assume that both ends of the rod are simply supported. Take E = 212 GPa.

Answer: 212 mm.

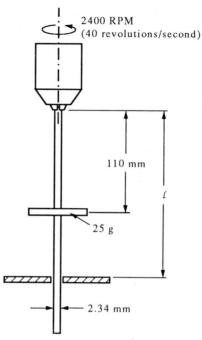

2.2   The diagram shows a stirrer for use in a chemical plant. The shaft of the stirrer is made from a steel tube having an outside diameter of 50 mm and a wall thickness

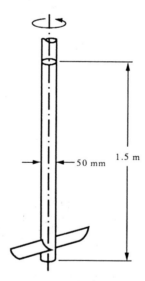

of 2 mm. The shaft is 1.5 m long. The bottom of the shaft is unsupported but the top of the shaft is rigidly fixed to the solid shaft of the drive motor. Estimate the rotational speed at which the stirrer will fail by whirling. Neglect the effect of the blades. Take $E = 212$ GPa. The density of steel is 7.8 Mg m^{-3}.

Answer: 1376 rpm.

3.1 The diagram shows the interference fit between a driving wheel and its tyre. Further information is given in Fig. 3.4. When the tyre is shrunk in place on the

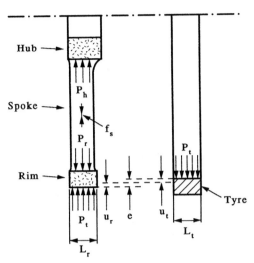

rim of the wheel there is a contact pressure p_t between the two components. The spokes exert an equivalent pressure p_r on the inside of the rim. Each spoke is subjected to a compressive force f_s. Show that

$$e = f_s \left\{ \left(\frac{E_r}{E_t} \right) \left(\frac{t_r}{t_t} \right) \left(\frac{a_t}{a_r} \right)^2 \left(\frac{l_s}{t_s w_s E_s} + \frac{a_r n}{2\pi t_r L_r E_r} \right) + \frac{l_s}{t_s w_s E_s} \right\}$$

where the subscripts t, r and s refer to tyre, rim and spoke respectively. Assume that the compliance of the hub is small.

3.2 Using information from Example 3.1 and Chapter 3 estimate the equivalent pressure p_h that the tyred wheel exerts on the outside of the hub. Make the simplifying assumptions that $t_r = t_t$, $a_r = a_t$ and $E_r = E_s$. Estimate by how much the internal radius of the hub will decrease because of this pressure.

Answers: 35 MPa; 70 µm.

4.1 A standard demonstration in thermodynamics is the Collapsible Can Experiment. The apparatus consists of a cylindrical can with an integral base and lid. The lid has a short spout which can be sealed with a rubber bung. A small amount of water is poured into the can which is then heated over a gas flame until the water is boiling and the space above the water is filled with steam. The can is removed and the spout is sealed off. As the water cools, the pressure of the saturated steam decreases, generating a partial vacuum in the can. Eventually the pressure differential increases to the point at which the can buckles inwards.

Fanta drinks cans are made of steel with an aluminium top. They are 100 mm long, 66 mm in diameter and have a wall 0.12 mm thick. You aim to use discarded cans for the Collapsible Can Experiment by sealing the top with a rubber disc.

Estimate the temperature at which the cans will collapse. Assume that $E = 212\,\text{GPa}$ and $v = 0.29$. Use the Steam Table in Appendix 1H. The yield stress of the cold-drawn steel is about 400 MPa.

Answer: 81 to 86°C.

4.2 A lightweight unmanned submersible vehicle for deep-sea operations has a frame made from extruded tubes of aluminium alloy. The tubes have a diameter of 200 mm, a wall thickness of 5 mm and are up to 4 m long. Each tube is sealed at both ends to keep out the water (this minimises the mass of the submersible and provides some buoyancy). After a dive to a depth of 200 m it is noticed that one of the tubes has become squashed flat for most of its length. Explain the failure. Would steel, GFRP or CFRP have offered a weight advantage over aluminium in this application? Data are given below.

Material	E (GPa)	ρ (Mg m^{-3})
Steel	212	7.8
Aluminium	69	2.7
GFRP	30	2.0
CFRP	120	1.5

Assume that $v = 0.3$ for all materials. The density of water is about 1 Mg m^{-3}. The yield stress of the extruded alloy is about 200 MPa.

Answers: p_{cr} (actual minimum) = 1.9 MPa; external pressure at 200 m = 2.0 MPa; a CFRP tube would be lighter than an aluminium tube by a factor of 2.2.

4.3 A short unconstrained ring of length L, mean radius r and thickness t ($r/t > 10$) is subjected to a uniformly distributed force F as shown in the diagram. Obtain an expression for the value of F that will give rise to elastic buckling.

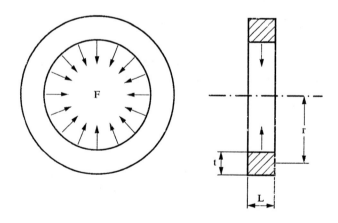

A similar ring has a constant but unspecified cross-section. The second moment of area of the section is I. Modify your expression for F_{cr} to describe this situation.

Answers: $F_{cr} = \dfrac{\pi EL}{2}\left(\dfrac{t^3}{r^2}\right)$ and $F_{cr} = \dfrac{6\pi EI}{r^2}$.

5.1 The observation windows in the side of a large public aquarium are made from injection-moulded polycarbonate. They take the form of a spherical cap clamped at the periphery, with dimensions $R = 500$ mm, $h = 200$ mm and $t = 5$ mm. The windows are convex to the water side. One of the large and dangerous fish in the tank gently bumps the centre of a window. The window promptly fails, causing Jaws-like alarm among the spectators. Given that $E \approx 2.6$ GPa, $v \approx 0.5$ and $\sigma_y \approx 55$ MPa estimate the depth of water in the aquarium at the level of the failed window. The density of water is about 1 Mg m^{-3}.

Answer: 5.1 m.

5.2 A steel aerosol can containing deodorant has a concave base shaped as a spherical cap with dimensions $r = 38.5$ mm, $R = 21$ mm and $t = 0.38$ mm. You can assume that the periphery of the cap is essentially clamped by the rolled-over edges. The propellant is probably butane or a mixture of butane and propane. If the maximum temperature that the can is likely to reach in domestic surroundings is taken as 90°C estimate the minimum safety factor that the base has against "flipping" when the propellant consists of (a) 100% butane, (b) 100% propane. Assume that $E = 212$ GPa, $v = 0.29$ and $\sigma_y = 400$ MPa.

Answers: (a) 6.0, (b) 2.0.

6.1 A cylindrical pressure vessel is rolled up from carbon–manganese steel plate 10 mm thick. Owing to a mistake in positioning one of the nozzles the wall of the vessel is left with a circular hole through it which measures 30 mm in diameter. It is decided to fill in the hole by machining up a circular disc from spare 10-mm plate so it will just fit into the hole. The edges of both the hole and the disc are then chamfered back by 30° and a full-penetration weld is run into the V-shaped groove. The heat from the arc warms up a circular area of plate which is concentric with the centre of the disc. As the area cools towards room temperature it contracts, generating a residual tensile stress which acts at right angles to the line of the weld bead. Estimate how big this stress will be if the temperature of the warm area is initially 100°C above that of the surrounding vessel. Comment on the practical implications of your answer. Data for the steel are as follows: $E = 212$ GPa, $\alpha = 12 \times 10^{-6}$ °C^{-1}, $\sigma_y = 250$ MPa.

Answer: 127 MPa.

6.2 A uniform bar of metal is subjected to a bending moment that is just sufficient to make the whole of the cross-section go plastic. When the bar is released the elastic "springback" leaves a residual elastic stress distribution in the bar.

Calculate the residual stresses for the following cases:

(a) A solid square cross-section with the neutral axis parallel to two opposite edges of the square. Residual stresses to be found at these opposite edges.
(b) A solid rectangular cross-section with the neutral axis parallel to two opposite edges of the rectangle. Residual stresses to be found at these opposite edges.
(c) A solid circular cross-section. Residual stresses to be found at the two points which are furthest away from the neutral axis.
(d) A thin-walled tube. Residual stresses to be found at the two points which are furthest away from the neutral axis.

Express each answer as a proportion of the uniaxial yield stress, σ_y. Use the results given in Appendix 1B for the elastic bending of beams. Second moments of area and fully plastic moments for the various sections are listed in Appendices 1B and 1C.

Answers: (a) ± 0.50, (b) ± 0.50, (c) ± 0.70, (d) ± 0.27.

7.1 Table 7.1 in Chapter 7 gives the dimensions of a number of annealed copper flue tubes. Calculate the actual minimum collapse pressure of each tube under conditions of external hydrostatic pressure. Use the material properties and equations given in Chapter 7 as appropriate. You will also need to refer to the Tube Buckling Map in Appendix 1B.

Answers: see Table 7.1 in Chapter 7.

7.2 Table 7.2 in Chapter 7 gives the dimensions of a number of tubular fireboxes made from annealed copper. Calculate the actual minimum collapse pressure of each firebox under external hydrostatic pressure. Use the material properties and equations given in Chapter 7 as appropriate. You will also need to use the Tube Buckling Map in Appendix 1B.

Answers: see Table 7.2 in Chapter 7.

8.1 The diagram gives the dimensions of a bicycle chain. The chain is driven by

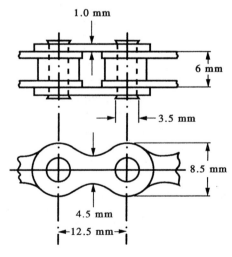

a chain wheel which has pitch diameter of 190 mm. The chain wheel is connected to the pedals by a pair of cranks set at 180° in the usual way. The centre of each pedal is 170 mm away from the centre of the chain wheel. The average Vickers hardness measured on the surface of a link is 532. Measured on the end of a pin it is 650. If the cyclist weighs 90 kg, estimate the factor of safety of the chain. You may assume that a link would fail in simple tension at the position of minimum cross-sectional area and that a pin would fail in double shear. You may assume that the chain is made from a ferritic steel because bicycle chains rust when they are not kept well oiled. Comment on your answer.

Answer: 9.7.

8.2   A workshop supervisor wants to move a milling machine which weighs about 1.5 tonnes from one side of a machine shop to the other. The shop is 5.5 m wide from wall to wall. The plan is to span the shop with a length of square hollow box section steel resting on top of the walls as shown in the diagram. The outside of

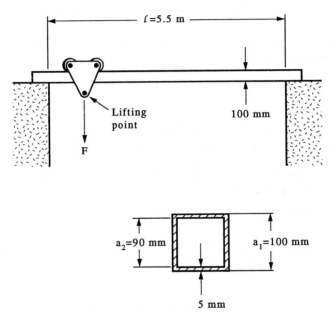

the box section measures 100 mm × 100 mm. The wall of the section is 5 mm thick. The machine is attached to the lifting point with a chain hoist and lifted off the floor. The lifting point is suspended from a wheeled trolley which is pushed along the beam with the intention of transporting the machine across the shop. As the trolley approaches the centre of the span the section bends. The machine drops to the floor, narrowly missing the feet of one of the assistants. Given that the yield strength of the steel is about 250 MPa, account for the failure.

9.1   A thin-walled tube is loaded in tension along its axis at the same time as it is subjected to an axial torque. The axial stress is σ and the shear stress due to the torque is τ. Assuming that the tube is on the point of yielding according to the Von

Mises yield criterion derive an expression relating σ and τ to the yield stress σ_y.

Answer: $\sigma^2 + 3\tau^2 = \sigma_y^2$.

9.2 The diagram shows a coupling between two rotating shafts designed to transmit power from a low-speed hydraulic motor to a gearbox. The coupling

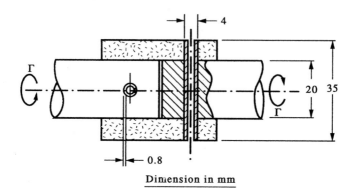

Dimension in mm

sleeve was a sliding fit on the shafts and the torque was taken by the two Bissell pins as shown in the diagram. Owing to a malfunction in the gearbox one of the pins sheared, disconnecting the drive. After the failure the assembly was dismantled. The hardness of the pin end was measured and was found to be 500 HV. Estimate the torque needed to cause the failure. Is there any evidence that the pin was at fault? (The normal material for Bissell pins is spring steel containing 0.75 to 0.82% carbon and 0.60 to 0.80% manganese.)

Answers: 16 kgf m. Material not at fault (see Solution for reason).

9.3 Using the information given in Example 9.2 work out the torque needed to bring the pin to the point of yielding in shear. Use the Tresca criterion and the correlation between Vickers hardness and yield stress given in Appendix 1C. Data from Smithells show that the flow stress of the spring steel rises by typically 100 MPa for a plastic strain of about 6%. Explain why your answer is likely to be slightly conservative.

Answer: 12 kgf m.

10.1 The diagram is a scale drawing of a metallurgical section taken from a welded connection between two bars of carbon–manganese structural steel. The left-hand

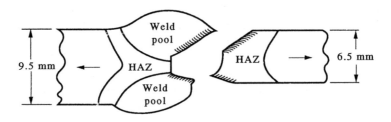

bar had a cross-section measuring 9.5 mm × 73 mm and the right-hand bar had a cross-section measuring 6.5 mm × 73 mm. The weld had fractured in response to a tensile load applied at right angles to the welded joint. The structure immediately next to the fracture surface showed evidence of plastic shearing as shown schematically on the diagram. The average Vickers hardness of the steel section and the weld pool were 215 and 193 HV respectively. Estimate the load at which the weld failed. What load would have been needed to make the fabrication break at a location away from the weld? What criticisms can you make of the weld geometry?

Answers: 20 tonnef and 33 tonnef.

10.2 When a uniform bar of metal is subjected to a progressively increasing bending moment the deformation ceases to be fully elastic when the outermost part of the cross-section reaches the yield stress. This condition defines the maximum elastic bending moment that the section can carry. As the bending moment is increased further the zone of plasticity gradually spreads in towards the neutral axis. The bending moment required to make the whole cross-section go plastic is called the plastic moment, M_p. Using results given in Appendices 1B and 1C calculate the ratio of the plastic moment to the maximum elastic moment for the following cross-sections: (a) solid square, (b) solid rectangle, (c) solid circle, (d) thin-walled tube. In cases (a) and (b) you may assume that the neutral axis is parallel to one edge of the cross-section.

Answers: (a) 1.50, (b) 1.50, (c) 1.70, (d) 1.27.

11.1 A thin walled tube made from 316 stainless steel contains gas under pressure inside a furnace at a temperature of 700°C. The pressure generates a tensile hoop stress of 40 MPa and a tensile axial stress of 20 MPa. The inspection procedure is to examine the tube every three months for signs of creep bulging and to pass it as safe if it does not show any significant deformation. You are not convinced that this is a safe procedure but all you have to hand are some fracture-mechanism and deformation-mechanism maps for several different grades of 316 steel (see the diagrams on pages 402 and 403). Explain whether or not you consider your reservations to be justified.

11.2 The heat exchanger in a reformer plant consisted of a bank of tubes made from $1\frac{1}{4}$Cr $\frac{1}{2}$Mo steel. The tubes contained hydrocarbon gas at a pressure of 4.3 MPa and were heated from the outside by furnace gases. The tubes had an internal diameter of 128 mm and a wall thickness of 6.6 mm. Owing to a temperature overshoot one of the tubes fractured and the resulting gas leak set the plant on fire.

   When the heat exchanger was stripped down it was found that the tube wall had fractured longitudinally over a distance of about 300 mm. At the edge of the fracture the wall had thinned down to about 2.9 mm. Metallurgical sections cut from the tube next to the failure showed a ferrite–pearlite microstructure with a slightly enlarged grain size and indications of recrystallisation. The iron carbide in the pearlite had not spheroidised. The diagrams on page 404 show the fracture-mechanism map for $2\frac{1}{4}$Cr 1Mo steel and the deformation-mechanism map for

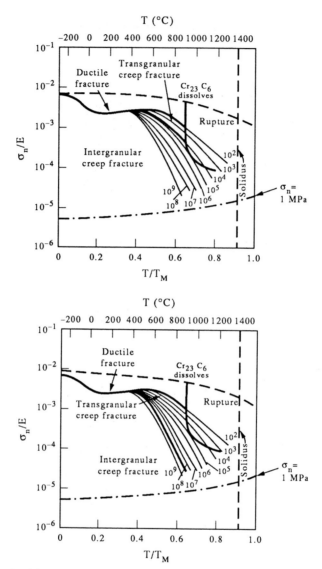

1Cr Mo V steel with a grain size of 0.1 mm. Using the information given estimate the minimum temperature at which the failure took place and the maximum time to failure.

12.1 A bar of hot material is subjected to a constant tensile load along its axis. The material creeps under the load according to the constitutive equation

$$\dot{\varepsilon} = B\sigma^n$$

where B and n are constants. The bar contains a length of slightly reduced cross-sectional area. Show how the value of n affects the shape of the bar as it creeps. Explain the relevance of your findings to

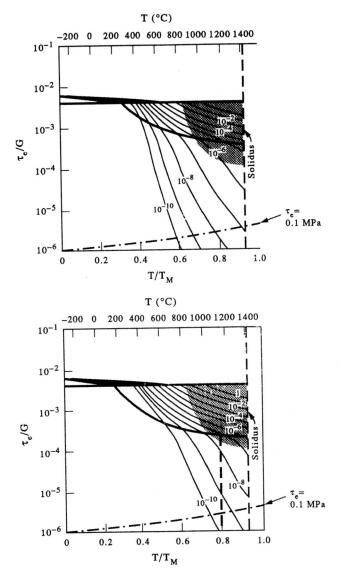

(a) the drawing of sections of a thick viscous liquid, such as hot glass;
(b) the superplastic forming of titanium alloy components.

Hint: alloys for superplastic forming have a very small grain size.

13.1 A large furnace flue operating at 440°C is made from a low-alloy steel with
the following composition in weight %.

C	Si	Mn	S	P	Cr	Ni	V	Cu
0.10	0.40	0.40	0.010	0.09	0.90	0.20	0.04	0.30

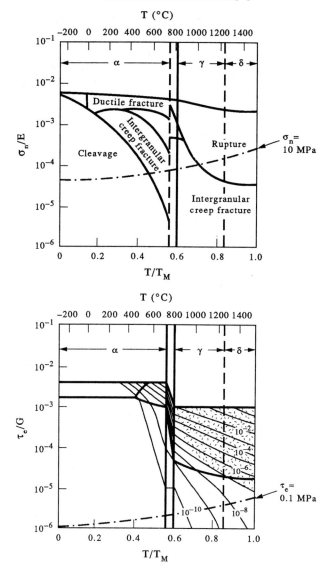

The chromium and nickel were added to increase the resistance of the steel to oxidation at the working temperature. The vanadium was added to obtain precipitates of vanadium carbide which increase the creep strength. The copper is present as an impurity from remelted scrap.

After 2 years in service, specimens were removed from the flue and Charpy V-notch impact tests were carried out at room temperature. The average value of impact energy was only about 4 J compared to a value when new of about 80 J. The loss in toughness was due to temper embrittlement caused by the impurity phosphorus: steels for high-temperature use normally contain molybdenum to reduce the rate of embrittlement and have lower levels of phosphorus.

The skin of the flue was made from plate 10 mm thick. Owing to the self-weight of the flue the plate has to withstand primary membrane stresses of up to 60 MPa. Estimate the length of through-thickness crack that will lead to the fast fracture of the flue when the plant is shut down. The yield stress of the steel at room temperature is about 440 MPa. How accurate is your answer likely to be? What steps would you take to increase the reliability of your answer? How would you search for possible cracks in the structure?

Answer: 160 mm.

13.2 A cylindrical pressure vessel in an ammonia plant was 7 m long, had an internal diameter of 1 m and had a wall thickness of 62 mm. The operating hoop stress was 285 MPa. The service temperature was 26°C. The material was a low-alloy steel in the quenched and tempered condition with a specified minimum yield stress of 570 MPa and a carbon equivalent of 0.56.

After 16 years in service the vessel exploded into a large number of fragments, some of which were hurled a distance of 1 km. All the fractures were flat; the micro-mechanism of fracture was cleavage. Semicircular "thumbnail" cracks, typically 4 mm deep, were found at the inner surface of the vessel; they had initiated at the edges of a series of fillet welds used to attach internal fittings to the vessel wall. The gas in the vessel contained 58% hydrogen. An analysis of the steel showed 0.4% free hydrogen in interstitial solid solution and the cracks were blamed on hydrogen cracking.

Tests on samples of the steel gave a value for K_{1c} of about 40 MPa $\sqrt{m}$. Account for the failure using fracture mechanics. Comment on the general advisability of making welded attachments to a pressure vessel.

From: J. D. Harrison, S. J. Garwood and M. G. Dawes, "Case Studies and Failure Prevention in the Petrochemical and Offshore Industries", in *Fracture and Fracture Mechanics: Case Studies*, edited by R. B. Tait and G. G. Garrett, Pergamon, 1985, p. 281.

14.1 A natural gas pipe made from a low-alloy steel with a yield strength of 350 MPa had a diameter of 600 mm and a wall thickness of 15 mm. It was rated to carry an internal pressure of 8 MPa. Individual runs of pipe (typically 6 to 10 m in length) were joined end-to-end with circumferential welds. When the pipe was two years old a fast fracture occurred under an operating pressure of 5.5 MPa. As a result of the failure the pipe split open like a peeled banana and the broken ends were flung 150 m.

The initial part of the fracture was a through-thickness crack that was more-or-less parallel to the axis of the pipe. It was flat and showed clear chevron marks. Once the crack had reached a length of about 0.6 m it changed over to a shear lip fracture. After the crack had travelled another few metres the fracture edge had drawn down to about 7 mm. Areas of the fracture surfaces showed a step several mm high which followed the centre-line of the plate.

Explain the changes in fracture geometry as the crack propagated and account for the step on the fracture surface.

Hint: Steel plates can exhibit a plane of weakness mid way between the two surfaces of the plate. When the plate is subjected to a tensile stress in the "through-thickness" direction a "delamination" failure can result. This effect used to be quite common when plates were hot rolled from cast ingots of steel. Impurities segregated to the centre of the ingot during solidification: when the ingot was passed through the hot mill the segregated impurities were rolled out into a plane on the plate centre-line, resulting in the plane of weakness. This is less of a problem now because modern steel making practice uses the continuous casting process and exercises much tighter control over impurities.

14.2   The shell of a heat exchanger consisted of a long tube having an external diameter of 328 mm and a wall thickness of 10 mm. The shell was rolled up from low-alloy steel plate and the ends of the plate were joined together with a longitudinal seam weld. Tests on spare pieces of plate gave a yield strength of 400 MPa. The shell operated at an internal pressure of 5.9 MPa and a temperature of 275°C.

A longitudinal crack initiated in the heat-affected zone at the inner wall of the shell and grew into the thickness of the plate by corrosion fatigue. At the point of failure the crack had grown until it was a very long thumbnail. At this stage the centre of the crack just breached the outer wall of the shell; at the inner wall the crack had a length of 400 mm. The crack then propagated by fast fracture and ran along the length of the shell. For most of its length the running crack deviated slightly from the heat affected zone and ran through the parent plate close to the weld.

What geometry would you expect for the running crack? Support your answer with an appropriate calculation.

15.1   The photographs show the end of a bicycle crank. The crank has fractured on either side of the thread hole through which the pedal spindle was screwed.

The thickness of the crank measured parallel to the axis of the hole is 10 mm. When the pedal was at its lowest position the downwards force on the centre of the pedal would have applied a bending moment to the crank cross-section, putting the outer face of the crank into tension. There is a dark fatigue crack on either side of the hole which initiated at the outer face of the crank in this region of tension and spread into the centre of the cross-section with time. The final fracture on each side of the hole is bright and "crystalline" and is a cleavage failure. Final fracture presumably occurred when the pedal was at its lowest position and was subjected to a single overload which made the fatigue crack go unstable. By approximating the final fracture to a Charpy V-notch test with a suitable correction for the area of the fracture estimate the energy required to cause the final fracture. If the overload was caused by dropping a 1 kg weight onto the end of the pedal estimate the height through which the weight was dropped in order to break the crank.

Answers: 2.5 to 5 J; 0.25 to 0.5 m.

15.2   You have been asked to design a hydraulic rock splitter for excavating rock in situations where blasting is not permitted (e.g. where the rock face is unstable, or where there are buildings). The procedure is to drill a hole deep into the rock and to insert into it a hydraulically operated device capable of exerting an outwards radial pressure on the wall of the hole. Estimate the maximum pressure required to do the job if the rock has a fracture strength of 20 MPa. If the diameter of the hole is 50 mm and Young's modulus for the rock is 60 GPa estimate the radial expansion of the hole when fracture occurs.

Answers: 20 MPa, 0.01 mm.

16.1   The photograph shows a specimen of grey cast iron which fractured in axial

torsion. Identify as closely as possible the point where the failure initiated and account for the shape of the fracture.

16.2 The photograph shows how a wooden rod failed in torsion. Explain why the shape of the fracture is very different from that seen when cast iron fails in torsion.

16.3 A prop shaft is made from a thin-walled tube of epoxy resin reinforced with carbon fibres. Along which directions would you set the carbon fibres in order to resist both torsional fracture and whirling deflection?

17.1 The diagram shows a strip of wood of thickness d being removed from a larger piece by splitting it along the grain. The sketch shows two ways of obtaining

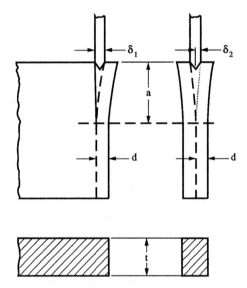

the strips. The first is to remove them from the edge of a wide plank of wood. The second is to split much narrower planks in half. For a crack of length a the critical values of the displacement δ at which the crack goes unstable are δ_1 and δ_2 for the two situations. By using an energy approach to the linear elastic fracture mechanics problem find the ratio of δ_1/δ_2.

When slabs of slate are split into thin plates for roofing purposes they are first split in half, then each half is halved again, and so-on until every piece has the required thickness. If, instead, thin plates are split from the face of the slab one after another it is found that most of them snap off at the tip of the crack. This can also happen when separating pieces of frozen sliced bread just after the loaf has been removed from a deep-freeze. Why is this the case?

Answer: $\sqrt{2}$.

17.2 The diagram shows the lower end of a laboratory storage bottle made from soda glass. The bottle was used to dilute sulphuric acid in the approved manner

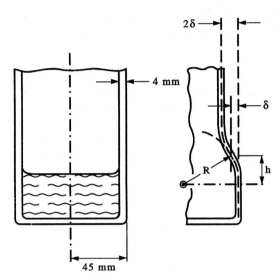

by putting some water into the bottle first and then slowly adding the acid. The dissolution reaction generated a significant amount of heat but there was no spitting or other side effect until, after a minute or two, a crack formed in the glass at the level of the meniscus and the top of the bottle became separated from the base. The failure was probably caused by thermal mismatch between the base and the top. Do a rough calculation to see whether this scenario gives the right sort of numbers. Model the thermal distortion as shown. Assume that the initial temperature of the bottle was 20°C and that the liquid ended up near 100°C. A rough guess for h might be 10 mm. For soda glass $E = 74$ GPa, $\alpha = 8.5 \times 10^{-6} \, °C^{-1}$ and $\sigma_{TS} = 50$ MPa. How might you avoid this problem in future?

18.1 After 3 years in service a boiler started to leak steam from a position below the water line. The leak was traced to a fine crack 20 mm long in the wall of the boiler barrel alongside the entry position of the water feedpipe. When the boiler had been made the water pipe had been inserted into a hole in the wall of the barrel and had been welded in place with a full-penetration weld. An internal inspection showed severe cracking of the boiler plate. It was concluded that the cracks had initiated at the inner surface of the barrel and had then propagated through the wall. There were several curved cracks which formed arcs of a circle concentric with the axis of the pipe. The longest cracks went about half way around a circle. The radial distance between the cracks and the weld was approximately equal to the plate thickness. A baffle had been fitted in the water space a short distance in from the feed pipe in order to deflect the cold feed water sideways along the boiler barrel. There was noticeable pitting corrosion of the plate in the region of the water inlet. Microsections cut from the cracked areas showed fine unbranched cracks running in a transgranular fashion through the plate at right angles to the surface. Indicate the likely mechanism of failure and support it with a simple calculation.

From: B. G. Whalley, "The Analysis of Service Failures", *Metallurgist and Materials Technologist*, **15**, 21 (1983).

19.1   An electric iron for smoothing clothes was being used when there was a loud bang and a flash of flame from the electric flex next to the iron. An inspection showed that the failure had occurred at the point where the flex entered a polymer tube which projected about 70 mm from the body of the iron. There was a break in the live wire and the ends of the wire showed signs of fusion. The fuse in the electric plug was intact. Explain the failure. Relevant data are given below.

Rating of appliance: 1.2 kW.
Power supply: 250 V, 50 Hz AC.
Fuse rating of plug: 13 A.
Flex: 3 conductors (live, neutral and earth) each rated 13 A.
Individual conductor: 23 strands of copper wire in a polymer sheath; each strand
    0.18 mm in diameter.
Age of iron: 14 years.
Estimated number of movements of iron: 10^6.

19.2   The photograph shows the fracture surfaces of two broken tools from a pneumatic drill. The circular fracture surface is 35 mm in diameter and the

"rectangular" surface measures 24 mm × 39 mm. Discuss the features of each fracture surface in relation to the mechanisms by which the failures took place.

20.1 Referring to the bicycle crank of Example 15.1, estimate the stress cycle at the origin of the fatigue cracks and hence find the approximate age of the bicycle. You may apply simple elastic beam theory to the cross-section of the crank at the location of the failure. Ignore any compressive load transfer between the pedal spindle and the side of the hole. Ignore stress concentrations as well. Additional data are given below.

> Horizontal distance between centre of crank cross-section and centre of pedal = 65 mm.
> Estimated vertical force on pedal = 45 kgf.
> Diameter of rear wheel = 690 mm.
> Gear ratio = 2.5.
> Estimated distance travelled = 15 km per day.
> Vickers hardness of crank = 234.

The specification of the crank is not known, but Smithells gives a tensile strength of 780 MPa for a heat treated steel containing 0.4% carbon and 0.8% manganese. The fatigue limit of this steel, measured on smooth specimens in uniaxial tension/compression, is ±278 MPa at 10^7 cycles.

Answers: ±298 MPa; roughly 10 years.

20.2 A rotating steel shaft from a high-speed textile machine had been repaired with a circumferential surface weld. The weld bead had been laid immediately next to a sharp change in the diameter of the shaft. The weld had been turned flush with the original surface but the lathe tool had left a coarse finish. The shaft failed after being put back into service. Indicate the likely mechanism and characteristics of the failure and identify the factors that would have contributed to it. Data are as follows: hardness of shaft away from repair = 220 HV; maximum hardness of HAZ = 560 to 580 HV; analysis of shaft material in weight %:

C	Si	Mn	S	P	Cr	Mo	Ni
0.40	0.27	0.72	0.041	0.012	0.96	0.22	0.23

The carbon equivalent of a steel is given by

$$CE = C + \frac{Mn}{6} + \frac{Cr + Mo + V}{5} + \frac{Ni + Cu}{15},$$

where the chemical symbols represent the concentration in weight % of the element concerned.

From: B. G. Whalley, "The Analysis of Service Failures", *Metallurgist and Materials Technologist*, **15**, 21 (1983).

21.1   The diagram shows the crank-pin end of the connecting rod on a large-scale miniature steam locomotive. The locomotive weighs about 900 kg and is designed

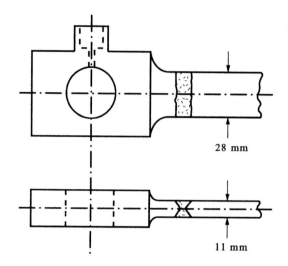

28 mm

11 mm

for hauling passengers around a country park. In the full-size prototype the connecting rod and big-end were forged from a single billet of steel. However, to save cost in building the miniature version, it is intended to weld the two parts together with a full-penetration double-sided weld, grinding the surface flush to hide the joint. Do you think that this design solution will have the required fatigue properties? Design and operational data are given below.

Diameter of cylinder = 90 mm.
Diameter of driving wheel = 235 mm.
Steam pressure inside cylinder at point of admission = 7 bar gauge.
Estimated annual distance travelled = 6000 km.
Design life = 20 years minimum.

21.2   Vibrating screens are widely used in the mining industry for sizing, feeding and washing crushed mineral particles. A typical screen consists of a box fabricated from structural steel plate which contains a mesh screen. During operation the box is shaken backwards and forwards at a frequency of up to 20 Hz. Although the major parts of the box are often fixed together using bolts or rivets, individual sub-assemblies such as a side of the box frequently have welded joints, particularly where frame stiffeners or gussets are added. Owing to the inertial forces generated by the rapid shaking the stresses in the unit can be significant and the fatigue design of the welded connections has to be considered carefully.

The side of one box consists of a relatively thin plate which is stiffened with triangular gussets as shown in the diagram on page 413. Strain gauges attached to the plate during operation show that the maximum principal stress range in the plate near the end of the gusset is 8 N mm^{-2}. Given that the screen is expected to work for 12 hours per day and 6 days per week estimate the time that it will take for there to be a 50% chance of a crack forming in the plate at the end of each

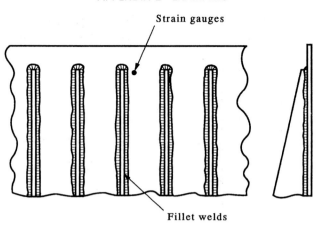

gusset. You should assume that the screens operate in a moderately corrosive environment. What would be the time for a 2.3% chance of cracking?

Answers: 11 years; 4 years.

From: P. R. Fry and M. E. Greenway, "An Approach to Assessing Structural Integrity and Fatigue Failures in Vibrating Equipment", in *Fracture and Fracture Mechanics: Case Studies*, edited by R. B. Tait and G. G. Garrett, Pergamon, 1985, p. 159.

22.1   A marine harbour wall was cased with sheet piling made from carbon–manganese structural steel. Five months after the piles had been driven a flaking corrosion product typically 0.8 mm thick had formed on the surface of the piles where they went through the tidal zone. Eight separate flakes were analysed for iron with the following results.

Sample no.	Area of flake (mm²)	Mass of iron (mg)
1	174	207.9
2	68	43.0
3	68	42.8
4	56	42.6
5	73	39.4
6	149	146.4
7	78	52.2
8	107	92.4

The piling was originally covered with a dense layer of iron oxide mill scale from the hot-rolling operation. The scale was estimated to have been 0.10 mm thick. It is assumed that the scale was incorporated in the flakes of corrosion product. Estimate the thickness of steel lost by wet corrosion in the marine environment assuming that the mill scale layer is composed of (a) Fe_2O_3, (b) Fe_3O_4, (c) FeO. The densities of the three oxides are respectively 5.26, 5.18 and 5.7 g cm^{-3}. The atomic weight of oxygen is 16.00. Other data are given in Appendix 1F.

Answers: (a) 0.053 mm, (b) 0.052 mm, (c) 0.043 mm. In other words the thickness of metal lost is of the order of 0.05 mm irrespective of the precise composition of the mill scale.

22.2 The low-alloy steel propeller shaft of a sea-going ship was covered with a tubular bronze sleeve with the aim of protecting it against corrosion. The sleeve was made up from two lengths of tube which were joined end-to-end with soft solder. The solder did not fill the gap properly and sea water found its way onto the surface of the shaft underneath. As a result the surface of the shaft suffered from pitting corrosion which in turn initiated fatigue cracks and led to the eventual breakage of the shaft. Explain why the pitting corrosion took place.

From: A. Ball, "The Fracture of a Ship's Propeller Shaft", in *Fracture and Fracture Mechanics: Case Studies*, edited by R. B. Tait and G. G. Garrett, Pergamon, 1985, p. 265.

23.1 When a car was 7 years old a water leak appeared in one of the core plugs fitted to the cylinder block. The plug was a dish-shaped pressing made from mild steel plate 1.5 mm thick. The plug was pressed into the core hole in the side of the block so that the flat central area of the pressing was in direct contact with the cooling water space. On closer inspection it could be seen that the centre of the plug had perforated. The water in the cooling system was reddish-brown in colour and it emerged that the engine coolant had not been changed since the car was made. Why did the failure take place?

23.2 Water taps with brass bodies frequently start to drip after a few years of use. The standard culprit is the rubber washer. Indeed, renewing the washer often stops .the leak, but usually only for a few weeks. If the brass seating is examined at this stage it will appear red in colour, in contrast to the original bright yellow. If the seating is machined back with a tap-seat cutter it is found that a depth typically of 0.5 mm has to be removed before the whole seating is coloured yellow again. If this is done the tap will function perfectly even with an old washer. The red metal that is removed is in fact porous copper which allows water to seep through underneath the washer. Why does this type of corrosion take place?

24.1 A pipe 300 mm in diameter was made by forming a long strip of austenitic stainless steel into a helix and butt welding the helical joint. The pipe was used to carry hot dilute acid across a river estuary. The temperature of the acid was 80°C and the pipe was therefore lagged with fibre insulation which was clad with a waterproof skin of aluminium. After several years in service the pipeline supports suffered significant corrosion. In order to repair them the cladding and lagging were removed and later renewed. Several months later leakage of acid was discovered at a number of locations. Numerous cracks were found in the wall of the pipe. Deposits on the external surface of the pipe were rich in chloride ions. It was also found that rain water had been seeping under the cladding. Give a likely explanation for the failure.

From: B. G. Whalley, "The Analysis of Service Failures", *Metallurgist and Materials Technologist*, **15**, 193 (1983).

24.2 A model steam boiler was made from copper parts which were silver soldered together. A blow-down valve was fitted to the lowest point in the boiler. The valve body was made of bronze and was silver soldered through the boiler wall. The valve spindle was made from austenitic stainless steel rod 4 mm in diameter. The end of the spindle was made conical so that it could be screwed down hard onto the bronze seating. After each run the blow-down valve was opened fully. As a result the residual steam pressure forced all the water out of the boiler. The valve was left open so that the residual heat would dry the boiler out completely. The boiler was steamed on five occasions using distilled water as the feed. The total steaming time was about 10 hours at an operating pressure of 6 bar gauge. After this time the valve started to leak and it was found that the end of the spindle that had been in contact with the water space had corroded to a depth of of 1 mm on one side. Explain the mechanism of corrosion.

25.1 A boiler for heating water runs off gas produced from sewage sludge. The gas has a composition of 70% by volume methane. The boiler is made from sections of cast iron. After running non-stop for three months a deposit about 3 mm thick had formed on the fire-side surfaces of the boiler. Flakes of the deposit were detached by tapping the boiler with a hammer. The flakes had a red-yellow colour consistent with the presence of iron compounds and elemental sulphur. On heating, the yellow particles melted, turned black and gave off sulphur dioxide. This confirmed the presence of elemental sulphur: sewage sludge gas contains hydrogen sulphide which burns to give sulphur in an oxygen-deficient atmosphere. One of the flakes weighed 7.81 g and had an area of 11.1 cm^2. An analysis of the flake gave weight percentages of 22.1 iron, 16.3 sulphur and 12.0 moisture.

Using the data estimate the loss in thickness of the cast iron wall. What assumptions would you make when estimating the loss in thickness over a 5-year period? How accurate do you think such an estimate might be?

Answer: 0.2 mm.

25.2 The diagram on page 416 is a schematic view of the lower end of a tube-and-shell heat exchanger made from mild steel. The unit was designed to heat oil in a chemical process plant. The oil was passed through the small tubes and the heat was supplied from steam which was injected into the shell. The unit had been in operation for only $2\frac{1}{2}$ years when one of the tubes perforated. When the tubes were extracted from the shell it was found that they had all corroded on the outside over a distance of about 160 mm from the lower tube plate. On the worst-affected areas attack had occurred to a depth of about 1.5 mm over regions measuring typically 10 mm by 20 mm. The corroded areas were light brown in colour.

The heat exchanger was operated on a cyclic basis as follows. First, saturated steam was admitted to the shell at a temperature of 180°C in order to heat a new batch of oil. The steam condensed on the surfaces of the tubes and the condensed water trickled down to the bottom of the shell where it was drawn off the via the condensate drain. When the oil was up to temperature the steam supply was cut off and the pressure in the shell was dropped to atmospheric. The cycle was repeated when it was time to heat up a new batch of oil.

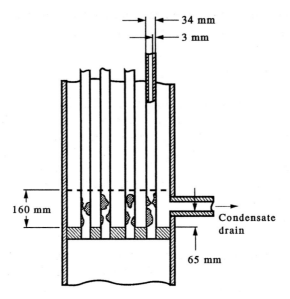

Account for the severity and location of the corrosion on the small tubes. Suggest a simple design modification which might reduce the rate of attack when the replacement tubes are installed.

26.1 The diagram shows a compression joint for fixing copper water pipe to plumbing fittings. When assembling the joint the gland nut is first passed over the

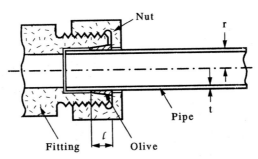

pipe followed by a circular olive made from soft copper. The nut is then screwed onto the end of the fitting and the backlash is taken up. Finally the nut is turned through a specified angle which compresses the olive on to the surface of the pipe. The angle is chosen so that it is just sufficient to make the cross-section of the pipe yield in compression over the length which is in contact with the olive. Show that the water pressure required to make the pipe shoot out of the fitting is given approximately by

$$p_w = 2\mu\sigma_y \left(\frac{t}{r}\right)\left(\frac{l}{r}\right)$$

where μ is the coefficient of friction between the olive and the outside of the pipe.

Calculate p_w given the following information: $t = 0.65$ mm, $l = 7.5$ mm, $r = 7.5$ mm, $\mu = 0.15$, $\sigma_y = 120$ MPa. Comment on your answer in relation to typical hydrostatic pressures in water systems.

Answer: 3.1 MPa, or 31 bar.

27.1   Find the axial compressive load which will make one of the diagonal bracing bars of a Tay Bridge pier fail by Euler buckling. Assume that the ends of the bar are simply supported. The working drawings given in Fig. 27.4 indicate that a diagonal bar was in fact made from *two* separate bars each of which had a cross-section measuring 0.5 in × 4 in. Consider both this situation and also the case of a one-piece bar measuring 1 in × 4 in. Assume that the bar is 13 ft long and take $E \approx 13{,}000$ tsi for wrought iron.

Answers: 0.44 ton, 1.74 tons.

28.1   (a) Find the maximum bending moment which can be supported by a cast-iron pipe from one of the 20-in columns of the Tay Bridge.
   (b) Estimate the bending moment needed to make a flanged connection in the column fail by yielding of the wrought-iron bolts.

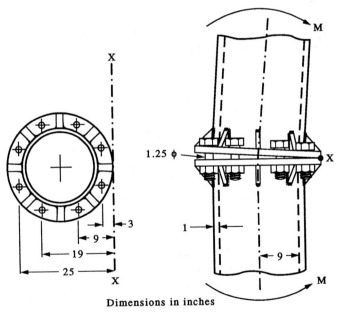

Dimensions in inches

Ignore the strengthening effect of the cement filling in both cases. The relevant dimensions are shown in the diagram. Assume that $\sigma_{TS} \approx 10$ tsi for cast iron and that $\sigma_y \approx 11$ tsi for wrought iron.

Answers: (a) 225 ton ft, (b) 126 ton ft.

SOLUTIONS TO EXAMPLES

1.1 For the present situation we have

$$\varepsilon_1 = 0, \quad \varepsilon_2 = 0, \quad \sigma_1 = \sigma_2 \quad \text{and} \quad E_{\text{ax}} = \frac{\sigma_3}{\varepsilon_3}.$$

Using the elastic stress–strain equations given in Appendix 1A we have

$$0 = \frac{\sigma_1}{E} - v\frac{\sigma_1}{E} - v\frac{\sigma_3}{E},$$

$$\varepsilon_3 = \frac{\sigma_3}{E} - v\frac{\sigma_1}{E} - v\frac{\sigma_1}{E}.$$

Rearranging equations then gives us

$$\sigma_1 = v\frac{\sigma_3}{(1-v)},$$

$$\varepsilon_3 = \frac{\sigma_3 - 2v\sigma_1}{E},$$

$$\varepsilon_3 = \frac{\sigma_3}{E}\left(\frac{1 - v - 2v^2}{1-v}\right) = \frac{\sigma_3}{E}\left\{\frac{(1-2v)(1+v)}{1-v}\right\},$$

$$E_{\text{ax}} = \frac{\sigma_3}{\varepsilon_3} = \frac{\sigma_3 E(1-v)}{\sigma_3(1-2v)(1+v)}$$

$$= \frac{E(1-v)}{(1-2v)(1+v)}.$$

1.2 As a first approximation we assume that the rubber is linear elastic for small strains. The strain energy stored in unit volume is then equal to

$$U^{\text{el}} = \frac{\sigma^2}{2E}.$$

When the solid surface is loaded in compression the rubber cannot expand sideways to any great extent because it is constrained by the sole of the shoe to which it is stuck. The appropriate modulus for calculating the strain energy is thus the axial modulus as defined in Example 1.1. On the other hand, provided the moulded surface is not compressed too much (which would close up the gaps) the protrusions are free to expand laterally. So the modulus to use for the moulded surface is simply Young's modulus.

Since the energy input is the same in both cases, we can now write

$$\sigma(\text{solid}) = \sigma(\text{moulded}) \times \sqrt{\frac{E_{\text{ax}}}{E}}$$

$$= \sigma(\text{moulded}) \times \sqrt{\frac{(1 - 0.49)}{(1 - 2 \times 0.49)(1 + 0.49)}}$$

$$= \sigma(\text{moulded}) \times \sqrt{17}$$

$$= 4.1 \times \sigma(\text{moulded}).$$

1.3 The unidirectional protrusions are prevented from expanding along their lengths. This is a situation of plane strain and Appendix 1A shows that the appropriate modulus in this case is given by

$$E_{\text{ps}} = \frac{E}{(1 - v^2)} = \frac{E}{(1 - 0.49^2)} = 1.3E.$$

We can then write

$$\sigma(\text{solid}) = \sigma(\text{moulded}) \times \sqrt{\frac{E_{\text{ax}}}{E_{\text{ps}}}}$$

$$= \sigma(\text{moulded}) \times \sqrt{\frac{17E}{1.3E}}$$

$$= 3.6 \times \sigma(\text{moulded}).$$

2.1 From Appendix 1B the appropriate expression is

$$f_{\text{cr}} = 1.103 \sqrt{\frac{EI}{Ml^3}}.$$

In addition,

$$I = \frac{\pi r^4}{4}.$$

Combining these results gives

$$f_{\text{cr}} = 1.103 \left\{ \frac{E}{M} \times \frac{\pi}{4} \left(\frac{r}{l} \right)^3 r \right\}^{1/2}.$$

Thus

$$l = \left\{ \frac{E}{M} \times \frac{\pi}{4} \times r^4 \times \frac{(1.103)^2}{f_{\text{cr}}^2} \right\}^{1/3}$$

$$= \left\{ \frac{\pi}{4} \times \frac{212 \times 10^9 \text{ N m}^{-2}}{0.025 \text{ kg}} \left(\frac{2.34 \times 10^{-3}}{2} \right)^4 \text{ m}^4 \times \frac{(1.103)^2}{(40)^2 \text{ s}^{-2}} \right\}^{1/3}.$$

Using $N = kg\ m\ s^{-2}$ we have

$$l = 0.212\ m = \underline{\underline{212\ mm}}.$$

This value of l places the mass almost exactly half-way along the free length of the rod, which is the situation assumed in deriving the expression for f_{cr}.

2.2   From Appendix 1B the appropriate expression is

$$f_{cr} = 0.560\sqrt{\frac{EI}{Ml^3}}$$

with

$$I = \pi r^3 t \qquad \text{and} \qquad M = 2\pi r t l\rho.$$

Thus

$$\frac{I}{M} = \frac{r^2}{2l\rho}$$

and

$$f_{cr} = 0.560\left\{\frac{1}{2}\left(\frac{r}{l}\right)^2 \frac{1}{l^2} \times \frac{E}{\rho}\right\}^{1/2}$$

$$= 0.560\left\{\frac{1}{2}\left(\frac{25\ mm}{1500\ mm}\right)^2 \times \frac{1}{(1.5^2)\ m^2} \times \frac{212 \times 10^9\ N\ m^{-2}}{7.8 \times 10^3\ kg\ m^{-3}}\right\}^{1/2}.$$

Using $N = kg\ m\ s^{-2}$ we have

$$f_{cr} = 22.9\ s^{-1} = \underline{\underline{1376\ rpm}}.$$

3.1   From the diagram,

$$e = u_t + u_r.$$

Also

$$\sigma_t = \frac{p_t a_t}{t_t}; \qquad \varepsilon_t = \frac{u_t}{a_t} = \frac{\sigma_t}{E_t}.$$

$$u_t = a_t \varepsilon_t = \frac{a_t \sigma_t}{E_t}$$

$$= \frac{a_t}{E_t} \times \frac{p_t a_t}{t_t} = \frac{p_t}{E_t} \times \frac{a_t^2}{t_t}.$$

Similarly,

$$\sigma_r = \frac{(p_t - p_r)a_r}{t_r},$$

and

$$u_r = \frac{(p_t - p_r)}{E_r} \times \frac{a_r^2}{t_r} = \frac{p_t a_r^2}{E_r t_r} - \frac{p_r a_r^2}{E_r t_r}.$$

Now

$$p_r = \frac{nf_s}{2\pi a_r L_r}$$

so

$$u_r = \frac{p_t a_r^2}{E_r t_r} - \frac{a_r^2}{E_r t_r} \times \frac{n f_s}{2\pi a_r L_r}.$$

$$\varepsilon_s = \frac{u_r}{l_s} = \frac{\sigma_s}{E_s}.$$

$$u_r = \frac{\sigma_s l_s}{E_s} = \frac{f_s l_s}{t_s w_s E_s}.$$

Thus

$$u_r = \frac{p_t a_r^2}{E_r t_r} - \frac{a_r^2}{E_r t_r} \times \frac{n f_s}{2\pi a_r L_r} = \frac{f_s l_s}{t_s w_s E_s}$$

giving

$$p_t = \frac{f_s E_r t_r}{a_r^2} \left(\frac{l_s}{t_s w_s E_s} + \frac{a_r n}{2\pi t_r L_r E_r} \right).$$

Finally,

$$u_t = \frac{p_t a_t^2}{E_t t_t} = f_s \left(\frac{E_r}{E_t} \right) \left(\frac{t_r}{t_t} \right) \left(\frac{a_t}{a_r} \right)^2 \left(\frac{l_s}{t_s w_s E_s} + \frac{a_r n}{2\pi t_r L_r E_r} \right)$$

which gives

$$e = f_s \left\{ \left(\frac{E_r}{E_t} \right) \left(\frac{t_r}{t_t} \right) \left(\frac{a_t}{a_r} \right)^2 \left(\frac{l_s}{t_s w_s E_s} + \frac{a_r n}{2\pi t_r L_r E_r} \right) + \frac{l_s}{t_s w_s E_s} \right\}.$$

3.2 Using the simplifying assumptions the result for e reduces to

$$e = f_s \left\{ \left(\frac{E_r}{E_t} \right) \left(\frac{l_s}{t_s w_s E_s} + \frac{a_r n}{2\pi t_r L_r E_r} \right) + \frac{l_s}{t_s w_s E_s} \right\}$$

$$= f_s \left\{ \frac{l_s}{t_s w_s E_s} \left(1 + \frac{E_r}{E_t} \right) + \frac{a_r n}{2\pi t_r L_r E_t} \right\}.$$

Now

$$\frac{E_r}{E_t} = \frac{E_s}{E_t} = \frac{152 \text{ GPa}}{212 \text{ GPa}} = 0.72.$$

Thus

$$e = \frac{f_s}{E_s} \left\{ \frac{1.72 l_s}{t_s w_s} + \frac{0.72 a_r n}{2\pi t_r L_r} \right\}.$$

Now

$$e \approx 1.3 \times 10^{-3} \times a_t \approx 1 \text{ mm}.$$

Thus

$$f_s = \frac{1 \text{ mm} \times 152 \times 10^3 \text{ MPa}}{\left\{ \dfrac{1.72 \times 550 \text{ mm}}{50 \text{ mm} \times 105 \text{ mm}} + \dfrac{0.72 \times 780 \text{ mm} \times 20}{2\pi \times 80 \text{ mm} \times 140 \text{ mm}} \right\}}$$

$$= 4.5 \times 10^5 \text{ MPa mm}^2.$$

$$p_h = \frac{nf_s}{2\pi A L_h} = \frac{20 \times 4.5 \times 10^5 \text{ MPa mm}^2}{2\pi \times 230 \text{ mm} \times 180 \text{ mm}}$$

$$\approx 35 \text{ MPa}.$$

$$\Delta B = \left(\frac{p_h}{E}\right)\left(\frac{2A^2 B}{A^2 - B^2}\right) = \left(\frac{p_h}{E}\right) B \left(\frac{2A^2}{A^2 - B^2}\right)$$

$$= \left(\frac{35 \text{ MPa}}{152 \times 10^3 \text{ MPa}}\right) 115 \text{ mm} \left(\frac{2 \times 230^2}{230^2 - 115^2}\right)$$

$$\approx 70 \text{ } \mu\text{m}.$$

4.1

$$\frac{L}{r} = \frac{100 \text{ mm}}{33 \text{ mm}} = 3.0$$

and

$$\frac{r}{t} = \frac{33 \text{ mm}}{0.12 \text{ mm}} = 275.$$

The Tube Buckling Map (Appendix 1B) shows that $n = 6$ for these values. The buckling equation is then

$$p_{cr}(\text{theoretical}) = \frac{0.86E}{(1 - v^2)^{3/4}} \left(\frac{t}{L}\right)\left(\frac{t}{r}\right)^{3/2}$$

$$= \frac{0.86 \times 212 \times 10^3 \text{ N mm}^{-2}}{(0.936)} \left(\frac{0.12 \text{ mm}}{100 \text{ mm}}\right)\left(\frac{1}{275}\right)^{3/2}$$

$$= 51.3 \text{ kPa gauge}.$$

Atmospheric pressure $= 101.3$ kPa, so the absolute pressure inside the can at the onset of buckling is $101.3 - 51.3 = 50$ kPa. The saturation temperature at this pressure is 81°C (see the Steam Table in Appendix 1H).

Note: at the onset of buckling, the compressive hoop stress in the can is given conservatively by

$$\sigma = \frac{p_{cr}(\text{theoretical})r}{t} = 0.051 \text{ MPa} \times 275 = 14 \text{ MPa}.$$

This is much less than the yield stress of the steel and the elastic buckling equation is valid. Since p_{cr} (actual minimum) $\approx 0.8 p_{cr}$(theoretical) the can could buckle at a gauge pressure as low as 0.8×51.3 kPa $= 41.04$ kPa. This is achieved when the absolute pressure inside the can is $101.3 - 41.0 = 60.3$ kPa, corresponding to a saturation temperature of 86°C.

4.2

$$\frac{L}{r} = \frac{4000 \text{ mm}}{100 \text{ mm}} = 40$$

and

$$\frac{r}{t} = \frac{100 \text{ mm}}{5 \text{ mm}} = 20.$$

The Tube Buckling Map (Appendix 1B) shows that $n = 2$ for these values. The appropriate buckling result is:

$$p_{cr}(\text{theoretical}) = \frac{E}{4(1 - v^2)} \left(\frac{t}{r}\right)^3 = \frac{69 \times 10^3 \text{ MPa}}{4 \times 0.91} \left(\frac{1}{20}\right)^3 = 2.37 \text{ MPa}.$$

$$p_{cr}(\text{actual minimum}) \approx 0.8 \times 2.37 \text{ MPa} = \underline{1.9 \text{ MPa}}.$$

$$p = h\rho g = 200 \text{ m} \times 10^3 \times 9.81 \text{ N m}^{-3} = \underline{2.0 \text{ MPa}}.$$

The pressure at a depth of 200 m is slightly more than p_{cr} (actual minimum), so the tube failed by external-pressure buckling.

The compressive hoop stress at failure is given by

$$\sigma = \frac{pr}{t} = 2.0 \text{ MPa} \times 20 = 40 \text{ MPa}.$$

This is much less than the yield stress of the tube and the elastic buckling equation is valid.

The mass of the tube is given by

$$m = 2\pi rt L\rho.$$

Now

$$p_{cr} = \frac{E}{4(1 - v^2)} \left(\frac{t^3}{r^3}\right)$$

which can be rearranged to give

$$t^3 = \frac{p_{cr} 4(1 - v^2)r^3}{E}$$

and

$$t = \{p_{cr} 4(1 - v^2)r^3\}^{1/3} \frac{1}{E^{1/3}}.$$

Thus

$$m = 2\pi r L\rho \{p_{cr} 4(1 - v^2)r^3\}^{1/3} \frac{1}{E^{1/3}}$$

$$= 2\pi L \{p_{cr} 4(1 - v^2)r^6\}^{1/3} \left(\frac{\rho}{E^{1/3}}\right).$$

Material	E (GPa)	ρ (Mg m^{-3})	$(\rho/E^{1/3})$	m (aluminium)/m
Steel	212	7.8	1.31	0.5
Aluminium	69	2.7	0.66	1.0
GFRP	30	2.0	0.64	1.0
CFRP	120	1.5	0.30	2.2

4.3 Since ring is unconstrained use the equation for the external-pressure buckling of a long tube (see Appendix 1B) and omit the plane-strain elastic constraint factor of $(1 - v^2)$. The buckling equation then becomes

$$p_{cr} = \frac{E}{4}\left(\frac{t}{r}\right)^3.$$

Now

$$p_{cr} = \frac{F}{2\pi r L}, \quad \text{or} \quad F = p_{cr} 2\pi r L.$$

Thus

$$F_{cr} = 2\pi r L \times \frac{E}{4}\left(\frac{t}{r}\right)^3 = \frac{\pi E L}{2}\left(\frac{t^3}{r^2}\right).$$

For the ring as given

$$I = \frac{L t^3}{12}.$$

Thus

$$F_{cr} = \frac{\pi E L t^3}{2 r^2} = \frac{\pi E}{2 r^2} \times 12 I = \frac{6 \pi E I}{r^2}.$$

5.1 From Appendix 1B

$$p_{cr}(\text{theoretical}) = \frac{2E}{\{3(1 - v^2)\}^{1/2}}\left(\frac{t}{r}\right)^2.$$

Now

$$r = \frac{R^2 + h^2}{2h} = \frac{500^2 \text{ mm}^2 + 200^2 \text{ mm}^2}{2 \times 200 \text{ mm}} = 725 \text{ mm}.$$

$$p_{cr}(\text{theoretical}) = \frac{2 \times 2.6 \times 10^3 \text{ MPa}}{(3 \times 0.75)^{1/2}}\left(\frac{5 \text{ mm}}{725 \text{ mm}}\right)^2$$

$$= 0.165 \text{ MPa gauge}.$$

$$p_{cr}(\text{actual minimum}) \approx 0.3 \times 0.165 \text{ MPa}$$

$$= 0.050 \text{ MPa} = 50 \text{ kPa gauge}.$$

$$p = h\rho g = h \times 10^3 \times 9.81 \text{ N m}^{-3} = 9.81 h \text{ kPa gauge}.$$

Finally,

$$h = \frac{50 \text{ kPa}}{9.81 \text{ kPa}} = 5.1 \text{ m}.$$

Note that the compressive hoop stress is

$$\frac{p_{cr}(\text{actual minimum})r}{2t} = \frac{0.050 \text{ MPa} \times 725 \text{ mm}}{5 \text{ mm}} = 7.25 \text{ MPa}.$$

This is much less than the yield stress so the elastic buckling equation is valid.

From Appendix 1B,

$$\lambda = \{12(1 - v^2)\}^{1/4}\left(\frac{r}{t}\right)^{1/2} \alpha.$$

Now

$$\sin \alpha = \frac{R}{r} = \frac{500 \text{ mm}}{725 \text{ mm}} \qquad \text{so } \alpha = 0.75 \text{ rad}.$$

Thus

$$\lambda = \{12 \times 0.75\}^{1/4}\left(\frac{725 \text{ mm}}{5 \text{ mm}}\right)^{1/2} 0.75 = 15.6$$

which is within the range of validity quoted in Appendix 1B.

5.2   We substitute the values for E, v, t and r into the expression given in Solution 5.1 for p_{cr} (theoretical) and obtain a figure of 24.9 MPa. p_{cr} (actual minimum) is then approximately equal to 0.3×24.9 MPa $= 7.5$ MPa $= 7500$ kPa. This is, of course, the gauge pressure and because atmospheric pressure is about 1 bar the absolute pressure is 7600 kPa.

Appendix 1H gives saturation temperatures and pressures for butane and propane. At 90°C the saturation pressures are 1265 and 3760 kPa absolute respectively. The factors of safety are therefore:

(a) 7600 kPa/1265 kPa = 6.0,
(b) 7600 kPa/3760 kPa = 2.0.

The biaxial compressive stress in the cap at 7500 kPa is

$$(7.5 \text{ MPa} \times 38.5 \text{ mm})/(2 \times 0.38 \text{ mm}) = 380 \text{ MPa}.$$

This is comparable to the yield stress assumed for the steel. However, at the assumed maximum operating pressure of 3760 kPa the cap will be well below yield.

The value of the parameter λ can be calculated from the expression given in Solution 5.1 by inserting the appropriate parameter values. Note that the angle α is given by $\sin \alpha = (R/r) = (21 \text{ mm}/38.5 \text{ mm}) = 0.545$, so $\alpha = 0.57$ radian. The value found for λ is 10.5, which is well within the required range for the analysis to be valid.

6.1   When the circular area has cooled down to the temperature of the rest of the vessel it will be in a state of equi-biaxial tension. We call the biaxial stress p. If we were able to cut the circular area out of the wall of the vessel then we would have to apply a uniform radial stress p to the circumference in order to stop it contracting radially. We would also have to apply a balancing radial stress p to the circumference of the hole to stop it expanding radially. The behaviour of both the circular area and the plate surrounding it can in fact be modelled by the formulae for a thick-walled tube given in Appendix 1A.

The increase in the external radius of a thick-walled tube subjected to a *negative* external pressure p is given by

$$\Delta a = \left(\frac{p}{E}\right) a \left\{\left(\frac{a^2 + b^2}{a^2 - b^2}\right) - v\right\}.$$

Since $b = 0$ for a solid disc the equation can be simplified and re-arranged to give the following expression for the radial strain:

$$\frac{\Delta a}{a} = \left(\frac{p}{E}\right)(1 - v).$$

The decrease in the internal radius of a thick-walled tube subjected to a *negative* internal pressure p is given by

$$\Delta B = \left(\frac{p}{E}\right)B\left\{\left(\frac{A^2 + B^2}{A^2 - B^2}\right) + v\right\}.$$

Since $A = \infty$ for a hole in an infinite plate the radial strain is:

$$\frac{\Delta B}{B} = \left(\frac{p}{E}\right)(1 + v).$$

The two radial strains sum to give the thermal strain, so

$$\frac{\Delta a}{a} + \frac{\Delta B}{B} = \alpha \, \Delta T,$$

$$\left(\frac{p}{E}\right)(1 - v) + \left(\frac{p}{E}\right)(1 + v) = \alpha \, \Delta T,$$

$$p = \frac{E\alpha \, \Delta T}{2}.$$

Finally, inserting the given values for E and α and setting $\Delta T = 100°C$ produces a value for p of 127 MPa.

The residual stress is about half the yield stress and yet it has been set up by a comparatively small temperature differential. Temperature differentials in welding operations are often much greater than the one assumed here, and it is common to have residual stresses that are equal to the yield stress. This has the following practical consequences.

(a) Depending on the environment welds are susceptible to stress corrosion cracking. Residual stresses also increase the driving force for hydrogen cracking.

(b) The fatigue properties of welds are generally poor compared to those of the parent plate.

(c) Cracks in welds go critical at lower values of operating stress.

(d) If the stress acts at right angles to the surface of the plate (in the "through-thickness" direction) the plate can "delaminate" (see Example 14.1).

(e) Welded fabrications can show distortion.

6.2 The residual stress distribution is found from the equations

$$M_{el} = \frac{\sigma_{el} I}{z},$$

$$M_{el} + M_p = 0,$$

$$\sigma_{res} = \sigma_y + \sigma_{el}.$$

For case (a)

$$I = \frac{a^4}{12}, \qquad z = \frac{a}{2} \qquad \text{and} \qquad M_p = \frac{\sigma_y a^3}{4}.$$

The equations gives residual stresses of $0.5\sigma_y$ in compression at one edge and $0.5\sigma_y$ in tension at the other edge. The required answer is thus $\underline{\pm 0.50}$.

The answers to cases (b), (c) and (d) are obtained in an exactly analogous way.

7.1

L/r	r/t	n	Equation	t/L	Actual minimum p_{cr} (MPa)
36.9	6.04	2	7.8		8.09
47.1/27.0	7.75/10.6	2	7.8		4.91/2.63
50.5/25.0	6.92/10.4	2	7.8		6.16/2.73
69.9/44.0	8.79/7.79	2	7.8		3.82/4.86
66.5/33.0	6.92/10.4	2	7.8		6.16/2.73
42.3	6.76	2	7.8		6.46
50.9	9.82	2	7.8		3.06
5.1	14.3	2 or 3	7.9	0.0137	2.28
8.5	14.0	2	7.9	0.00843	1.84

The Tube Buckling Map in Appendix 1B is used to find the value of n and to select the correct buckling equation. Equation (7.8) is developed as shown in Chapter 7 to give

$$p_{cr}(\text{actual minimum}) \approx \frac{0.8a}{2.83(1 - v^2)^{1/2}} \left(\frac{t}{r}\right)^2.$$

Since $a = 0.98 \times 10^3$ MPa and $v = 0.34$, this becomes

$$p_{cr}(\text{actual minimum}) \approx 2.95 \times 10^2 \text{ MPa} \left(\frac{t}{r}\right)^2.$$

The predicted collapse pressures are then derived using the values of r/t given in the table above.

Equation (7.9) is developed as shown in Chapter 7 to give

$$p_{cr}(\text{actual minimum}) \approx \frac{0.528a}{(1 - v^2)^{3/8}} \left(\frac{t}{L}\right)^{1/2} \left(\frac{t}{r}\right)^{5/4}$$

which leads to

$$p_{cr}(\text{actual minimum}) \approx 5.42 \times 10^2 \text{ MPa} \left(\frac{t}{L}\right)^{1/2} \left(\frac{t}{r}\right)^{5/4}.$$

L/r	r/t	t/L	Actual minimum p_{cr} (MPa)
3.65	14.8	0.0186	2.55
1.84	16.2	0.0334	3.05
8.68	20.9	0.00552	0.90
3.56	15.3	0.0183	2.42
3.56	15.8	0.0178	2.30
2.02	20.8	0.0239	1.87
1.49	21.8	0.0308	2.02

The Tube Buckling Map in Appendix 1B shows that Eqn. (7.9) is appropriate for all the fireboxes. This is developed as shown in the solution to Example 7.1 to give

$$p_{cr}(\text{actual minimum}) \approx 5.42 \times 10^2 \text{ MPa} \left(\frac{t}{L}\right)^{1/2} \left(\frac{t}{r}\right)^{5/4}.$$

8.1   We can estimate the tensile strength of the steel used to make the link by using the correlation between tensile strength and Vickers hardness given in Appendix 1C. Then,

$$\sigma_{TS} \approx 3.2 \times 532 \text{ MPa} \approx 1702 \text{ MPa}.$$

The breaking load of each link plate in tension is given approximately by the minimum cross-sectional area multiplied by the tensile strength. The total breaking load of the two links in parallel is double this figure and is given by

$$T = 2(1702 \text{ N mm}^{-2} \times 1 \text{ mm} \times 4.5 \text{ mm})$$

$$= 1.53 \times 10^4 \text{ N}.$$

The tensile strength of the pin is given by

$$\sigma_{TS} \approx 3.2 \times 650 \text{ MPa} \approx 2080 \text{ MPa}.$$

We can estimate the shear failure stress of the pin using the correlation with tensile strength given in Appendix 1C. Then

$$k_u \approx \frac{2080 \text{ MPa}}{1.6} \approx 1300 \text{ MPa}.$$

The failure load of the pin in double shear is obtained by multiplying this shear failure stress by twice the cross-sectional area of the pin to give

$$T = 2\left\{1300 \text{ N mm}^{-2} \times \pi \left(\frac{3.5}{2}\right)^2 \text{ mm}^2\right\} = 2.50 \times 10^4 \text{ N}.$$

This load is 1.63 times greater than the load needed to break the links and the strength of the chain is therefore given by the lower figure of 1.53×10^4 N.

To estimate the tension produced in the chain during use we take moments about the centre of the chain wheel to give

$$90 \text{ kgf} \times 170 \text{ mm} \approx T \times \frac{190 \text{ mm}}{2},$$

$$T \approx 161 \text{ kgf} \approx 1.58 \times 10^3 \text{ N}.$$

The factor of safety is then given by

$$\frac{1.53 \times 10^4 \text{ N}}{1.58 \times 10^3 \text{ N}} = \underline{\underline{9.7}}.$$

Comments

(a) The factor of safety is calculated assuming static loading conditions. The maximum loadings experienced in service might be twice as much due to dynamic effects.

(b) The chain must also be designed against fatigue and this is probably why the factor of safety is apparently so large.

8.2 As shown in Appendix 1B, suspending the machine from the beam at mid span would produce a bending moment at this point given by

$$M = \frac{Fl}{4}.$$

The load required to make the beam go fully plastic at mid span is then given by

$$F = \frac{4M_p}{l}.$$

From Appendix 1C we see that

$$M_p = \frac{\sigma_y}{4}(a_1^3 - a_2^3).$$

Thus

$$F = \frac{\sigma_y}{l}(a_1^3 - a_2^3) = \frac{250 \text{ N mm}^{-2}}{5500 \text{ mm}}(100^3 - 90^3) \text{ mm}^3$$

$$= 12{,}318 \text{ N} = \underline{\underline{1.26 \text{ tonnef.}}}$$

This is about $\frac{1}{4}$ tonne less than the weight of the milling machine, which explains why the failure occurred before the trolley reached mid span.

9.1   Refer to Appendix 1A for initial information. The stress tensor is

$$\sigma_{ij} = \begin{pmatrix} 0 & \tau & 0 \\ \tau & \sigma & 0 \\ 0 & 0 & 0 \end{pmatrix}.$$

We form the determinant

$$\begin{vmatrix} (0-\lambda) & \tau & 0 \\ \tau & (\sigma-\lambda) & 0 \\ 0 & 0 & (0-\lambda) \end{vmatrix} = 0$$

which has the characteristic equation

$$\lambda^3 - \sigma\lambda^2 - \tau^2\lambda = 0.$$

This has roots given by

$$\lambda = 0,$$

$$\lambda = \frac{\sigma + \sqrt{\sigma^2 + 4\tau^2}}{2},$$

$$\lambda = \frac{\sigma - \sqrt{\sigma^2 + 4\tau^2}}{2},$$

which are the three principal stresses. If we substitute the principal stresses into the Von Mises equation (see Appendix 1C) we get the result

$$\underline{\underline{\sigma^2 + 3\tau^2 = \sigma_y^2.}}$$

9.2   The correlations given in Appendix 1C give

$$\sigma_{TS}(MPa) \approx 3.2 \times HV = 3.2 \times 500 = 1600$$

and

$$k_u \approx \frac{\sigma_{TS}}{1.6} = \frac{1600 \text{ MPa}}{1.6} = 1000 \text{ MPa}.$$

The cross-sectional area of the pin is

$$A = \pi(2^2 - 1.2^2) \text{ mm}^2 = 8.0 \text{ mm}^2.$$

The force needed to shear this area is

$$f_s = k_u A = 1000 \text{ N mm}^{-2} \times 8.0 \text{ mm}^2 = 8000 \text{ N}.$$

Finally, the failure torque is

$$\Gamma = 2(f_s \times 10 \text{ mm}) = 2 \times 8000 \text{ N} \times 10 \text{ mm} = 1.6 \times 10^5 \text{ N mm}$$

$$= 160 \text{ N m} \approx 16 \text{ kgf m}.$$

Appendix 1G shows that martensite containing 0.75% carbon has a hardness of about 880 HV. The hardness of the pin is only 500 HV. The steel appears to have been tempered properly and there is no evidence of a fault in the material.

9.3   The correlation given in Appendix 1C gives

$$\sigma_y (MPa) \approx \frac{H (MPa)}{3} = \frac{500 \times 9.81}{3} = 1635.$$

This is the yield stress of the pin after it has been subjected to the additional plastic strain of 8% caused by the hardness test. The increase in the flow stress produced by this strain is given by

$$100 \text{ MPa} \times \tfrac{8}{6} = 133 \text{ MPa}.$$

The yield stress of the pin in the as-received condition is therefore

$$(1635 - 133) \text{ MPa} = 1502 \text{ MPa}$$

and the shear yield stress according to the Tresca criterion is given by

$$k = \frac{1502 \text{ MPa}}{2} = 751 \text{ MPa}.$$

We then have

$$f_s = kA = 751 \text{ N mm}^{-2} \times 8.0 \text{ mm}^2 = 6008 \text{ N}$$

which gives

$$\Gamma = 2 \times 6008 \text{ N} \times 10 \text{ mm} = 1.2 \times 10^5 \text{ N mm}$$

$$= 120 \text{ N m} \approx 12 \text{ kgf m}.$$

For accurate results the Von Mises yield criterion should be used. As shown in Appendix 1C this assumes that

$$k = \frac{\sigma_y}{1.732}$$

so using the Tresca criterion will be conservative by a factor of

$$\frac{2}{1.732} = 1.16.$$

10.1   Using the empirical correlation between Vickers hardness and tensile strength given in Appendix 1C we estimate that the tensile strength of the weld pool material is 3.2×193 MPa $= 618$ MPa. The shear stress at failure is found by using the empirical correlation between tensile strength and shear failure stress given in Appendix 1C: it is 618 MPa$/1.6 = 386$ MPa.

By scaling measurements from the diagram we find that the upper fracture surface has a length of 5 mm and makes an angle of 35° with the axis of loading. The shear force in the upper surface at failure is approximately equal to $386 \text{ N mm}^{-2} \times 5 \text{ mm} \times 73 \text{ mm} = 14$ tonnef. The component of this force which acts along the loading direction is 14 tonnef $\times \cos 35° = 11$ tonnef.

The lower fracture surface in the diagram is 3 mm long and makes an angle of 11° with the axis. The shear force in the lower surface at failure is $386 \text{ N mm}^{-2} \times 3 \text{ mm} \times 73 \text{ mm} = 9$ tonnef. The force component along the tensile axis is 9 tonnef $\times \cos 11° = 9$ tonnef.

The total force along the loading axis is then given by $11 + 9 = 20$ tonnef.

To make the assembly break away from the welds we would have to make the smaller of the two cross-sections fail in tension. The tensile strength of the bar is given by 3.2×215 MPa $= 688$ MPa. The breaking load is then equal to $688 \text{ N mm}^{-2} \times 6.5 \text{ mm} \times 73 \text{ mm} = 33$ tonnef.

The joint has a lack of penetration at the root of each weld bead and this has made a shear failure mechanism possible. With full penetration the joint would have had a strength comparable to that of the smaller bar. The joint is badly aligned and the axes of the two bars do not coincide. There will thus be a bending moment in the joint, which will decrease its strength in tension. Finally, the lower weld bead has insufficient overlap with the smaller bar.

10.2 The maximum elastic bending moment is given by

$$M_{el} = \frac{\sigma_y I}{c}.$$

For case (a) we have

$$I = \frac{a^4}{12}$$

and

$$c = \frac{a}{2}.$$

Inserting these results into the first equation then gives

$$M_{el} = \frac{\sigma_y a^3}{6}.$$

Since

$$M_p = \frac{\sigma_y a^3}{4}$$

we find that

$$\frac{M_p}{M_{el}} = \left(\frac{\sigma_y a^3}{4}\right)\left(\frac{6}{\sigma_y a^3}\right) = \underline{\underline{1.5}}.$$

The answers to cases (b), (c) and (d) are obtained in an exactly analogous way.

11.1 The principal stresses in the tube wall are

$$\sigma_1 = 40 \text{ MPa}, \quad \sigma_2 = 20 \text{ MPa}, \quad \sigma_3 = 0.$$

The maximum tensile stress is 40 MPa and this is marked on the fracture-mechanism maps (see the diagrams on page 433) at the operating temperature of 700°C. The maps give failure times of approximately 10^8 and 4×10^8 seconds respectively (40 and 160 months).

As shown in Appendix 1D, the equivalent shear stress for the present stress state is equal to half the hoop stress and is therefore 20 MPa. This is marked on the deformation-mechanism maps (see the diagrams on page 434) at the operating temperature of 700°C. The maps give equivalent shear strain rates of approximately 4×10^{-9} and 10^{-10} per second.

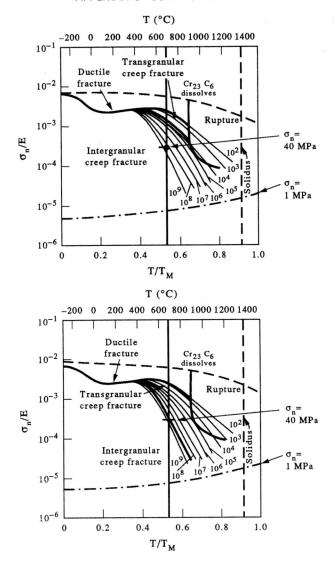

The Levy–Mises equations for strain rate given in Appendix 1D can be used to derive the hoop strain rate from the equivalent shear strain rate. Since

$$\sigma_2 = \frac{\sigma_1}{2}, \qquad \sigma_3 = 0$$

the Levy–Mises equations give

$$\dot{\varepsilon}_1 = \frac{3C\sigma_1}{4}, \qquad \dot{\varepsilon}_2 = 0, \qquad \dot{\varepsilon}_3 = -\frac{3C\sigma_1}{4}.$$

In other words,

$$\dot{\varepsilon}_1 = -\dot{\varepsilon}_3, \qquad \dot{\varepsilon}_2 = 0.$$

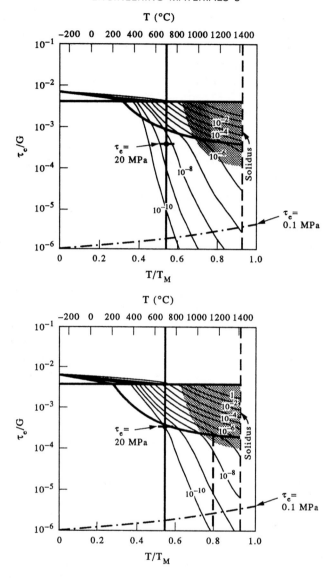

Substituting these relations into the expression for the equivalent shear strain rate gives us

$$\dot\gamma_e = [\tfrac{2}{3}\{(\dot\varepsilon_1 - 0)^2 + (0 + \dot\varepsilon_1)^2 + (-\dot\varepsilon_1 - \dot\varepsilon_1)^2\}]^{1/2}$$

so the hoop strain rate is given by

$$\dot\varepsilon_1 = \frac{\dot\gamma_e}{2}.$$

Using the values obtained above for the equivalent shear strain rates we find hoop strain rates of about 2×10^{-9} and 5×10^{-11} per second respectively. The worst case is to multiply the minimum time to failure of 10^8 seconds by the minimum hoop strain rate of 5×10^{-11} per second to give a strain increment at failure of about 0.5%.

If the tube has an initial diameter of 100 mm a 0.5% hoop strain will be equivalent to an increase in diameter of 0.5 mm. This is a small change to measure accurately in the confines of a furnace, and it is also likely to be of the same order as the manufacturing tolerance on the tube. On this basis your reservations seem justified. However, as shown in Appendix 1D, creep rates depend critically on structure. Suppliers' data for the exact grade of stainless steel used for the tubes might in fact show that the procedure was safe.

11.2 The iron–carbon phase diagram shows that pearlite is stable only up to 723°C. If the steel had been held just below this temperature for a reasonable length of time the iron carbide in the pearlite would have coarsened. This process is driven by the energy of the interfaces between the ferrite and carbide phases. The mechanism is one of diffusion. Smaller carbides dissolve whilst carbides grow at their expense. At the same time the growing carbides change from the plate-like form that they had in the pearlite to more spherical shapes in order to minimise the interfacial energy. This process is called spheroidisation. Since spheroidisation was not observed it is unlikely that the tube failed below 723°C.

When steel is taken above 723°C the nodules of pearlite are replaced by grains of austenite. As the temperature increases the remaining ferrite transforms progressively to austenite. Above the A_3 temperature the structure is 100% austenite. If the steel is held at a high enough temperature the austenite grains will coarsen. When the steel is cooled back below 723°C the austenite will revert to a mixture of ferrite and pearlite. The grain coarsening is consistent with this scenario and suggests that the tube was heated to well above 723°C.

The hoop stress in the tube wall is given by

$$\sigma = \frac{pr}{t} = \frac{4.30 \text{ MPa} \times 64 \text{ mm}}{6.6 \text{ mm}} = 42 \text{ MPa}.$$

Following the solution to Example 11.1 we see that the maximum tensile stress is 42 MPa and the equivalent shear stress is 21 MPa. These stresses are marked on the fracture-mechanism and deformation-mechanism maps respectively (see the diagrams on page 436).

The 42 MPa line on the fracture map lies in an intergranular field from 400°C to 700°C, just nips a transgranular field between700°C and 750°C, passes through another intergranular field between 750°C and 900°C and finally passes into the rupture field. The considerable ductility observed at the edge of the fracture suggests that the failure took place in the rupture field at a temperature of at least 900°C.

The 21 MPa line on the deformation map intersects with an equivalent shear strain rate of about 10^{-3} per second at 900°C. As shown in the solution to Example 11.1 this will give a hoop strain rate of 5×10^{-4} per second. If we assume a strain to failure of say 50%, or 0.5, then failure would take 1000 seconds, or 17 minutes. This is a maximum value since failure will occur more quickly at a higher temperature.

At 900°C the 21 MPa line lies inside the region of dynamic recrystallisation on the deformation map, which again is consistent with the metallurgical observations.

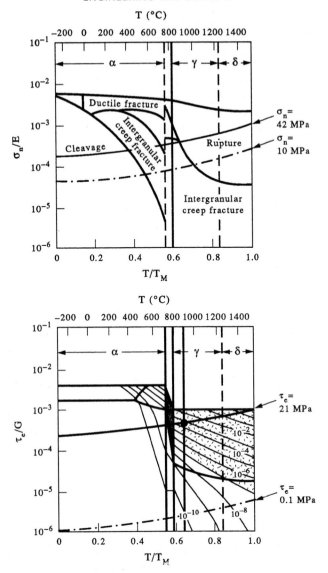

12.1 We denote the length and cross-sectional area of the reduced section by l and A respectively. The tensile force is F. Since volume is conserved during creep we can then write equations as follows:

$$Al = \text{constant},$$

$$A\,dl + l\,dA = 0,$$

$$\frac{dA}{A} = -\frac{dl}{l} = -d\varepsilon,$$

$$\frac{1}{A}\frac{dA}{dt} = -\frac{d\varepsilon}{dt},$$

$$\frac{\dot{A}}{A} = -\dot{\varepsilon} = -B\sigma^n,$$

$$\frac{\dot{A}}{A} = -B\left(\frac{F}{A}\right)^n = \frac{\text{constant}}{A^n}.$$

If $n = 0$, then $\dot{A}/A$ is constant: different sections of the bar draw down in the same proportion and the shape of the bar remains unchanged. If $n = 1$, as in diffusion creep or linear viscous flow, then

$$\frac{\dot{A}}{A} \propto \frac{1}{A}.$$

This means that the reduced cross-section draws down slightly faster than the rest of the bar. If $n = 5$, a value typical of power-law creep, then

$$\frac{\dot{A}}{A} \propto \frac{1}{A^5}$$

so the reduced section will draw down much faster than the rest, leading to rapid failure.

(a) Since $n = 1$ for a linear viscous liquid, drawing operations can be carried out without much risk of unstable necks forming. This is why molten glass can be drawn down into thin fibres or tubes.

(b) As shown in Appendix 1D, a material with a small grain size has a very large diffusion creep field. Diffusion creep, with $n = 1$, occurs up to relatively high stresses giving strain rates that are sufficient for economic forming operations. The risk of necking instability is much less than if the material had a normal grain size and exhibited power-law creep.

13.1 See Appendix 1E for appropriate results.

$$K_{1c} \approx 15.5(\text{CVN})^{1/2}.$$

$$\text{CVN} = 4 \times 0.737 \text{ lbf ft} = 2.95 \text{ lbf ft}.$$

$$K_{1c} \approx 15.5(2.95)^{1/2} \text{ ksi }\sqrt{\text{in}} = 26.6 \text{ ksi }\sqrt{\text{in}} \approx 30 \text{ MPa }\sqrt{\text{m}}.$$

$$K_{1c} = Y\sigma\sqrt{\pi a},$$

$$Y = 1 \text{ (Case 1)},$$

$$K_{1c} = \sigma\sqrt{\pi a},$$

$$a = \frac{1}{\pi}\left(\frac{K_{1c}}{\sigma}\right)^2 = \frac{1}{\pi}\left(\frac{30 \text{ MPa }\sqrt{\text{m}}}{60 \text{ MPa}}\right)^2 = 0.080 \text{ m} = 80 \text{ mm}.$$

$$2a = \underline{160 \text{ mm}}.$$

Check on validity of plane strain LEFM:

$$2.5\left(\frac{K_{1c}}{\sigma_y}\right)^2 = 2.5\left(\frac{30 \text{ MPa }\sqrt{\text{m}}}{440 \text{ MPa}}\right)^2 = 11.6 \text{ mm}.$$

This is much less than either a or $(W - a)$. It is slightly more than the value given for t of 10 mm, but plane strain LEFM should still be pretty accurate. The main uncertainty lies in the use of the correlation between Charpy energy and fracture toughness. In order to increase our confidence in the result we should obtain direct measurements of fracture toughness using compact tension specimens made from samples of plate removed from the flue. Cracks can be found by using any of the standard methods such as dye-penetrant testing, magnetic-particle inspection or ultrasonics.

13.2   Relevant results are given in Appendix 1E.

$$K_{1c} = Y\sigma\sqrt{\pi a}.$$

From Case 10 of Appendix 1E we have

$$Y = 1.12 \times 0.64 = 0.72$$

for a thumbnail crack lying at right angles to the surface of the plate. The critical radius of the crack is given by

$$a = \frac{1}{\pi}\left(\frac{K_{1c}}{Y\sigma}\right)^2 = \frac{1}{\pi}\left(\frac{40 \text{ MPa }\sqrt{m}}{0.72 \times 285 \text{ MPa}}\right)^2$$

$$= 0.012 \text{ m} = 12 \text{ mm}.$$

Check on validity of plane strain LEFM:

$$2.5\left(\frac{K_{1c}}{\sigma_y}\right)^2 = 2.5\left(\frac{40 \text{ MPa }\sqrt{m}}{570 \text{ MPa}}\right)^2 = 12 \text{ mm}.$$

This is much less than the dimensions of the surrounding plate. It is not greater than a so the analysis is valid.

The critical crack size determined above is greater than the size of the cracks found in the vessel by a factor of about 3. There are two main reasons why this could be the case.

(a) The steel at the edge of the welds could have been subjected to a residual tensile stress left by the welding operations. However, the vessel had been subjected to a post-welding heat treatment so the stresses should have been relaxed. Depending on the time and temperature of the annealing treatment the residual stress would probably have been of the order of 200 MPa. The residual stress field would have decayed away rapidly from the toe of the fillet weld but it might still have been 100 MPa at the centre of the 4-mm cracks. The total stress on the crack would then have been 100 MPa + 285 MPa = 385 MPa. This gives a critical crack depth of 6.6 mm, much closer to that observed. Note that this is less than the minimum crack size for which plane strain LEFM is strictly valid. However, the error is probably not very much. The plane-stress plastic zone radius is 0.8 mm, which is still small compared to the crack depth.

(b) The fracture toughness of the heat-affected zone next to the weld bead was probably less than that of the main plate. If the fracture toughness was reduced from 40 MPa $\sqrt{m}$ to say 35 MPa $\sqrt{m}$ at the assumed total stress of 385 MPa then the critical crack depth would have been reduced to 5 mm, which is close enough!

Making small attachment welds on the surface of a thick plate of steel which has a high carbon equivalent is likely to give small, hard, heat-affected zones with poor toughness. Many pressure vessel failures have been traced to such welds and they are best avoided if at all possible.

14.1 Appendix 1E shows that, as the ratio of plate thickness to plastic zone size decreases, the geometry of fast fracture changes from flat through shear lip to shear lip with necking. This is what we see in the pipe. The radius of the plastic zone in plane stress is given by

$$r_y = \frac{1}{2\pi}\left(\frac{K_1}{\sigma_y}\right)^2 = \frac{1}{2\pi}\left(\frac{Y\sigma\sqrt{\pi a}}{\sigma_y}\right)^2$$
$$= \frac{a}{2}\left(\frac{Y\sigma}{\sigma_y}\right)^2$$

where σ is the hoop stress in the wall of the pipe. We can see from this result that, as the crack grows, the plastic zone increases in size in direct proportion to the crack length. This is why the ratio of thickness to zone size decreases as the crack runs even though the thickness of the plate is constant.

The hoop stress is given by

$$\sigma = \frac{pr}{t} = \frac{5.5 \text{ MPa} \times 300 \text{ mm}}{15 \text{ mm}} = 110 \text{ MPa}.$$

The size of the plane stress plastic zone at the end of the flat part of the fracture is then

$$r_y = \frac{a}{2}\left(\frac{Y\sigma}{\sigma_y}\right)^2 = \frac{0.3 \text{ m}}{2}\left(\frac{1 \times 110 \text{ MPa}}{350 \text{ MPa}}\right)^2 = 15 \text{ mm}.$$

This is the same as the plate thickness. This means that the crack tip will be in plane stress, allowing a shear lip to form right through the plate.

As shown in Appendix 1E, there is a through-thickness tensile stress at the tip of a crack owing to the triaxiality of the stress field. This would have been relieved by the formation of the large plastic zone but was evidently enough to have caused delamination in the wall of the pipe. The delamination would have divided the crack in half, allowing the step to form.

14.2 The hoop stress in the shell is given

$$\sigma = \frac{pr}{t} = \frac{5.9 \text{ MPa} \times 164 \text{ mm}}{10 \text{ mm}} = 97 \text{ MPa}.$$

Using the result derived in Solution 14.1 the radius of the plastic zone in plane stress is

$$r_y = \frac{a}{2}\left(\frac{Y\sigma}{\sigma_y}\right)^2 = \frac{400 \text{ mm}}{2}\left(\frac{1 \times 97 \text{ MPa}}{400 \text{ MPa}}\right)^2 = 12 \text{ mm}.$$

This is comparable to the thickness of the plate. The crack tip is therefore in a state of plane stress and the running crack should have a slant geometry as shown in Appendix 1E.

15.1 The diagram shows the fracture surface with dimensions scaled from the photograph. To make the calculation of area easier the dimensions may be

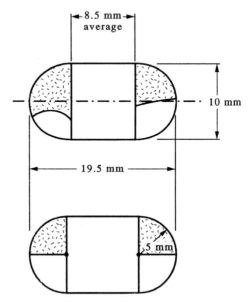

approximated as shown. The total area of the cleavage fracture is then $(\pi/2) \times$ $(5 \text{ mm})^2 = 39 \text{ mm}^2$. The area of the fracture surface in a standard Charpy V-notch test is 80 mm^2 (see Appendix 1E). Since the fracture was by cleavage this is equivalent to being on the lower shelf of the Charpy test which should give a standard Charpy impact energy of roughly 5 to 10 J depending on the steel used. The energy required to break the cracked crank is therefore about $(39/80) \times 5 \text{ J}$ to $(39/80) \times 10 \text{ J} = \underline{2.5 \text{ to } 5 \text{ J}}$. The potential energy released when a mass of 1 kg falls from a height of $\overline{0.25 \text{ m}}$ is $1 \times 9.81 \times 0.25 \text{ J} = 2.5 \text{ J}$ so dropping the weight a distance of $\underline{\underline{0.25 \text{ m to } 0.5 \text{ m}}}$ should have broken the crank.

Note that it is not appropriate to make area corrections to standard Charpy impact energies other than on the lower shelf because of the pronounced effect of specimen dimensions on plastic deformation.

15.2 Since rock is a brittle material, fracture generally starts where the tensile stress is a maximum and leads to a crack which is aligned at right angles to the direction of the initiating tensile stress. The principal stresses around the hole can be found from the equations for a thick-walled tube given in Appendix 1A. The only tensile principal stress is the hoop stress, given by

$$\sigma_1 = \frac{pb^2(a^2 + r^2)}{r^2(a^2 - b^2)}.$$

Since a is much greater than b in the present situation this reduces to

$$\sigma_1 = \frac{pb^2}{r^2}.$$

The stress is a maximum when $r = b$ (at the bore of the hole) when it is equal to p. The internal pressure needed to crack the rock is therefore the same as the fracture strength of 20 MPa. Fracture will start at the bore of the hole and the crack will run outwards along a radial plane.

The strain is given by

$$\frac{\Delta b}{b} = \left(\frac{p}{E}\right)\left\{\left(\frac{a^2 + b^2}{a^2 - b^2}\right) + v\right\}.$$

Since a is large compared with b this becomes

$$\frac{\Delta b}{b} = \left(\frac{p}{E}\right)(1 + v).$$

Assuming that Poisson's ratio is about 0.3 we find that the strain at the fracture pressure is $(20\ \text{MPa}/60\ \text{GPa}) \times 1.3 = 4 \times 10^{-4}$. This gives an expansion in the radius of the hole of $25\ \text{mm} \times 4 \times 10^{-4} = 0.01\ \text{mm}$.

16.1 Since cast iron is essentially a brittle material fracture generally starts where the tensile stress is a maximum and leads to a crack which is aligned at right angles to the direction of the initiating tensile stress. As shown in Appendix 1A, when a circular tube is subjected to torsion the principal stresses lie at $\pm 45°$ to the axis. We would therefore expect the initial crack to lie at $45°$ to the axis of the specimen and to be at right angles to the principal *tensile* stress. In addition, the crack should have started at the outside of the bar because this is where the stresses generated by the torque are a maximum. The probable point of initiation is marked with an arrow on the photograph.

As soon as the initial crack has grown to a reasonable size its presence modifies the original stress state and the direction of crack propagation changes in order to keep the crack more-or-less perpendicular to the changing direction of maximum tensile stress. This feature is obvious in the photograph. Also clear from the photograph is the absence of any ductility in the fracture. The edges of the fracture surfaces are sharp (no shear lips) and there has not been any drawing down of the section.

16.2   Wooden sections are always made with the grain parallel to the axis. This is done for obvious reasons: not only are trees long in the direction of the grain but a section machined with the grain running at right angles to the axis would snap very easily in bending because wood splits very easily across the grain. This, of course, reflects the big anisotropy in the mechanical properties of wood which results from its aligned structure. In torsion, however, a wooden section is very weak because the component of tensile stress that acts at right angles to the torsional axis, although much less than the principal stress, is still large enough to pull the wood apart across the grain, resulting in the delamination type of failure seen in the photograph.

16.3   When a tube made from un-reinforced epoxy resin is subjected to a torque it will break in a brittle fashion, just like the cast iron in Example 16.1. To prevent this happening carbon fibres should be incorporated into the structure as contra-helical windings making angles of $\pm 45°$ with the axis of the shaft. In order to resist lateral bending deflection we must also incorporate fibres parallel to the axis. In practice layers of longitudinal stiffening are alternated with layers of helical strengthening in a way governed by the need to optimise the mix of properties and ensure elastic compatibility between the layers.

17.1   For fast fracture at fixed displacements the energy equation is

$$-\frac{dU^{el}}{da} = G_c t.$$

For the single cantilever situation the stored strain energy is

$$U^{el} = \int_0^{\delta_1} F \, d\delta.$$

From Appendix 1B we have

$$\delta = \frac{Fa^3}{3EI}.$$

Thus

$$U^{el} = \frac{3EI}{a^3} \int_0^{\delta_1} \delta \, d\delta = \frac{3}{2}\left(\frac{EI\delta_1^2}{a^3}\right),$$

$$-\frac{dU^{el}}{da} = \frac{9}{2}\left(\frac{EI\delta_1^2}{a^4}\right) = G_c t,$$

$$\delta_1 = \frac{\sqrt{2}}{3}\left(\frac{\sqrt{EG_c}}{E}\right)\left(\frac{t}{I}\right)^{1/2} a^2.$$

For the double cantilever situation the stored strain energy is

$$U^{el} = 2 \int_0^{\delta_2} F \, d\delta.$$

Thus

$$\delta_2 = \frac{1}{3}\left(\frac{\sqrt{EG_c}}{E}\right)\left(\frac{t}{I}\right)^{1/2} a^2,$$

$$\frac{\delta_1}{\delta_2} = \sqrt{2}.$$

The maximum tensile stress in the beam occurs right next to the crack tip. Both slate and ice are brittle materials. If the maximum tensile stress is big enough a new crack will initiate at the tip of the main crack and will run across the root of the beam at right angles to the plane of the main crack. The maximum tensile stress is linear in δ. Thus the stress at the root of the single cantilever is $\sqrt{2}$ times the stress at the roots of the double cantilevers and fracture is more likely to occur in the single cantilever.

17.2 The radial mismatch is $2\delta = 45 \text{ mm} \times 8.5 \times 10^{-6}\,°C^{-1} \times 80°C = 3.1 \times 10^{-2} \text{ mm}$. For $\delta/2R \ll 1$, $R = h^2/2\delta = 10^2 \text{ mm}^2/(3.1 \times 10^{-2} \text{ mm}) = 3.2 \times 10^3 \text{ mm}$. From Appendix 1B,

$$M = \frac{EI}{R},$$

$$\sigma_{max} = \frac{Mc}{I} = \frac{Ec}{R},$$

$$\sigma_{max} = \frac{74 \times 10^3 \text{ MPa} \times 2 \text{ mm}}{3.2 \times 10^3 \text{ mm}} = 46 \text{ MPa}.$$

This is of the same order as the tensile strength so the scenario is quite reasonable. Note that the stress depends on h^{-2} so changing our original guess for h will have a strong effect on our estimate of the stress. To get an accurate answer we would have to run a computer model of the time-dependent heat flow along the wall and use this in turn to model the stress pattern. To be really smart we should also use Weibull statistics. Such labour—and expense—are hardly justified in the present case, but there are critical industrial applications (e.g. ceramic engine parts) where it is necessary to do a rigorous analysis.

The failure could have been avoided by using a pyrex glass flask and cooling the outside of the beaker with cold water during the mixing operation. The coefficient of thermal expansion of pyrex is about one-half that of soda glass. As the beam-bending equations show the stress is linear in the wall thickness, so the flask should be as thin as possible consistent with mechanical strength—the opposite of what one might expect.

18.1 A typical operating pressure for an industrial steam boiler is 10 bar gauge. The corresponding water temperature is about 180°C (see the steam table in Appendix 1H). If the feed water is at 10°C then the thermal shock that the plate experiences when feed water is put into the boiler could be as much as 170°C. The coefficient of thermal expansion of steel is about $12 \times 10^{-6}\,°C^{-1}$ (see Example 6.1) so the thermal strain in the surface of the plate, $\alpha \Delta T$, is about 2×10^{-3}. The strain required to give yield is

$$\varepsilon_y = \frac{\sigma_y}{E} = \frac{250 \text{ MPa}}{212 \text{ GPa}} = 1.2 \times 10^{-3},$$

using data from Example 6.1 for yield strength and Young's modulus. The thermal strain is therefore sufficient to cause plastic deformation. Since the thermal shock is cyclic (it occurs every time water is put into the boiler) it is likely that the cracks formed and grew by low-cycle fatigue. When the plastic strain amplitude is equal to the yield strain typically 10^4 cycles are required for failure. Over a three-year period we would only have to put water into the boiler ten times a day to achieve this number of cycles. In practice the number of cycles could have been appreciably greater.

The baffle would have made the cold water sweep sideways over the inner surface of the plate, ensuring efficient cooling. The problem would have been less acute if a baffle had not been fitted. The corrosion indicates that the environment around the feed pipe was aggressive and this would have increased the rate of crack growth. The transgranular unbranched cracking is consistent with fatigue. Branched or intergranular cracks would have indicated stress corrosion cracking (mild steel is liable to SCC if the water contains sodium hydroxide).

19.1   Each time the iron was moved backwards and forwards the flex would have experienced a cycle of bending where it emerged from the polymer sheath. The sheath is intended to be fairly flexible to avoid concentrating the bending in one place. Possibly the sheath was not sufficiently flexible and the flex suffered a significant bending stress at the location of failure. The number of cycles of bending is well into the range for high-cycle fatigue and fatigue is the likely cause. The scenario is that the individual strands in the live conductor broke one by one until the current became too much for the remaining strands to carry. At this stage the last strands would have acted as a fuse and melted, causing the fire. If 23 strands are rated to carry 13 A, then a single strand should carry about 0.57 A safely. The iron draws 4.8 A, which is 8.4 times the safe capacity of one strand. It is therefore not surprising that, when only a few wires were left intact, the flex was no longer able to take the current without overheating. Failures of this sort have also occurred with appliances such as vacuum cleaners. However, these tend to have a smaller current rating and failure does not always result in a fire.

19.2   A labelled photograph is given on page 445.

Circular Cross-section
   The failure was caused by fatigue. A marks the point of initiation. The early part of the fatigue crack is very smooth, but as the crack grows we can see clam-shell marks, e.g. B, which mark the tip of the crack at a given instant. The early clam-shell marks are parts of a circle concentric with the point of initiation. As the crack progresses the clam-shell marks become less curved. There are clam-shell marks almost all the way to the back face of the cross-section and the final fracture seems to have occurred as a small ligament C terminating in two shear lips D and E.

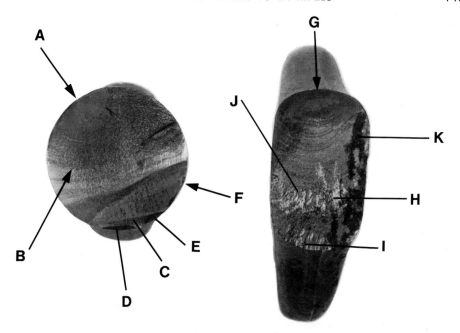

There is another shear lip F at the end of the last-but-one clam-shell mark which indicates appreciable plane-stress plasticity at the crack tip. The small area of the final fracture indicates that the tensile stress in that area was small.

Rectangular Cross-section
The failure was again caused by fatigue which initiated at G. The bright areas H and I are not cleavage facets but high spots where the two broken parts of the tool hammered into one another immediately after the failure. This fretting has obscured the later stages of the failure but the final fracture probably started at the clam-shell line marked J: this marks the transition from a smooth surface typical of fatigue to a rough grey surface typical of fibrous fast fracture. The large area of the final fracture indicates that the remaining ligament was subjected to a fairly high stress. There are areas of rust on the surface, e.g. K.

20.1 The diagram shows the approximate cross-section of the crank at the level of the fatigue cracks. When the pedal is at the bottom of its stroke the vertical

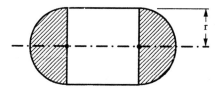

downwards force on the pedal generates a bending moment about the neutral axis of the cross-section of $45 \times 9.81 \times 65$ mm = 28,694 N mm. This moment gives rise to a bending stress distribution across the section. The maximum stress occurs at the surface and is given by

$$\sigma = \frac{Mc}{I}.$$

The second moment of area of the section about the neutral axis is

$$I = \frac{\pi r^4}{4} = \frac{\pi}{4} (5 \text{ mm})^4 = 491 \text{ mm}^4.$$

The surface stress is therefore

$$\sigma = \frac{28,694 \text{ N mm} \times 5 \text{ mm}}{491 \text{ mm}^4} = 292 \text{ N mm}^{-2}.$$

This stress is tensile at the origin of the fatigue cracks.

The downwards force on the pedal also generates a uniform direct tensile stress over the whole cross-section. The area of the cross-section is approximately $\pi(5 \text{ mm})^2 = 79 \text{ mm}^2$, so the direct stress is $(45 \times 9.81 \text{ N})/79 \text{ mm}^2 = 6 \text{ MPa}$.

The total tensile stress at the crack origin is thus $292 + 6 = 298 \text{ MPa}$.

When the pedal is at the top of its stroke the situation is reversed and the total stress at crack origin is 298 MPa in compression.

As shown in Appendix 1C, the tensile strength of the crank in MPa is given approximately by $3.2 \times$ Vickers hardness $= 3.2 \times 234 = 750 \text{ MPa}$. The steel quoted by Smithells therefore has a similar strength to that of the crank; it probably has a similar fatigue strength as well.

In ten years the rear wheel makes an estimated number of revolutions equal to $(10 \times 365 \times 15 \times 1000 \text{ m})/(\pi \times 0.69 \text{ m}) = 2.52 \times 10^7$. The number of revolutions made by the crank is less than this by a factor of 2.5, and is approximately 10^7. This is the same as the number of cycles used to determine the fatigue life of the smooth specimens. The stress cycle in the fatigue tests of $\pm 278 \text{ MPa}$ is very similar to the one estimated for the crank of $\pm 298 \text{ MPa}$, indicating that the bicycle was roughly 10 years old.

In practice the area of the cross-section in the compressive part of the stress distribution will be greater than assumed because there will be load-bearing contact between the spindle and the hole in compression. As a result the actual stress range will differ somewhat from our estimate. The assembly of crank and pedal spindle is statically indeterminate so an exact analysis is impossible, but approximate analyses in such cases are often surprisingly close to reality. Our calculation also ignores the stress concentrations caused by the complex changes of section in the region of the failure: these will increase the stress range available for fatigue initiation but their calculation is difficult.

20.2 The mechanism is one of fatigue. Fatigue cracks grew from the HAZ at a number of positions on the circumference and spread in towards the centre of the shaft. Where adjacent fatigue cracks ran sideways into one another they left radial "ratchet" markings on the fracture surface. Microsections cut from the HAZ showed cracks. As Appendix 1G shows, a steel with 0.40% carbon should have a Vickers hardness of about 670 in the fully martensitic condition. The measured maximum hardness of 560 to 580 HV indicates a high proportion of martensite. The HAZ would have cracked when the weld cooled because of the low toughness of the HAZ and the high residual stresses. The HAZ cracks would have been ideal starters for the fatigue cracks.

The carbon equivalent of the shaft can be calculated from the composition given by using the formula. The answer is 0.78. This is a high value. If the CE of a steel is much above 0.45 a hard HAZ is likely unless the parent metal is first preheated and the heat input from the weld electrode is sufficiently large relative to the thermal mass of the component. These conditions were not satisfied in the repair operation and a hard HAZ was a foregone conclusion.

The failure was encouraged by the surface stress concentration caused both by the coarse turning marks and the sudden change in diameter. Appendix 1A lists SCF values for a stepped shaft in pure bending. For a change in diameter of 10% (similar to that of the failed shaft) and a fillet radius of 0.05 × (shaft diameter) the SCF is nearly 2.

21.1 The maximum pressure in the cylinder occurs at the point of admission. The maximum force acting on the piston is therefore given by

$$0.7 \text{ N mm}^{-2} \times \pi(45 \text{ mm})^2 = 4453 \text{ N.}$$

The stress in the connecting rod next to the joint is

$$4453 \text{ N}/(28 \text{ mm} \times 11 \text{ mm}) = 14.5 \text{ N mm}^{-2}.$$

Since the locomotive is double-acting the stress *range* is twice this value, or 29 N mm^{-2}.

The number of revolutions that the driving wheel is likely to make in 20 years is $(20 \times 6000 \times 1000 \text{ m})/(\pi \times 0.235 \text{ m}) = 1.6 \times 10^8$.

Data for the fatigue strengths of welded joints are given in Appendix 1E. The type of weld specified is a Class C. The design curve shows that this should be safe up to a stress range of 33 N mm^{-2}, which is slightly more than the figure calculated above. If a fatigue limit operates after 10^7 cycles, the design curve gives a stress range of 75 N mm^{-2}. This generates an additional safety factor of about 2 and also means that the joint should last indefinitely. In practice, as discussed in Chapter 21, the fatigue curve probably falls somewhere between these two limiting situations: the joint may fail eventually but will probably last much more than 20 years. Of course, our calculations have ignored dynamic effects due to the reciprocating masses of the piston and connecting rod and these should be investigated as well before taking the design modification any further.

21.2 Data for the fatigue strengths of welds are given in Appendix 1E. The weld is a surface detail on the stressed plate and the weld classification is Class F2. As explained in Chapter 21, because the environment is corrosive there is probably no change in the slope of the fatigue curve above 10^7 cycles. We therefore extrapolate the curve following the dashed line for Class F2 until we hit the stress range of 8 N mm^{-2}.

The mean-line fatigue curve gives the data for a 50% chance of cracking. For the stress range of 8 N mm^{-2} the cycles to failure are 3×10^9. The time to failure is $(3 \times 10^9)/(20 \times 60 \times 60 \times 12 \times 6 \times 52) = 11$ years.

The design curve gives the data for a 2.3% chance of cracking. The number of cycles to failure is 10^9 and the time to failure is therefore 4 years.

22.1 The following table gives the mass of iron per unit area of corrosion product for each flake.

Sample no.	Mass of iron (mg mm^{-2})
1	1.195
2	0.632
3	0.629
4	0.761
5	0.540
6	0.983
7	0.670
8	0.864

The average value is 0.784 mg mm^{-2}.

(a) The mill scale that covers an area of 1 mm^2 has a volume of 1 mm$^2 \times 0.10$ mm $=$ 0.1 mm^3. The mass of Fe$_2$O$_3$ in this volume is 5.26 g cm$^{-3} \times 0.1$ mm$^3 = 0.526$ mg. Given that the atomic weight of iron is 55.85 the mass of iron in the volume of scale is

$$\left(\frac{2 \times 55.85}{2 \times 55.85 \times 3 + 16}\right) \times 0.526 \text{ mg} = 0.368 \text{ mg}.$$

A parallel procedure can be used for Fe$_3$O$_4$ and FeO to give masses of (b) 0.375 mg and (c) 0.443 mg. The masses of iron taken up by the wet corrosion process are thus given by:

(a) $0.784 - 0.368 = 0.416$ mg.
(b) $0.784 - 0.375 = 0.409$ mg.
(c) $0.784 - 0.443 = 0.341$ mg.

Each mass of iron is then divided by the density of iron (7.87 g cm^{-3}) to give the volume of iron removed by the wet corrosion. The volumes for the three cases are:

(a) 0.053 mm^3.
(b) 0.052 mm^3.
(c) 0.043 mm^3.

The surface area over which the volume of iron is lost is 1 mm^2 in each case, so the thicknesses of steel lost by wet corrosion are:

(a) 0.053 mm.
(b) 0.052 mm.
(c) 0.043 mm.

This degree of precision is meaningless in view of the uncertainty over the thickness of the original mill scale layer and the assumption that the corrosion product did indeed include all the mill scale. It is enough to say that the loss in thickness is of the order of 0.05 mm, irrespective of the precise composition of the mill scale. If the mill scale had not been incorporated in the corrosion product the volume of steel lost by wet corrosion would simply have been $0.784 \text{ mg}/7.87 \text{ g cm}^{-3} = 0.10 \text{ mm}^3$, giving a thickness loss of 0.10 mm.

22.2 The table of standard electrode potentials given in Appendix 1F shows that iron is electrochemically active with respect to copper. The same will be true for a low-alloy steel relative to a copper alloy such as bronze. Because of the physical contact between the shaft and the sleeve the two alloys were in good electrical contact. There was also excellent electrochemical contact between the two metals via the water because of the high conductivity of sea water. There would therefore have been a tendency for the steel to corrode by galvanic action. The dissolved oxygen which was initially present in the entrained water would have been used up quite soon by the cathodic reaction on the surface of the bronze. Fresh oxygen would have diffused in from the sea water outside, but only slowly, and this would have restricted the cathodic current available to drive the anodic reaction at the surface of the steel. This would have encouraged limited, selective corrosion rather than wholescale uniform attack. The variable geometry of the crevice between the shaft and the sleeve would also have encouraged selective attack. The steel would have corroded most rapidly where the crevice was at its narrowest: the concentration of oxygen in such places would have been a minimum and this would have discouraged the formation of stable surface films. The rate at which the pits developed would have been increased by the presence of chloride ions owing to their role in the autocatalytic processes that take place inside a pit.

23.1 The coolant in a new car usually consists of a mixture of water and "anti-freeze". The anti-freeze is generally ethylene glycol but it also contains corrosion inhibitors (e.g. sodium nitrite) to develop and maintain a passivating film on the metal surfaces in the cooling system. The concentration of inhibitor needs to be well above the minimum level needed for full protection. If the concentration is insufficient the metal will be liable to selective attack and will tend to pit. In addition the inhibitor solution will have a high ionic conductivity which will allow the pitting reactions to proceed at a high rate. Indeed the rate of penetration with insufficient inhibitor will be far greater than if no inhibitor were present at all. Since the inhibitor gets used up over a period of time it is normal practice to change the anti-freeze every two years in order to replenish the inhibitor. Since this was not done in the present instance it is not surprising that the steel plug rusted through. The rusty colour of the coolant also indicates large-scale corrosion of the interior of the block and it would only have been a matter of time before the block itself perforated, with expensive consequences.

23.2 Because brass is an alloy of copper and zinc it consists of two metals which have a very large difference in electrode potential. Pure zinc is highly active with respect to pure copper and is attacked preferentially in water. In fact, as the table of standard electrode potentials in Appendix 1F shows, the difference in potentials between the two metals is 1.1 volts. The situation is more complicated in brass, of course: in single-phase alpha brass (which contains less than 32 weight % Zn) the two elements are associated as a solid solution; and in two-phase alpha-beta brass (e.g. 60Cu–40Zn) there are also precipitates of the intermetallic beta phase (which has the approximate formula CuZn). The precise mechanism by which the zinc is attacked so as to leave a porous residue of copper is not fully understood, but the phenomenon of "dezincification" is common whenever ordinary brasses are exposed to corrosive environments. Because of this special brasses have been developed which contain arsenic as a dezincification inhibitor.

24.1 At temperatures above 70°C austenitic stainless steel is susceptible to stress corrosion cracking in the presence of stress and chloride ion. Failures have been reported even with very low levels of stress and chloride concentration. In the present failure the stress was probably provided by the residual stresses from the welding process. In addition large concentrations of chloride were found. There seems little doubt that this was the cause of the failure. The chloride presumably originated from the marine environment and was carried under the cladding by the leaking rain water. Chemical analysis of some unused lagging also revealed soluble chloride which would have provided an extra source of chloride ion.

24.2 The table of standard electrode potentials given in Appendix 1F shows that iron is electrochemically active with respect to copper. The same will be true of iron alloys such as stainless steel relative to copper alloys such as bronze. Because of the physical contact between the spindle and the valve body the two alloys were in excellent electrical contact. There was a large area of copper and bronze available to support the cathodic reaction but only a small area of stainless steel to provide the balancing anodic reaction. This combination is conducive to the intense galvanic attack of the stainless steel. Oxygen dissolved in the feed water would have provided a ready source of oxygen for the cathodic reaction. Although distilled water should not contain any soluble salts it is nevertheless aggressive, with a pH as low as 4.5. This is usually due to dissolved carbon dioxide which forms carbonic acid. The electrical conductivity of the water would therefore have been sufficient to couple the end of the spindle to a large area of cathode through the water side of the electro-chemical circuit. Oxygen would have been excluded from the crevice formed between the spindle and the seating. This would have encouraged the breakdown of the passive film of chromium oxide, allowing the attack to take place. This is a good example of the strong tendency that stainless steels have to suffer from crevice corrosion. The operating pressure of 6 bar gauge gives a pressure of 7 bar absolute. The Steam Table in Appendix 1H shows that this will give a saturation temperature of 165°C. The high temperature will accelerate the rates of the thermally activated corrosion reactions. In view of the above it is not surprising that the stainless steel corroded at a rate of 0.1 mm/hour!

25.1 The mass of iron in the flake is 0.221×7.81 g $= 1.73$ g. The density of iron is 7.87 g cm^{-3} (Appendix 1F) so this mass would have occupied a volume of iron of

$1.73 \text{ g}/7.87 \text{ g cm}^{-3} = 0.22 \text{ cm}^3$. Given that the area of metal from which this volume was lost is 11.1 cm^2 the thickness lost is $0.22 \text{ cm}^3/11.1 \text{ cm}^2 = 0.020 \text{ cm} = \underline{\underline{0.20 \text{ mm}}}$. If we assume that the rate of attack is constant with time the thickness lost over five years would be about 4 mm. This estimate involves a large extrapolation without any physical justification. In corrosion testing it is essential to conduct long-term tests to obtain data for long-term applications. The statistical weight of the data obtained from a single flake is also very small. For example, the analysis of another flake from the boiler gave a section loss of only 0.11 mm. We would also need to be satisfied that the flakes did not pick up iron from any other source (e.g. carried over with the gas) and that only one layer of corrosion product had been produced. The five-year estimate is therefore not very accurate but it does show that there is a potential problem which needs to be monitored by periodic inspection.

25.2   The red colour of the corrosion product indicates that it is red rust, hydrated Fe_2O_3. Of the three forms of iron oxide (Fe_2O_3, Fe_3O_4 and FeO) this has the highest ratio of oxygen to iron. It is the form of oxide that is produced when there is an adequate supply of oxygen. When the oxygen concentration is low, the corrosion deposit consists instead of hydrated magnetite (Fe_3O_4) which is black. But there was no evidence that this was present in the corrosion product. There is evidence therefore of oxygen in the condensate which presumably came from air dissolved in the make-up feed water to the boiler. This would have provided the oxygen needed for the cathodic reaction.

The design of the unit allows condensate to build up to the level of the drain. It is interesting that corrosion has only occurred in, or just above, the pool of condensate; it has not taken place further up the tubes even though they would have been dripping with condensed steam. A likely scenario is that when the shell was let down to atmosphere the water at the bottom of the shell was boiled off by the residual heat in the tube plate. This would have left either a concentrated solution or a solid residue containing most of the impurities that were originally dissolved in the condensate pool. With each cycle of operation the concentration of impurities in the pool would have increased. A prime suspect is carbonic acid, derived from carbon dioxide dissolved in the feed water. This would have made the liquid in the pool very acidic and given it a high ionic conductivity both of which would have resulted in rapid attack. As the electrochemical equilibrium diagram for iron shows (see Appendix 1F), iron does not form a surface film in acid waters. Finally, the temperature is high so the rates of the thermally activated corrosion processes should be high as well.

The design modification is to provide a condensate drain at the lowest point of the shell to stop water accumulating there.

26.1   If p is the radial pressure that the olive exerts on the outside of the pipe then we can write

$$\sigma_y = \frac{pr}{t}$$

provided we neglect the strengthening effect of the sections of pipe that lie outside the olive. If we assume that the end of the pipe far away from the fitting has an end cap (or a bend that functions as an end cap) then the force trying to push the pipe out of the fitting is $p_w \pi r^2$. This force is balanced by the frictional force between the olive and the pipe so we can write

$$p_w \pi r^2 = \mu p 2\pi r l.$$

Combining the two equations to eliminate p gives

$$p_w = 2\mu\sigma_y \left(\frac{t}{r}\right)\left(\frac{l}{r}\right).$$

Using the data given we get

$$p_w = 2 \times 0.15 \times 120 \text{ MPa} \left(\frac{0.65}{7.5}\right)\left(\frac{7.5}{7.5}\right) = 3.1 \text{ MPa}.$$

The hydrostatic head of water in a seven-floor building is about 2 bar, so the joint could actually cope with pumping water to the top of a seventy-floor skyscraper and still have a factor of safety of 1.5. However, the pressure in water systems frequently exceeds the static head, often substantially, because of "water hammer". This is the dynamic overpressure that arises when taps are suddenly shut off.

27.1 From Appendix 1B the buckling force is given by

$$F_{cr} = 9.87\left(\frac{EI}{l^2}\right).$$

From Appendix 1B the second moment of area of the 0.5 in × 4 in section is

$$I = \frac{bd^3}{12} = \frac{4 \times 0.5^3}{12} \text{ in}^4 = 0.042 \text{ in}^4.$$

$l = 156$ in, so $F_{cr} = 0.22$ ton. The buckling load for the two-piece bar is thus 0.44 ton.
  The second moment of area of the 1 in × 4 in section is 0.33 in^4, giving a buckling load for the one-piece bar of 1.74 tons.

28.1 (a) Following Eqn. (16.4), the maximum moment is given by

$$M_{max} = \frac{\sigma_{TS} \pi (r_1^4 - r_2^4)}{4c}.$$

$r_1 = c = 10$ in, $r_2 = 9$ in and $M_{max} = 225$ ton ft.
  (b) When the bolts yield, the connection can be approximated as a mechanism which hinges at X. The cross-sectional area of one bolt is $\pi(1.25/2)^2$ in$^2 = 1.23$ in^2. The yield load of one bolt is 1.23 in$^2 \times 11$ tsi $= 13.5$ tons. The moment at yield is given by

$$M \approx (2 \times 13.5 \times 25) + (2 \times 13.5 \times 19) + (2 \times 13.5 \times 9) + (2 \times 13.5 \times 3)$$

$$= 1512 \text{ ton in} = 126 \text{ ton ft.}$$

The hinge must react the total load from the bolts, which is 108 tons. This means that in practice the hinge will extend over a finite area of contact. X will lie in from the outer edge of the flange by about 1 in to 2 in but the effect on the bending moment will be small.

LIST OF PRINCIPAL SYMBOLS

a	acceleration
	atomic weight
	constant in strain-hardening law (small strains)
	crack length
	outer radius
	side of square
A	area
	constant in constitutive equation for creep
	constant in fatigue crack-growth law
	constant in strain-hardening law (large strains)
A_1	eutectoid temperature
A_3	ferrite start temperature
b	breadth
	inner radius
B	constant
c	maximum distance from neutral axis
C	constant
CE	carbon equivalent
CVN	Charpy V-notch impact energy
d	depth
	diameter
D	diameter
e	radial misalignment
E	Young's modulus
E_{ax}	axial modulus
E_t	tangent modulus
f	force
	frequency
f_{cr}	critical whirling speed/mode 1 natural vibration frequency
F	force
F_{cr}	Euler buckling load
g	acceleration due to gravity on the Earth's surface
	effective throat dimension of fillet weld
G	shear modulus
G_c	toughness (critical strain-energy release rate)

G_{1c}	plane-strain toughness
h	height
H	hardness
HAZ	heat-affected zone
HB	Brinell hardness
HRC	Rockwell C-scale hardness
HV	Vickers hardness
i	current density
I	second moment of area of structural section
I_x, I_1	second moments of area in parallel-axis theorem
k	shear yield stress
	spring constant
k_e	Von Mises equivalent shear stress
k_u	shear failure stress
K	bulk modulus
	thermal conductivity
K_c	fracture toughness
K_1	mode 1 stress intensity factor
K_{1c}	plane-strain fracture toughness
ΔK	K range in fatigue cycle
l	direction cosine
	length
L	length
m	direction cosine
	exponent in fatigue crack-growth law
	exponent in strain-hardening law
	mass
	Weibull modulus
M	bending moment
	mass
M_f	martensite finish temperature
M_p	fully plastic moment of structural section
M_s	martensite start temperature
n	direction cosine
	exponent in constitutive equation for creep
	mode number
N	number of fatigue cycles
N_f	number of fatigue cycles to failure
p	distance from neutral axis in parallel-axis theorem
	pressure
p_{cr}	critical pressure for external-pressure buckling
P	load
P_s	survival probability
q	distance from neutral axis to elemental area
Q	activation energy for creep
r	radius
r_y	radius of crack-tip plastic zone

R	gas constant
	normal reaction force
	radius
S	stiffness
SCF	elastic stress-concentration factor
t	thickness
	time
t_f	time to failure
T	temperature
	tension
T_g	glass transition temperature
T_M	melting temperature
ΔT	temperature differential
	thermal shock resistance
u	displacement
U^{el}	elastic strain energy
v	velocity
V	volume
	volume fraction
V_0	volume of test specimen
w	weight loss
	width
W	weight
	width of cracked component
Y	crack geometry factor
z	charge on metal ion
	distance from neutral axis
α	angle
	linear coefficient of thermal expansion
β	angle
δ	deflection
Δ	dilatation
ε	strain
ε_e	Von Mises equivalent tensile strain
ε_f	strain to failure
ε_{ij}	general strain tensor
ε_p	principal strain tensor
ε_y	yield strain
$\varepsilon_1, \varepsilon_2, \varepsilon_3$	principal strains
$\delta\varepsilon$	increment of plastic strain
$\Delta\varepsilon^{pl}$	plastic strain range in fatigue
γ	angle
	engineering shear strain
γ_e	Von Mises equivalent shear strain
Γ	torque
η	viscosity
μ	coefficient of static friction

v	Poisson's ratio
ϑ	angle
ρ	density
σ	stress
σ_e	Von Mises equivalent tensile stress
σ_f	fracture stress
σ_{ij}	general stress tensor
σ_{max}	maximum stress
σ_n	nominal tensile stress
σ_0	reference stress in Weibull equation
σ_p	principal stress tensor
σ_{TS}	tensile strength
σ_y	yield stress
$\sigma_1, \sigma_2, \sigma_3$	principal stresses
$\Delta\sigma$	stress range in fatigue
τ	shear stress
τ_e	Von Mises equivalent shear stress
ω	angular frequency
$1, 2, 3$	right-handed orthogonal axes
$1_p, 2_p, 3_p$	principal directions

INDEX